Discovering Physical Science

PRENTICE-HALL INTERMEDIATE SCIENCE SERIES
SERIES EDITOR:
William A. Andrews
Professor and Chairman of Science Education
Faculty of Education, University of Toronto

Understanding Science 1
Understanding Science 2
Physical Science: An Introductory Study
Biological Science: An Introductory Study
Discovering Physical Science
 — Student Text
 — Teacher's Guide
Discovering Biological Science

William A. Andrews, Faculty of Education, University of Toronto
T. J. Elgin Wolfe, Faculty of Education, University of Toronto
Donna K. Moore, David and Mary Thomson, C.I. Scarborough, Ontario

CLARKE ROAD SECONDARY SCHOOL
SCIENCE DEPARTMENT

Discovering Physical Science

PRENTICE-HALL CANADA INC., Scarborough, Ontario

CANADIAN CATALOGUING IN PUBLICATION DATA

Andrews, William A., 1930-
   Discovering physical science

For use in schools
Includes index.
ISBN 0-13-215749-7

1. Physics. 2 Chemistry. I. Wolfe, T. J. Elgin (Thomas John Elgin), 1938- II. Moore, Donna K., 1944- III. Title.
QC23.A56   530   C81-094957-1

© 1982 by Prentice-Hall Canada Inc., Scarborough, Ontario

All Rights Reserved.
No part of this book may be produced in any form or by any means without permission in writing from the publisher.

Prentice-Hall Inc., Englewood Cliffs, New Jersey
Prentice-Hall International, Inc., London
Prentice-Hall of Australia, Pty., Ltd., Sydney
Prentice-Hall of India, Pvt., Ltd., New Delhi
Prentice-Hall of Japan, Inc., Tokyo
Prentice-Hall of Southeast Asia (PTE) Ltd., Singapore

ISBN 0-13-215749-7

Metric Commission Canada has granted permission for use of the National Symbol for Metric conversion.

Production editor: Iris Skeoch
Designer: Maher & Murtagh Inc.
Illustrators: Acorn Technical Art Inc.
Production: Sharon Lindala
Compositor: Canadian Composition Limited

Printed and bound in Canada by The Bryant Press Limited
  2 3 4 5 6 7 8 9   90 89 88 87 86 85 84 83 82

Contents

Unit One: The Nature of Science
Chapter 1: What is Science? *4*
Chapter 2: Doing Science *13*
Chapter 3: Building Models *21*

Unit Two: Measurement
Chapter 4: Measuring Length *32*
Chapter 5: Measuring Area *45*
Chapter 6: Measuring Volume *49*
Chapter 7: Measuring Mass *61*

Unit Three: The Physical Properties of Matter
Chapter 8: Solids, Liquids, and Gases *70*
Chapter 9: Changes of State and the Particle Theory *80*
Chapter 10: Using the Particle Theory *92*
Chapter 11: Everyday Uses of the Particle Theory *106*

Unit Four: Solutions and Mixtures
Chapter 12: Solutions and Other Mixtures *120*
Chapter 13: Separation of Substances: Filtration and Distillation *135*
Chapter 14: Separation of Substances: Sedimentation and Floc Formation *150*

Unit Five: The Chemical Properties of Matter
Chapter 15: The Classification of Matter *160*
Chapter 16: Chemical Change *173*
Chapter 17: The Structure of Matter *187*
Chapter 18: Atoms and the Periodic Table *198*

Unit Six: Applying Your Knowledge of Chemistry
Chapter 19: Air Pollution *214*
Chapter 20: Water Pollution *227*
Chapter 21: How to be a Thinking Consumer *238*

Unit Seven: Mechanics and Machines
Chapter 22: Force and Pressure *248*
Chapter 23: Work, Energy and Power *263*
Chapter 24: Machines *274*

Unit Eight: Vibrations and Waves: Sound
Chapter 25: The Nature of Sound *292*
Chapter 26: Transmission of Sound *308*
Chapter 27: Characteristics of Sound *324*

Unit Nine: Heat Energy
Chapter 28: Heat Transfer, Expansion and a Theory *342*
Chapter 29: Quantity of Heat *355*
Chapter 30: Heat and Change of State *367*

Unit Ten: Electricity
Chapter 31: Electrostatics *380*
Chapter 32: Sources of Current Electricity *391*
Chapter 33: Series and Parallel Circuits *404*

Unit Eleven: Energy: A Critical Resource
Chapter 34: The Nature of Energy *420*
Chapter 35: Non-Renewable Energy Sources *432*
Chapter 36: Renewable Energy Sources *445*
Chapter 37: Conserving Energy for Heating *458*
Chapter 38: Conserving Electricity and Gasoline *471*

Acknowledgements

The authors wish to acknowledge the competent professional help received from the staff of Prentice-Hall Canada Inc., in the production of this text. In particular, we extend our thanks to Iris Skeoch for her skillful editorial work and to Sharon Lindala for her dedicated coordination of the production aspect of this text. We also thank Steve Lane for his untiring and valuable assistance in the planning and development of this text. This book owes its final shape and form largely to the efforts of Peter Maher and Rand Paterson.

We wish, further, to thank the many teachers who reviewed the manuscript and offered their constructive criticisms. Many students volunteered to serve as models in the photographs. We thank them and the people mentioned in the photo credits below for their assistance in making this book more appealing and useful.

We would be remiss if we did not express our appreciation of the imaginative, attractive, and accurate art work of Acorn Technical Art Inc. Finally, we extend a special word of appreciation to Barbara Wolfe for her assistance in the typing of the manuscript and to Lois Andrews for her skillful and dedicated preparation of the final manuscript.

W. A. Andrews
Editor and Senior Author

Photo Credits

Cover Photo: The Canadian Government photo processing laboratory, housed in the National Personnel Records Centre in Ottawa. About 1/3 of the laboratory's hot water requirements are supplied by 64 liquid flat-plate collectors mounted on the fourth floor wall. Individual collectors measure 213 cm by 92 cm. Courtesy of John Tailkur, Public Works of Canada, Solar Program. Figs. 1-5; 23-5: courtesy of Ontario Ministry of Transportation and Communications. Fig. 4-1: courtesy of the Ford Motor Company of Canada, Limited. Fig. 8-4: courtesy of Dofasco Inc. Fig. 14-3: courtesy of the Ontario Ministry of the Environment. Fig. 19-0: courtesy of the Ontario Ministry of Agriculture and Food. Fig. 19-3: Donna K. Moore. Figs. 22-0; 24-6,A; 32-10; 38-10: courtesy of Ontario Hydro. Figs. 22-1; 23-6; 24-1; 24-5,B; 24-6,C; 24-9,B; 24-11,C; 31-0; 31-1; 31-2; 31-8; 32-2; 32-4; 33-2; 37-1; 37-3; 37-4; 37-5; 37-7; 37-13; 37-14; 40-4; 41-2: T. J. Elgin Wolfe. Figs. 22-2; 23-7; 24-5,C: courtesy of the Ontario Ministry of Industry and Tourism. Fig. 22-6: Tony Evangelista. Fig. 24-5: courtesy of the Canadian Rehabilitation Council for the Disabled. Fig. 24-11: courtesy of Miles S. Nodal, Action Photographics. Fig. 28-0: courtesy of the Ontario Ministry of Natural Resources. Fig. 30-4: courtesy of the Horticultural Institute of Ontario, Ontario Ministry of Agriculture and Food. Fig. 38-2: courtesy of Northern Miner Press Ltd. Fig. 39-1: courtesy of the Ontario Ministry of Municipal Affairs and Housing. Fig. 40-3: Polyflex® and Fin-Seal® are registered trade marks of Schlegel Canada Inc. All other photos by W. A. Andrews.

To the Student

This is a first course in physical science. Its aim is to give you knowledge and skills that you can use in your day-to-day life. Physical science is part chemistry and part physics. The first half of this book is mainly chemistry. The last half is mainly physics.

One of the first things you do is review metric measurement. We all need to know that today. Then you begin your study of chemistry. You first look at solids, liquids, and gases. In doing so, you will find out many interesting things. For example, why does salt melt ice? How does antifreeze protect a car radiator? Why do hail stones form on hot days?

The chemistry part of the book has many interesting experiments. You also learn what atoms and molecules are. This part of the book also shows you how to use your chemistry knowledge. For example, you will find out what you can do to help stop pollution.

The physics half of the book begins with mechanics. This is the study of forces, work, power and energy. What is gravity? How do car brakes work? Do seat belts really help? How much does it cost to run a 1000 W (watt) electric heater for a day? These are some of the interesting questions you will study in mechanics.

In this book you learn about things two ways. Sometimes you read about them. Other times you do experiments. We call the experiments activities. The readings and activities give you important knowledge. They also train you to use the scientific method. Can you believe television commercials? Is brand X really twice as good as brand Y? You can usually find out by using the scientific method.

How To Use This Book

This book was written to make the learning of physical science as easy as possible for you. The following are some of the ways the book helps make learning easy. To get the most out of this book, you should keep these points in mind.

1. Reading Sections

Each chapter has some reading sections. Read them carefully. They provide the basic information you need to do activities. They also summarize and explain activities. As well, they provide interesting examples from the world around us.

Always do the questions in the *Section Review* at the end of a reading section. They aren't hard. You can find the answers to all the questions in the section. Finding these answers will help you understand what you have read. These questions will also help you make notes that summarize the important points of the reading section.

2. Laboratory Activities

Always read the procedure before you begin the activity. Then follow the steps carefully. Finally, try all the *Discussion* questions. They are included to help you find out if you understand the activity. Some of these questions are hard. Don't be discouraged if you can't do all of them. But make sure you understand the answers when they are discussed in class.

3. **Chapter Overview**

Each chapter begins with an overview of the sections, both narratives and activities, in it. Read the overview before you start the chapter. It will give you a feeling for what the chapter is about.

4. **Chapter Summary**

Every chapter ends with a list of *Main Ideas*. Always read them. If you don't understand any of them, ask your teacher for help. A *Glossary* is also included at the end of each chapter. It gives the meanings of key terms. When necessary, it also gives their pronunciations. Use the glossary to review the key terms.

5. **Chapter Questions**

The end of each chapter includes five kinds of questions:
a) *True or False Items.* These questions are pure recall. That means they test what you have remembered from the chapter. However, don't look back until you have tried to answer the questions.
b) *Completion Items.* These questions serve the same purpose as the true or false items.
c) *Multiple Choice Items.* Some of these are recall questions. Like the two previous types, they test only how well you remember the material. However, other items will help you find out how well you *understand* the material.
d) *Using Your Knowledge.* These questions give you a chance to use the knowledge you have gained in the chapter. Usually they deal with practical applications that are interesting and useful.
e) *Investigations.* These include further classroom activities, home projects, library projects, debates, and so on. The investigations give you a chance to work on your own.

It's fun to know why things happen as they do, and it's useful too. We hope you enjoy this course and learn many helpful things.

W.A. Andrews
Editor and Senior Author

Prentice-Hall Canada Inc., Educational Book Division and the authors of DISCOVERING PHYSICAL SCIENCE are committed to the publication of instructional materials that are as bias-free as possible. This text was evaluated for bias prior to publication.

The authors and publisher of this book also recognize the importance of appropriate reading levels and have therefore made every effort to ensure the highest possible degree of readability in the text. The content has been selected, organized, and written at a level suitable to the intended audience. Standard readability tests have been applied at several stages in the text preparation to ensure an appropriate reading level.

Readability tests, however, can only provide a rough indication of a book's reading level. Research indicates that readability is affected by much more than word or sentence length; factors such as presentation, format and design, none of which are considered in the usual readability tests, also greatly influence the ease with which students read a book.

One other important factor affecting readability is the extent to which the text is motivational for students. Thus the following features were incorporated into this book to increase reader comprehension further. Page references are given to provide examples of most features.

Real World Examples:
Wherever possible, the text relates the theory to the everyday world. This motivates students and makes the content more meaningful by showing that applications of physical science are all around us. In addition, some of the applications of physical science are developed in major sections or entire chapters. Examples include: everyday use of the particle theory in Chapter 11 (evaporation of water, how salt melts ice, the use of antifreeze in a car), air pollution (Ch. 19), water pollution (Ch. 20), how to be a thinking consumer (Ch. 21), noise pollution (p. 332), heat energy, (Unit 9), food and exercise (p. 374), conserving energy for heating (Ch. 40), conserving electricity and gasoline (Ch. 41).

Student Involvement:
The text recognizes that students learn best if they have concrete, hands-on experiences. Hence the book features over 100 **laboratory activities** integrated with the narratives. Instructions for each lab activity guide students to understand exactly what they are to do and observe in order to complete the activity successfully. Examples of typical laboratory activities can be found on pp. 142 and 475. Each activity consists of a short **paragraph** spelling out the purpose of the activity, a statement of the **problem** to be investigated, a list of required **materials**, the **procedure** carefully and thoroughly laid out in a step-by-step fashion, and **questions** at the conclusion of the activity. In addition, safety in the laboratory is stressed at all times. The word "CAUTION" in brown ink draws the student's attention to specific precautions whenever necessary.

Questions:
The text features a wealth of questions. **Review questions** after each

narrative section (pp. 22, 229, 442) are at the recall level and assist students with their reading and understanding of the narratives. These questions also help students to prepare notes that summarize the material of the narratives. **Discussion** questions after each lab activity (pp. 170, 346) are at the recall and comprehension level. In addition, some items require application of knowledge and skills from the activity to real-life situations. Furthermore, **study questions** at the end of each chapter help students recall, apply and extend important content (pp. 48, 306). These consist of: *true or false* (recall); *sentence completion* (recall); multiple choice (recall and comprehension); *using your knowledge* (some application questions); and *investigations* (a few lab activities, home projects, debates, library projects, etc.).

Attractive Format:
The book employs an attractive format to appeal to students and encourage the use of the text. It features an uncluttered, single-column format. In addition, generous use is made of headings and subheadings as well as second colour.

Numerous Illustrations:
Over 400 line drawings and photographs provide visual reinforcement of the printed word.

Topic Development:
Recognizing that most students using the text have short attention spans, the book is broken up into units, each consisting of several short chapters. Each chapter contains some narrative and some activity sections.

New Terms:
The terms are printed in second colour and are clearly explained in context when first introduced. Many of the terms are phonetically pronounced (p. 151). The most important of these terms are then listed (with their meanings) at the end of the chapter (p. 156).

Chapter Openings:
Each chapter opens with a brief outline of content to enable students to see at a glance where the chapter is headed. These introductions consist of an overview of both the narrative and the activity sections (p. 263).

Chapter Conclusions:
Each chapter concludes with a short listing of the main ideas of the chapter and a glossary of the major terms of the chapter, including their pronunciations (when required) and meanings (p. 482).

End of Text:
The book concludes with a comprehensive index to help students use and find important physical science terms and information (p. 485).

Discovering Physical Science

The Nature of Science

CHAPTER 1
What is Science?

CHAPTER 2
Doing Science

CHAPTER 3
Building Models

What is science? Is it a collection of facts that scientists have discovered? Is it a collection of theories scientists have developed? Is it a method scientists use in their experiments? Or is it a combination of these things?

In this unit you will find out what science is. You will also learn what scientists do. But, most important, you will find out that you can do science, too. You will also find out that science is very useful to you.

Fig. 1-0 What is science? Are these people doing science?

1 What is Science?

1.1 Why Study Science?
1.2 The Scientific Method

This chapter describes what science is. It also explains why science is important to all of us. Most important, it outlines the scientific method. You will be using this method in this course.

1.1 Why Study Science?

What is science? What good is science to me? Why should I study science? You are probably asking yourself questions like these as you begin this course. Let's answer them before we go further.

What is Science?

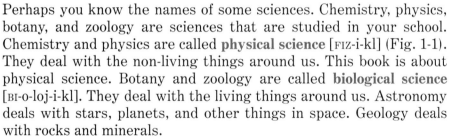

Perhaps you know the names of some sciences. Chemistry, physics, botany, and zoology are sciences that are studied in your school. Chemistry and physics are called **physical science** [FIZ-i-kl] (Fig. 1-1). They deal with the non-living things around us. This book is about physical science. Botany and zoology are called **biological science** [BI-o-loj-i-kl]. They deal with the living things around us. Astronomy deals with stars, planets, and other things in space. Geology deals with rocks and minerals.

Every science has two main parts. First, it has a body of knowledge. For example, certain facts, laws, and other information make up the science of physics. Second, each science has a method for discovering and using knowledge. This method is much the same for all sciences. Therefore it is called the scientific method.

This course is about physical science. Therefore you will be learning facts, laws, and other information about physics and chemistry. More important, you will also learn how to find out this information and how to use it.

Why Study Science?

Occasions often arise when you can use science. And usually you will be glad that you did. Here are two examples.

Example 1
A cigarette ad says that a certain brand has a "clean, fresh taste". It also shows healthy people smoking. But the package has a warning on it. "Smoking may be dangerous to your health" (Fig. 1-2). Is the warning true? Or is the ad true? Do all smokers stay healthy like the people in the ad?

A scientific way to answer the questions is as follows:

Fig. 1-1 Two sciences in your school

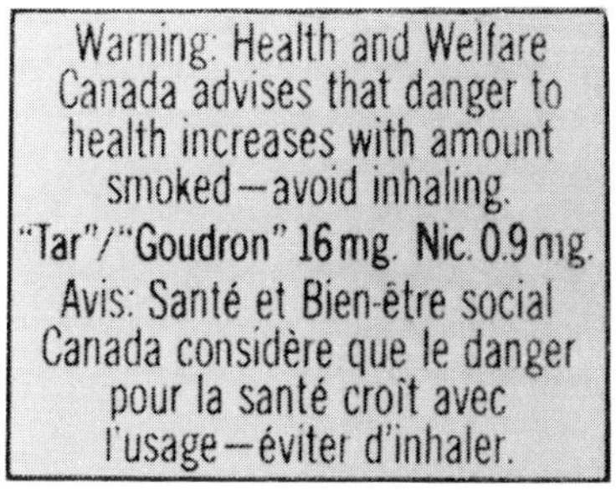

Warning: Health and Welfare Canada advises that danger to health increases with amount smoked—avoid inhaling. "Tar"/"Goudron" 16 mg. Nic. 0.9 mg.
Avis: Santé et Bien-être social Canada considère que le danger pour la santé croit avec l'usage—éviter d'inhaler.

Fig. 1-2 Is this warning true?

1. Find out what chemicals are in tobacco smoke.
2. Find out what those chemicals do to humans.
3. Think about the above two questions carefully.
4. Decide whether you should believe the warning or the ad.
5. Discuss the questions and your conclusion with other people.

Example 2

Most of us get a headache from time to time. And we may take pills to ease the pain. What brand of pills should we buy? There are so many brands in the drugstore. Some brands are expensive. Some are much cheaper. But are the cheap ones as good as the expensive ones?

You can often save money if you use the scientific method to help you decide. A scientist might do this by:

1. Reading the labels closely;
2. Making a list of the chemicals in each brand;
3. Writing down the recommended daily dose;
4. Figuring out what each daily dose will cost;
5. Reading the maker's claims carefully;
6. Deciding why some brands cost more than others;
7. Deciding if the extra cost of some brands will really give you "extra relief".

Table 1-1 shows the results of such a study. Four brands of headache tablets were studied. All contain ASA, a proven pain-reliever; two contain caffeine. Caffeine is also found in tea and coffee. A little caffeine "perks" you up. Too much makes you tense. In fact, it can give you a headache!

It is easy to see that brand A is a better buy than brand B. Both have the same amount of ASA; both have the same daily dose. But brand A is much cheaper.

Now, compare brands A and C. A daily dose of brand C costs more. What do you get for the extra money? A daily dose of brand A has 325 x 9 = 2925 mg of ASA. A daily dose of brand C has 454 x 6 = 2724 mg of ASA. So you pay more for brand C but get less ASA. You get some caffeine for the extra money, but do you really want it?

Make some more comparisons. Then decide what brand you would buy. Discuss your reasons with a group of your classmates.

Table 1-1 Which brand would you buy? Why?

Brand	Chemicals in one pill	Daily dose	Cost of a daily dose	Maker's claims
A	325 mg of ASA	9 pills	6¢	Fast relief of pain of headache...
B	325 mg of ASA	9 pills	13¢	Fast relief from pain of headache...
C	454 mg of ASA 28 mg of caffeine	6 pills	9¢	Fast pain relief of headache...
D	500 mg of ASA	4 pills	13¢	Strong; relieves headaches..

Why Should You Study Science?

You can likely see now why you should study science. There are two reasons. First, you will learn important information about the world in which you live. Second, you will learn how to think scientifically. Learning these two things will help you make many decisions in your life.

Section Review

1. What does physical science deal with?
2. What sciences make up physical science?
3. What does biological science deal with?
4. What sciences make up biological science?
5. What are the two main parts of any science?
6. There are two reasons why you should study science. What are they?

1.2 The Scientific Method

The last section mentioned the scientific method, but it did not say exactly what this method is. You will be using the scientific method throughout this course. Let us see, then, what it is.

There is no one scientific method that all scientists must use in their studies. However, most scientists tend to follow six steps. We will use those six steps in this book. Hopefully, you will learn how to use them in your daily life. The steps are:
1. Recognizing a problem;
2. Collecting information on the problem;
3. Making a hypothesis;
4. Doing an experiment;
5. Observing and recording results;
6. Making a conclusion.

Now, let us take a look at each of these steps.

1. Recognizing a Problem

All scientific studies begin with the **recognition of a problem**. Someone must recognize or see that there is a problem to be solved. That is, someone must see something worth studying. Here is an example.

Some people will not use seatbelts in cars. They say that seatbelts do not save lives. Rather, seatbelts may trap you in your car if you have an accident. Other people always use seatbelts (Fig. 1-3). They say seatbelts save lives and reduce injuries. Who is right? When you ask a question like that, you have recognized a problem.

Usually *curiosity* helps us recognize a problem. Most of us wonder about the hows and whys of things that happen around us; scientists wonder too. They are always asking questions about things that happen in their experiments. That's their job.

Fig. 1-3 The problem: Do seatbelts save lives and help reduce injuries?

2. Collecting Information on the Problem

Suppose you wanted to find out if seatbelts save lives. What should you do? You should first see if anyone else has studied this problem. Perhaps you can find out in a library (Fig. 1-4). Also the government will have information on the problem. Therefore you should begin your study by **collecting information** on the problem.

Scientists begin their studies the same way. They collect all the related information they can. They get it from books, journals, and other sources. Then they can build on the work of others. If they did not do so, science would never make any progress. We would be finding out the same things over and over again.

You should also do some reading before you do experiments. Before you do any experiments in this book, read the related material.

Fig. 1-4 A library has information on most problems.

3. Making a Hypothesis

Sometimes the information you collect solves the problem. If this happens, then the scientific method goes no further. For example, many studies have been done on seatbelts. The studies show that seatbelts save lives and prevent many injuries. If you accept this conclusion, the problem is solved. If you do not accept it, you should go on with the scientific method. You should do experiments to prove whether or not seatbelts are useful.

Before doing an experiment, a scientist usually makes a **hypothesis** [hy-POTH-uh-sis]. A hypothesis is a prediction of the results of the experiment. Sometimes we call it an educated guess. What good is a hypothesis? It gives the scientist something to work toward. It directs the scientist's work. Thus a scientist studying seatbelts might make this hypothesis: If seatbelts are used, the chance of being killed or hurt in a car accident will be lowered.

This hypothesis makes sense to most people, but that does not make it true. No hypothesis is true until experiments show it to be so.

4. Doing the Experiment

The next step, then, is to make up and do an **experiment** [ex-PER-i-ment] to test this hypothesis. To study the seatbelt problem, one could put dummies in cars. Some dummies would have seatbelts on and some would not. The cars would then be run into brick walls. The dummies would be checked for "injuries" after the collisions (Fig. 1-5).

An experiment must always be done under **controlled conditions**. All factors that can affect the results (**variables**) must be controlled (kept the same) except the one being studied. For example, the results will not mean much if the dummies with seatbelts are in small cars and those without seatbelts are in large cars. The size of the car is one **variable** [VAI-ri-a-bul] that must be controlled. All cars should

Fig. 1-5 A study of seatbelts and safety

Section 1.2   7

be the same size. Can you think of other variables that must be controlled?

5. Observing and Recording Results

A scientist always records all observations made during an experiment. This may be done in a table, graph, diagram, or written paragraph. Such information is called the **data** [DAT-tuh] of the experiments. Be sure that you record all your observations as you do experiments (Fig. 1-6). We will tell you how to do that later. Table 1-2 shows a data table for a seatbelt experiment.

Table 1-2 Seatbelts and Injuries

Car	Seatbelt On	Seatbelt Off	Mass of car	Speed of car	Effect on dummy
Car A					
Car B					
Car C					
Car D					

Fig. 1-6 It is important that you always record your data as you do the experiment.

You must always be careful not to confuse observations and conclusions. For example, if you say "seatbelts save lives" you have made a conclusion. If you simply state what happened in the experiment, you are giving your observations. *An observation is what you see, smell, feel, taste, or hear.* Therefore, when making observations, stick to what you see, smell, feel, taste, or hear. Do not draw conclusions until all the evidence is in.

Observations may be both qualitative and quantitative. **Qualitative observations** [KWAL-i-tat-iv ob-sur-VAY-shuns] are observations which describe such factors as colour, odour, shape, and taste. **Quantitative observations** [KWAN-ti-tat-iv] are observations which deal with measurements such as length, mass, and temperature. That is, they involve a number. Both types of observations are made by scientists. They usually make qualitative observations first, since they are the easiest to make. Quantitative observations, however, are often more useful. You will make both types in your studies.

6. Making a Conclusion

Once the observations have been made, the scientist draws them together and makes a **conclusion** [con-CLUE-zhun]. If the hypothesis was correct, the conclusion will be the same as the hypothesis. For example, the data from the seatbelt experiment would lead to this conclusion: The use of seatbelts reduces the chance of being hurt or killed in an accident.

Usually more experiments are done to check the conclusion. If many experiments support the conclusion, it may be called a **theory** [THEE-oh-ree].

If the experiments do not support the conclusion, the original hypothesis must be changed. It may even have to be discarded. Then the whole process must start over again.

Summary of Scientific Method

Figure 1-7 shows the six steps in the scientific method. The scientific method begins with the **recognition of a problem**. This usually happens when a person's curiosity is aroused. The next step is the **collection of information**. This is done by reading and talking to other people. This is followed by the formation of a **hypothesis**, or prediction. You try to guess the answer to the problem using the information you have collected. Next comes an **experiment** to test the hypothesis. The experiment must have controls. Then **observations** can be made and recorded. They may lead to a **conclusion** that solves the problem.

Section Review

1. a) How do all scientific studies begin?
   b) How does curiosity fit into the scientific method?
2. Why should you collect information on the problem before you start doing experiments?
3. a) What is a hypothesis?
   b) Of what use is a hypothesis to us?
   c) What is the difference between a hypothesis and a theory?
4. a) What is a variable?
   b) What does "controlled conditions" mean?
5. a) What are data?
   b) What is an observation?
   c) How does a conclusion differ from an observation?
6. List, in order, the six main steps in the scientific method.

Main Ideas

1. Physical science is made up of chemistry and physics.
2. If you study science you will learn useful things including how to think scientifically.
3. The scientific method is used for solving problems. It has six main steps: recognition of a problem, collection of information, making of a hypothesis, doing experiments, making observations, and drawing conclusions.

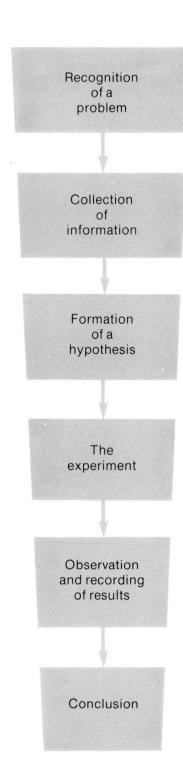

Fig. 1-7 The six main steps in the scientific method

Glossary

conclusion	con-CLUE-zhun	a general statement made by drawing observations together
data	DAT-tuh	the facts, figures, and other information gathered during experiments
experiment	ex-PER-i-ment	the action taken to test a hypothesis or gather data
hypothesis	hy-POTH-uh-sis	a prediction or educated guess that helps direct the course of an experiment
observation	ob-sur-VAY-shun	what a person discovers with one of the five senses during an experiment
theory	THEE-oh-ree	an explanation based on data from many experiments
variable	VAI-ri-a-bul	a factor that can affect the results of an experiment

Study Questions

A. True or False

Decide whether each of the following sentences is true or false. If the sentence is false, rewrite it to make it true. (Do not write in this book.)
1. Physical science deals with non-living livings.
2. A variable is any factor that can affect the results of an experiment.
3. Scientists usually begin the study of a problem by doing an experiment.
4. Physical science is made up of physics and chemistry.
5. If you measure the length of an object, you have made a qualitative observation.

B. Completion

Complete each of the following sentences with a word or phrase which will make the sentence correct. (Do not write in this book.)
1. After recognizing a problem you should ▉▉▉▉ on the problem before you begin to study it.
2. A hypothesis is a ▉▉▉▉ of the results of an experiment.
3. An experiment must always be done under ▉▉▉▉ conditions.
4. The ▉▉▉▉ of an experiment may be recorded in a table, graph, diagram, or written paragraph.
5. If the hypothesis is correct, the ▉▉▉▉ of the experiment will be the same as the hypothesis.

C. Multiple Choice

Each of the following statements or questions is followed by four responses. Choose the correct response in each case. (Do not write in this book.)

1. When ice is heated, it melts. This statement is best called
    a) an observation
    b) a guess
    c) a conclusion
    d) a hypothesis
2. When water is heated, it bubbles and gives off a colourless vapour. This statement is best called
    a) an observation
    b) a guess
    c) a conclusion
    d) a hypothesis
3. Which of the following is the most common order for the scientific method?
    a) collection of information, recognition of a problem, hypothesis, experiment, observations, conclusion
    b) collection of information, hypothesis, recognition of a problem, experiment, observations, conclusion
    c) recognition of a problem, collection of information, experiment, hypothesis, observations, conclusion
    d) recognition of a problem, collection of information, hypothesis, experiment, observations, conclusion
4. A scientist decided to study the effect of light on a new kind of plastic. Before doing any experiments she went to the library. She read about the topic and discussed it with other scientists. She then made this statement: "If light is shone on the plastic for several days, the plastic will crack". This statement is best called
    a) an observation
    b) a guess
    c) a conclusion
    d) a hypothesis
5. A toothpaste commercial states that "regular use of Brand X with fluoride reduces cavities by 28%." Assume that the statement is proven to be true. Which of the following best describes the statement?
    a) It is a conclusion based on controlled experiments.
    b) It is an observation based on many experiments.
    c) It is a hypothesis that will direct further experiments.
    d) It is a conclusion but is not based on controlled experiments.

D. Using Your Knowledge

1. A can of furniture polish has this label on it: "PICKS UP SIX TIMES MORE DUST". Think about the label like a scientist would. Then write down three questions you would like to ask the maker of the polish.
2. Are big cars safer than small cars? To find out, scientists put dummies in both sizes of cars. Then they ran the cars into brick walls. They observed that three times as many dummies were "hurt or killed" in the small cars. Could you conclude that big cars are safer? Explain your answer.
3. Suppose you are shopping for a box of cornflakes. You want the

best value for your money. What three variables should you consider before making your selection from the shelves?

E. Investigations

1. Collect at least ten cigarette ads from newspapers and magazines. What methods do the ads use to sell cigarettes? Make a list of problems that you recognize in these ads.
2. Get a copy of the "Fuel Consumption Guide" from Transport Canada. Study the data carefully. What do you observe about the ten makes of cars with the best ratings? What do you conclude from your observations?
3. Examine the labels of all the bottles of pain reliever pills in your home. Write the names of the brands in a column. Write the ingredients opposite each name. Visit the library, if necessary, and find out what each ingredient does. Calculate the cost of a daily dose of each brand. Which brand is the best buy? Why?

2 Doing Science

2.1 Activity: Making Observations
2.2 Activity: Making a Hypothesis
2.3 Activity: Testing a Hypothesis
2.4 Activity: Using the Scientific Method

In Chapter 1 you read about the scientific method. In this chapter you will use the scientific method in some simple activities.

2.1 ACTIVITY Making Observations

Observing is one of the most important steps in the scientific method. All of us need practice in observing. If we don't have practice, we often miss small but important things.

This activity gives you practice in observing. You will study a burning candle and write down your observations. This sounds simple, doesn't it? Well, let's see how good you are. A trained observer can make over 50 observations. How many can you make?

Points to Remember

Be sure to do these things as you observe:
1. Record each observation in your notebook as soon as you make it.
2. Record all observations, even if some do not seem important.
3. Make both qualitative and quantitative observations (see page 32).
4. Be careful not to confuse observations and conclusions.

Problem

How many observations can I make on a burning candle?

Materials

candle   matches   ruler

Procedure

a. Look closely at an unlit candle. Record your observations in a list. Be sure to number them.
b. Light the candle. Look at it closely again. Record your observations in your list. Keep trying for at least 20 min (Fig. 2-1).

Fig. 2-1 How many observations can you make?

Discussion

1. **a)** How many observations did you make?
   b) How many of these were qualitative? quantitative?
2. What could you have done to get more observations?
3. What is the difference between an observation and a conclusion? Make two conclusions from your observations.

2.2 ACTIVITY Making a Hypothesis

As you have already learned, we usually make a hypothesis before doing an experiment. A hypothesis is a prediction of the results of the experiment. Some people call it an educated guess. We make a hypothesis because it gives us something to work toward. It directs our work.

In this activity you will study a pendulum (Fig. 2-2). A pendulum is any device that swings back and forth in a regular way. A playground swing is an example. The one you will use is just a rubber stopper ("bob") on a string.

You are to make up a hypothesis that will help you solve this problem: **What variables (factors) affect how fast a pendulum swings back and forth?**

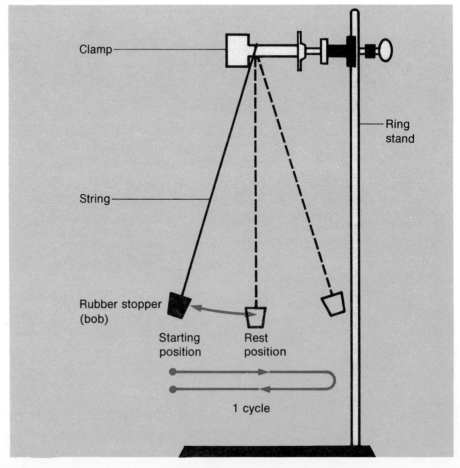

Fig. 2-2 A simple pendulum

Terms

You need to know three terms before you can do this activity and the next one.
1. *The **displacement** [dis-PLACE-ment] is the distance the bob is pulled to one side.*
2. *A **cycle** [SY-kel] is the motion of the bob from the starting point to the other extreme and back again.*
3. *The **period** is the time required for one cycle.*

Problem

What variables affect how fast a pendulum swings back and forth?

Materials

string   rubber stopper (bob)   ring stand   clamp

Procedure

a. Set up a pendulum as shown in Figure 2-2. Start it swinging. Study its movement closely.
b. Copy Table 2-1 into your notebook.
c. Ask yourself this question: What variables affect the period? That is, what factors affect how fast the pendulum swings back and forth? Write the variables in the left column of your table. You should be able to think of three variables.
d. Now ask yourself this question: How will the first variable affect the period? That is, will a certain change in the variable speed up or slow down the pendulum? Write your answer in your table.
e. Repeat step (d) for each of the other variables.

Table 2-1   Variables affecting period

Variable	Hypothesis (How does variable affect period?)

Discussion

1. The things you wrote in Table 2-1 are called hypotheses. Why?
2. Are your hypotheses true? How do you know?
3. What is the purpose in making hypotheses like these?

2.3   ACTIVITY   Testing a Hypothesis

In this activity you will test the hypotheses you made in the last activity. That is, you will do experiments to see if the hypotheses are true.

You likely decided that the distance the bob was pulled to one side, or the displacement, is one variable that affects the period. Perhaps you made a hypothesis like this: "The further out you pull the bob, the longer it will take to make one cycle." That is, the greater the displacement, the greater the period. That seems to make sense. But is it true? The only way to find out is to do an experiment.

In Procedure A of this activity we will help you do that experiment. Then in Procedure B you will test the other hypotheses on your own.

Problem

How do displacement and the other two variables affect the period?

Materials

string
rubber stopper (bob)
ring stand
clamp
ruler
watch or clock with second hand

Procedure A   Effect of Displacement on the Period

a. Set up a pendulum as shown in Figure 2-2. Make the length of the pendulum about 70 cm. The length should be measured from where the string is tied to the clamp to the middle of the bob.
b. Copy Table 2-2 into your notebook.
c. Pull the bob to one side a distance of 5 cm. This is the displacement. Let the bob go and time the bob for 5 cycles. Now find the period. (It is the time needed to complete 1 cycle.) Do this by dividing the total time for 5 cycles by 5. Record the time in seconds in your table. Here is an example. Suppose 9 s were needed for 5 cycles. Then the period is $\frac{9 \text{ s}}{5} = 1.8$ s. ("s" is the symbol for "second".)
d. Repeat step (c) 2 more times. Then average your 3 answers. This gives the average period for a displacement of 5 cm.
e. Repeat steps (c) and (d) using displacements of 10 cm and 20 cm. Record all your results in your table.

Procedure B   Effects of Other Variables on the Period

a. Write the name of another variable that might affect the period. (Look back to your notes on Activity 2.2.)
b. Write a hypothesis that tells how that variable might affect the period. (Look back to your notes on Activity 2.2.)
c. Make up a procedure for testing the hypothesis. Write it in your notebook. Be sure to include a data table for the results.
d. Do the experiment to test your hypothesis.
e. Repeat steps (a) to (d) for the other variable that might affect the period.

Table 2-2   Period of a Pendulum

Displacement	Trial	Time for 5 cycles (s)	Period (s)	Average period for 3 trials
5 cm	1			
5 cm	2			
5 cm	3			
10 cm	1			
10 cm	2			
10 cm	3			
20 cm	1			
20 cm	2			
20 cm	3			

Discussion A

1. Draw a conclusion from your results. It should say how the displacement affects the period.
2. Is your conclusion the same as your hypothesis? That is, was your hypothesis correct?
3. Do not worry if your hypothesis was wrong. Even scientists often have a wrong hypothesis. If they could always make a correct hypothesis, they would never have to do any experiments. A wrong hypothesis is still useful to us. In what way?
4. The other two variables were controlled while you studied the effect of displacement. What were these variables?
5. Why was each displacement tried three times?

Discussion B

1. Make up 2 conclusions from your results for Procedure B.
2. Are these conclusions the same as your hypotheses?
3. Why must variables be controlled in an experiment? Explain your answer by referring to this pendulum experiment.

2.4   ACTIVITY Using the Scientific Method

In this activity you will use the scientific method. You will try to solve a problem with it. Here is the problem:

Cobalt chloride crystals are pink. Something in the crystals helps make them pink. You are to find out what that is.

Problem

What makes cobalt chloride crystals pink?

Materials

Bunsen burner cobalt chloride crystals
clean dry test tube clamp
CAUTION: Wear safety goggles during this activity.

Procedure

a. Put a few cobalt chloride crystals in the test tube.
b. Describe the crystals (shape, colour, size, texture). Put your answer in your notebook.
c. Light the Bunsen burner. Your teacher will tell you how.
d. Heat the crystals gently with a low flame. Slant the test tube as shown in Figure 2-3. Heat only the bottom centimetre of the test tube. Record your observations.
e. Now heat the cobalt chloride strongly. Continue until no more changes occur. Then heat the whole test tube until no more changes occur. Write your observations in your notebook.

Fig. 2-3 Heat the cobalt chloride crystals like this.

Discussion

1. What do you think comes out of the crystals when they are heated?
2. What do you think helps make the crystals pink? Write your answer in sentence form. That is your hypothesis.
3. What can you do to test your hypothesis? Write this in your notebook. Now try it. This is your experiment.
4. What did you observe in your experiment? Record your observations.
5. Write a conclusion based on your observations.

Main Ideas

1. Careful observations must be made in all experiments.
2. Observations are what you see, smell, hear, feel, and taste.
3. Conclusions can often be made from observations.
4. A hypothesis helps direct your work in an experiment.
5. Variables must be controlled in an experiment.

Glossary

cycle	sy-kel	the motion of a pendulum from the starting point to the other extreme and back again

displacement	dis-PLACE-ment	the distance the bob of a pendulum is pulled to one side
period		the time required for one cycle

Study Questions

A. True or False

Decide whether each of the following sentences is true or false. If the sentence is false, rewrite it to make it true. (Do not write in this book.)

1. "The candle was 2 cm shorter after heating" is an observation.
2. A hypothesis helps direct your work in an experiment.
3. A hypothesis must always be correct.
4. Displacement is a variable that affects the period of a pendulum.
5. Cobalt chloride crystals contain water.

B. Completion

Complete each of the following sentences with a word or phrase which will make the sentence correct. (Do not write in this book.)

1. The colour of a candle flame is a ▓▓▓▓ observation.
2. The length of a candle is a ▓▓▓▓ observation.
3. Displacement and ▓▓▓▓ must be controlled when you are studying the effect of length on the period of a pendulum.
4. Sometimes many observations are needed before you can make a ▓▓▓▓ .
5. A substance can contain ▓▓▓▓ yet not appear wet.

C. Multiple Choice

Each of the following statements or questions is followed by four responses. Choose the correct response in each case. (Do not write in this book.)

1. Copper sulfate crystals, when heated, changed from a blue colour to a white colour. This statement is best called
    a) a conclusion
    b) an observation
    c) a hypothesis
    d) a prediction
2. A student measured a candle. He then wrote in his book: "The candle is 12 cm long." This statement is best called
    a) a hypothesis
    b) a conclusion
    c) a quantitative observation
    d) a qualitative observation
3. Cobalt chloride crystals turn blue when heated. This statement is best called
    a) a hypothesis
    b) a conclusion
    c) a quantitative observation
    d) a qualitative observation
4. Two variables affect the period of a pendulum. They are
    a) mass and length
    c) displacement and length

    b) displacement and mass   **d)** mass and the material of the bob
5. Water helps to give cobalt chloride crystals
    a) their shape and colour   **c)** their colour only
    b) their shape only   **d)** their size only

D. Using Your Knowledge

1. The pendulum of a grandfather clock is made of brass. Brass expands (gets longer) when it is heated. The clock loses time in warm weather. Why?
2. A scientist left a bottle of dry crystals on a window ledge. The window faced south. A few days later the crystals were gone. In their place was a wet solid. What happened?
3. Suppose you are trying to find out why cars rust so quickly in the winter. What hypothesis would you use to direct your work?

E. Investigations

1. What part of a candle actually burns? Is it the wick, the wax, or something else? Try to find out.
   Hint 1. Try burning a piece of wax.
   Hint 2. Try burning a piece of string (wick).
   Hint 3. Blow out a candle flame. Then quickly hold a lit match about 1 cm above the wick (Fig. 2-4).
2. Get some copper sulfate crystals (bluestone) from your teacher. Find out what gives them their blue colour.
3. Make up and try an experiment that answers this question: Will two bricks tied together fall faster than one brick? Include the following in your plan: a hypothesis, the procedure (include controlled conditions), observations, conclusion.
4. Make a candle in which the wick is a piece of woven wire.

Fig. 2-4 What is it that burns?

3 Building Models

3.1 Direct and Indirect Observations
3.2 Black Boxes and Models
3.3 Activity: Building a Model for a Black Box

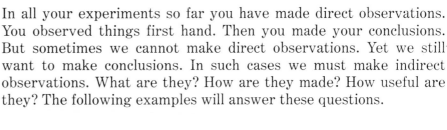

Fig. 3-1 The mechanic says you need spark plugs. But he never looked at the old ones. How good is his conclusion?

We often have to draw conclusions without making direct observations. For example, suppose the engine of your family car is not running smoothly. You take the car to a garage. The mechanic listens to the engine (Fig. 3-1). Then he says, "You need new spark plugs." He made this conclusion without looking at the spark plugs. That is, he made it without direct observations.

In this chapter you will learn how to draw conclusions without direct observations.

3.1 Direct and Indirect Observations

In all your experiments so far you have made direct observations. You observed things first hand. Then you made your conclusions. But sometimes we cannot make direct observations. Yet we still want to make conclusions. In such cases we must make indirect observations. What are they? How are they made? How useful are they? The following examples will answer these questions.

Example 1 What is in This Can?

Fig. 3-2 Can you find out what is in a can without opening it?

Imagine you have been given a sealed can (Fig. 3-2). It has no label on it. And you are not allowed to open it. Can you find out what is in the can?

You might begin by looking at the size and shape of the can. From your observations you might conclude that the can contains soup, juice, vegetables, or fruit. You would probably shake the can. Suppose the contents slosh easily. Then you might conclude that the contents are mainly liquid. You could also take the can to a store. There you could compare its size and shape to that of cans on the shelves. You could even weigh the can. Perhaps its weight is the same as that of one of the cans on the shelves.

All these observations are indirect. With just indirect observations you will never know for sure what is in the can. You could only be sure of the contents by opening the can and making direct observations.

Example 2 What is the Moon Made Of?

We know much about the moon today. **Astronauts** [as-TRUH-nots] brought back samples of moon rocks and dust. Then scientists made direct observations on them. However, we knew much about the moon long before Neil Armstrong first walked on the moon. Scientists used equipment on earth to make indirect observations of the moon. These observations gave scientists a good idea of what the moon is made of. They could not be sure though until moon material was brought back to earth.

Direct or Indirect?

Do you know the difference now? Direct observations are made first hand. You study what is to be observed directly. Yet sometimes you cannot make direct observations. You have to make indirect observations. Indirect observations are usually not as useful as direct observations. You have not made them first hand. But they are often all that you can make. If that is so, make lots of them. The more you make, the better your conclusion will be.

Section Review

1. What is a direct observation?
2. What is an indirect observation?
3. Why do direct observations usually lead to the best conclusions?
4. Why are indirect observations often used?

3.2 Black Boxes and Models

Indirect observations are used to study black boxes. They are also used to build models. This section explains what black boxes and models are. The next two sections let you study a black box and build models.

Learn this section well. Then you will be better prepared for the activities that follow.

What is a Black Box?

Many things can only be studied by indirect observations. They cannot be observed directly. Therefore much remains unknown about them. Scientists call such things black boxes. The word "black" suggests the unknown. A **black box**, then, is any object that can be studied only by indirect observations. Here are three examples of black boxes.

Example 1 A Sealed Can Without a Label

The can in Section 3.1 is a black box if it is not opened. You can only learn about its contents by indirect observations. You can never be sure what is in it unless you open it. The more experiments you do, the more you will learn about the contents.

Example 2 A Television Set

A television set is a black box for most of us (Fig. 3-3). We have never seen inside it. We don't know exactly what is in the set. Therefore we don't know exactly how it works.

We do, however, know some things about the inside of a television. This is because we have done "experiments" on its outside. These experiments give us indirect observations. These observations, in turn, suggest things that might be happening inside. For example, if we turn the "Volume" control, the sound gets louder or softer. If we turn the "Tint" control, faces may turn green. And if we press the "On-Off" control, the set goes on or off.

We can only guess how these controls work, but the more we try them, the better our guesses become.

Example 3 A Car Engine

A car engine is a black box if you have never seen inside one. You don't know exactly what is in it. Therefore you don't know exactly how it works. But you can do "experiments" on the outside of the engine. These experiments will give you indirect observations. They, in turn, may suggest what is going on in the engine. For example, you could turn off the gasoline supply. The engine will stop. What does this tell you about the inside of the engine? You could also unhook a spark plug wire. The engine will run roughly. If you unhook two wires, it may stop. What does this tell you about the inside of the engine?

Again, you may be guessing, but the more experiments you do, the better your guesses become. You can only find out about black boxes by collecting indirect observations on them.

What is a Model?

You can never know for sure what is in a black box, but you can get some ideas about the contents by doing experiments. The experiments give indirect observations about the contents. Then you can make conclusions from those observations. That is, you can guess what the contents are like. The more experiments you do, the better your guess will be.

You must not make wild guesses. Instead, make scientific guesses. Use your indirect observations to form a mental picture of what is in the black box. Such a mental picture is called a **model**. A model is a mental picture that you form to explain your observations. Let us form models for two of the black boxes we just discussed.

Example 1 A Sealed Can Without a Label

Suppose you shake the can and it sloshes. You could conclude that the

Fig. 3-3 Is this television a black box for the person fixing it?

can contains water. If you do, you have formed a model for the contents. But you have not actually observed the liquid. Your observations are indirect. The liquid may not be water. You don't know for sure, do you? Scientists would say your model is not very complete.

How could you make the model more complete? Water freezes at 0°C. Therefore you could put the can in a refrigerator at 0°C. If the liquid freezes, you can be more sure that it is water. You still cannot be certain however. Some other liquids freeze at 0°C.

You will find out later that there are other experiments you can do. Each experiment tells you more about the liquid in the can. Each experiment makes your model better; but, without opening the can, you will never have a complete model. That is, you will never know for sure what is in the can.

Example 2   A Car Engine

You likely have in your head a model for a car engine. You know that the engine needs fuel. You know that it gives off heat. Thus you may have a mental picture, or model, of fuel burning in the engine. But you have never seen the flames, have you? You are just guessing. Your guess is a scientific one. You based it on observations.

Here is another model for a car engine. The car builder put little creatures from a distant planet in the engine (Fig. 3-4). These creatures eat fuel. If you feed them well, they pedal rapidly. This makes the car go. When they work hard, they get hot. As a result, the engine gets hot. They even sweat. You may have seen their sweat. It comes out of the tailpipes of cars. What do you think of this model? Can you prove that it is wrong without taking the engine apart? How?

Fig. 3-4 A model for a car engine

Scientists and Models

Scientists use models most of the time. They do experiments and then make up a model to explain the results. Next they make predictions (hypotheses) based on the model. Finally they do experiments to test the predictions. Often this results in a changed model or even a new one.

A good example of a model is a description of an **atom**. Many books have been written on atoms. These books tell us much about the structure of the atom. Yet no one has observed an atom directly. What we read is just a model for the structure of the atom. This model is a mental picture of the atom. It is not a real picture (Fig. 3-5). It is a description of the atom as scientists think it must be. They got their ideas by doing experiments on atoms. As they do more experiments, they learn more about atoms. Then the model is changed.

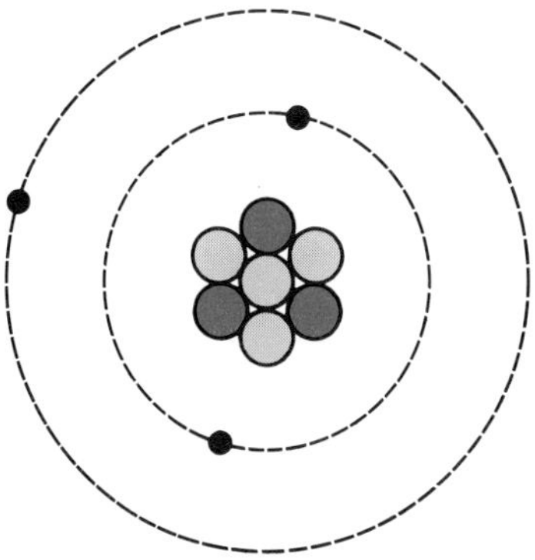

Fig. 3-5 This is one model for an atom. Scientists have better ones today. Why did the model change?

Section Review

1. What is a black box?
2. Explain why a sealed can without a label is a black box.
3. What is a model?
4. Can a model ever be complete? Why?
5. Why are pictures of atoms in books called models?

3.3   ACTIVITY   Building a Model for a Black Box

Your teacher will give you a box with an object in it. You must not open the box (Fig. 3-6). Therefore it is a black box. You can only learn about the object inside by indirect observations. Try some experiments on the box. Then make up a model for the contents.

Problem

To develop a model for the contents of a black box.

Materials

black box   probe

Notes

1. Do not open the box at any time.
2. Do not make the holes larger.

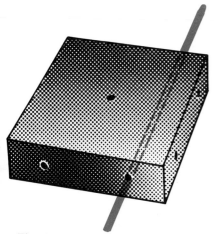

Fig. 3-6 A black box. Can you build a model for it?

Procedure

a. Make a table in your notes like Table 3-1.
b. Move the box in any way you wish. Tilt it. Move it back and forth.

Section 3.3   25

Turn it over. You will learn most by using carefully planned movements. Record everything you do, observe, and conclude in your table.
- **c.** Write down what you think is in the box. This is your model for the object in the box.
- **d.** Now try to improve your model by doing more experiments. Probe it from all sides. Try to find out the size and shape of the object. Change your model, if necessary.

Table 3-1 A Model for a Black Box

Procedure	Observation(s)	Conclusion(s)

Discussion

1. Why is the box and its contents a black box?
2. Write your best description of the object in the box. This is your final model.
3. Why is your description called a model?
4. List 3 or 4 things you don't know about the object in the box.

Main Ideas

1. There are two kinds of observations, direct and indirect.
2. Direct observations usually tell us the most. But sometimes we can only use indirect observations.
3. A black box is something that can be studied only by indirect observations.
4. A model is a mental picture that we form to explain observations.
5. We can build better models by doing more experiments.

Glossary

astronaut	as-TRUH-not	person travelling in outer space
atom		small particle of matter
black box		something that can be studied only by indirect observations
model		mental picture formed to explain observations

Study Questions

A. True or False

Decide whether each of the following sentences is true or false. If the sentence is false, rewrite it to make it true. (Do not write in this book.)
1. Direct observations are made first hand.
2. Indirect observations are usually more useful than direct observations.
3. It is possible to know everything about a black box.
4. A model is a mental picture formed to explain observations.
5. Indirect observations are often used to build models.

B. Completion

Complete each of the following sentences with a word or phrase which will make the sentence correct. (Do not write in this book.)
1. There are two kinds of observations, ▓▓▓ and ▓▓▓.
2. The more observations you make, the better your ▓▓▓ will be.
3. If you have never seen inside a television, it is a ▓▓▓ to you.
4. A description of an atom is best called a ▓▓▓ of an atom.
5. You can make a model better by doing more ▓▓▓.

C. Multiple Choice

Each of the following statements or questions is followed by four responses. Choose the correct response in each case. (Do not write in this book.)
1. No one has ever seen an atom. Yet many books contain descriptions of atoms. These descriptions are best called
   a) conclusions
   b) models
   c) black boxes
   d) indirect observations
2. A mechanic listened to the clacking sound of a car engine. Then he told the customer: "Your camshaft is worn." This statement is
   a) an indirect observation
   b) a direct observation
   c) a conclusion based on indirect observations
   d) a conclusion based on direct observations
3. Another mechanic tore down the engine and looked at the camshaft. Then she told the customer: "Your camshaft is worn." This statement is
   a) an indirect observation
   b) a direct observation
   c) a conclusion based on indirect observations
   d) a conclusion based on direct observations
4. Before astronauts visited the moon, astronomers believed that moon rocks contained iron. This belief is best described as
   a) an indirect observation

b) a direct observation
   c) a conclusion based on indirect observations
   d) a conclusion based on direct observations
5. Astronauts visited the moon and brought back moon rocks. Scientists studied them. Then they said that moon rocks contained iron. This statement is best described as
   a) an indirect observation
   b) a direct observation
   c) a conclusion based on indirect observations
   d) a conclusion based on direct observations

D. Using Your Knowledge

1. Name 3 things that are black boxes to you.
2. Is a motorbike engine a black box to you? Why?
3. A visitor to a laboratory asked a scientist this question: "Do you think you will get the right answer in your experiment?" The scientist replied: "No, I'll never get the right answer. But I am trying to get the best answer."

   Is the scientist making direct or indirect observations? How do you know?

E. Investigations

1. Visit a local car repair shop or your school shop (if the school has one). Study a car engine. Look at the inside, if possible. Talk to a mechanic. Then make up a model that tells how a car engine works.
2. A pop machine is a black box, unless you have seen inside it. You know that you must put the correct change in the slot. Then a can of pop drops down. Do you know how the machine works? Here is one model. A person inside the machine takes your money. He counts it. Then he releases a can of pop.

   Can you disprove this model without opening the machine? Perhaps you could try coaxing a free drink out of the person inside. The maker of the machine thought of that. He used a person who does not understand English. Perhaps the model is correct!

   Now, decide if you would like to defend the model or "shoot it down". If you want to "shoot it down", pair up with a student who wants to defend it. Say what you think is wrong with the model. Also, suggest experiments that could disprove the model. Your partner must defend the model against your sayings. May the best scientist win!

Measurement

CHAPTER 4
Measuring Length

CHAPTER 5
Measuring Area

CHAPTER 6
Measuring Volume

CHAPTER 7
Measuring Mass

Everyone has to be able to measure. We use a watch to measure time. We use a thermometer to measure temperature. We step on the bathroom scales to measure our weight. Someone measures us when we buy clothes.

Farmers measure the seed and the field before they sow crops. Cooks measure the ingredients of their mixes. Car makers measure the acceleration and fuel economy of their cars. Yes, everyone has to be able to measure. Can you?

In this unit you will learn how to measure many things. You will also learn how to measure accurately. Learn this unit well. It will help you in the rest of this course. It will also help you in your out-of-school life.

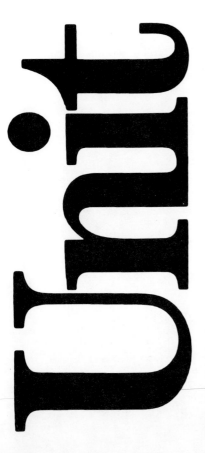

Fig. 4-0 How many things are measured by the gauges in this car?

4 Measuring Length

4.1 The Need for Measurement
4.2 History of the Metre
4.3 Metric Prefixes
4.4 Length Units and Symbols
4.5 Activity: Relationships among Length Units
4.6 Activity: Estimating and Measuring the Length of Lines
4.7 Activity: Drawing Lines of Given Length
4.8 Activity: Your Metric Measurements

Fig. 4-1 The Ford Escort is called a "world car". It is built and sold in many countries. All these countries must use the same measurement system.

Our country used the imperial system (foot, inch, mile, pound, etc.) until a few years ago. Now, we use the metric system. So do over 95% of the people on earth. This is because many countries found it necessary to use the same measurement system. It makes cooperation easier. For example, Ford produces what it calls a "world car" (Fig. 4-1). This car is built and sold in many countries. Also, parts built in one country are used by several other countries. Obviously these countries must use the same measurement system. Since the metric system is the easiest to use, most countries use it.

This chapter introduces the need for measurement. Then it explains the metric system. Finally, it applies the system to the measurement of length.

4.1 The Need for Measurement

In this course and in your everyday life you often have to describe things. Before you can describe something, you have to observe it. You have already made both qualitative and quantitative observations. You have also learned that quantitative observations are often more useful. Do you remember why?

This section reviews the terms qualitative and quantitative. Then it explains the importance of quantitative observations.

Qualitative Observations

When you make qualitative observations you may note such things as colour, odour, and taste. The following are qualitative observations: John has brown hair. A sewer has a bad odour. Water is colourless. Lemon juice tastes sour.

Qualitative observations also include a rough kind of measurement. "Tall" and "short" are rough measurements of a person's height. "Heavy" and "light" are rough measurements of a person's mass. "Hot" and "cold" are rough measurements of temperature.

Qualitative observations can be helpful. For example, after observing a rock, you could make this qualitative description: The rock is gray. It has no odour. It is heavy and smooth.

Such a description may help another person identify the rock. However, by themselves, qualitative observations are often not good enough. Many kinds of rocks are gray. Most rocks have no odour. And many rocks are heavy and smooth.

In addition, many people other than John have brown hair. Many things other than sewers have a bad odour. Many liquids are colourless, including poisonous acids. And many things besides lemon juice are sour. Further, how hot is "hot"? And how tall is "tall"?

In our day to day lives qualitative observations are often enough. But when we need to be exact, we must give quantitative observations.

Quantitative Observations

Quantitative observations involve actual measurements. Thus they are often more useful than qualitative observations. Suppose, for example, that you are going on a canoe trip. You know you will have to portage the canoe (Fig. 4-2). Would you rather have qualitative or quantitative facts when selecting a canoe?

Table 4-1 gives both qualitative and quantitative facts about a canoe. Which are the most useful facts about the canoe?

Table 4-1   Information about a Canoe

	Qualitative	Quantitative
Length	long	5.2 m
Width (at widest place)	fairly wide	1.0 m
Mass	light	28 kg
Load limit	greater than most canoes	450 kg

Fig. 4-2 Would qualitative or quantitative facts help you select the best canoe for a long canoe trip?

All things have characteristics that can be measured. Such characteristics are called **physical quantities**. Length, area, volume, mass, and temperatures are examples of physical quantities. That is, they are characteristics that can be measured. When a physical quantity is measured, the result is a **quantitative observation**. We also call it a **measurement**.

All measurements are made up of two parts, a **numeral** and a **unit**. For example, the length of a football is thirty centimetres (30 cm). The numeral is 30. The unit is centimetres. Each unit has a **symbol**. In this case **cm** is the symbol for **centimetres** [SEN-ti-me-ter].

The following are some examples of measurements:

The length of a pencil = twenty centimetres (20 cm)

The area of a floor = thirty square metres (30 m²)

The volume of gasoline purchased = forty litres (40 L)

The mass of a can of soup = three hundred grams (300 g)

Normal human body temperature = thirty-seven degrees Celsius (37°C)

Note that all these measurements have a **numeral** (20, 30, 40, 300, 37). Also, all have a **unit** (centimetre, square metre, litre, gram, degree Celsius). Note, too, that the units can be represented by **symbols** (cm, m², L, g, °C). You probably know the meanings of some of these units and their symbols. You will find out later in this unit what the others mean.

Measurements (quantitative observations) are usually more useful than qualitative observations. This is particularly so when you are identifying things. Many people have brown hair, but few people with brown hair are 181 cm tall and have a mass of 77.2 kg. Also, many liquids are colourless, but few besides water freeze at 0°C.

Section Review

1. What are qualitative observations?
2. What are quantitative observations?
3. Why are quantitative observations often more useful than qualitative observations?
4. What is a physical quantity?
5. What are the parts of a physical quantity?
6. What is the difference between a unit and a symbol?

4.2 History of the Metre

The metric system was developed toward the end of the eighteenth century. Scientists wanted a system that was easy to use. Therefore they developed a system with two important features:
1. It is based on standard units.
2. It uses multiples of ten.

To see why these features make the system easy to use, let us look at length measurement.

Primary Unit of Length

The **primary unit** of length in the metric system is the **metre**. "Primary" means first. Thus all other units of length are based on the metre. They are multiples or submultiples of it. For example, a kilometre is a multiple of a metre (one thousand metres). A centimetre is a submultiple of a metre (a hundredth). These multiples and submultiples are called **secondary units**.

The First Metre

In the 1790s France decided to make up a new measuring system. This system was called the metric system. It was based on a few primary units. The secondary units were related to the primary units by factors of 10.

The first primary unit to be selected was the metre. It was defined in 1799 as one ten-millionth of the distance from the equator to the

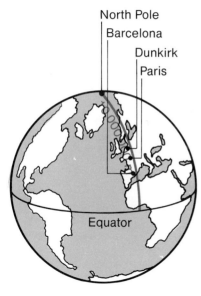

Fig. 4-3 The first metre. There are 10 000 000 m from the North Pole to the equator.

North Pole (Fig. 4-3). Two French scientists surveyed the meridian than runs through Paris — from Barcelona to Dunkirk — to find the length of a metre.

A Standard Metre

In 1889 the metre was re-defined. It was now based on the length of a standard platinum-iridium bar. A metre was defined as the distance between two scratches on the bar at 0°C. This bar was called the **International Prototype Metre**. It was kept near Paris. Copies were given to many other countries.

The Modernized Metric System (SI)

By the 1960s scientific work required a better standard. As a result, the metre was again re-defined. This definition is too complex to be included here. However, this new definition allows scientists to measure very accurately. They can now measure one metre to an accuracy of 1 part in 100 000 000!

This new metre is part of a modernized metric system. Many countries helped France modernize the system. As a result, the modernized metric system is called **Le Système International d'Unités**. It is called **SI** for short. Canada and most countries of the world now use this system.

Section Review

1. What does "primary" mean?
2. What is the primary unit of length in the metric system?
3. What is a secondary unit?
4. Name two secondary units of length.
5. What was the first definition of a metre?
6. Describe the International Prototype metre.
7. Why was the metre re-defined in the 1960s?
8. What does SI stand for?

4.3 Metric Prefixes

The metric system uses prefixes for the secondary units. Table 4-2 gives the prefixes you should know at this time. You should know these prefixes from memory.

The system is simple to use. **Kilo** means one thousand. If you put "kilo" in front of "metre" you get **kilometre** [KILL-o-me-ter]. A kilometre is one thousand metres. Or, in symbols, 1 km = 1000 m.

Milli means a thousandth, or 0.001. If you put "milli" in front of "metre", you get **millimetre** [MILL-i-me-ter]. A millimetre is a thousandth of a metre. Or, in symbols, 1 mm = 0.001 m.

The same prefixes are used for metric units other than length.

Thus one **kilogram** is one thousand grams (1 kg = 1000 g). One millilitre is a thousandth of a litre (1 mL = 0.001 L).

Table 4-2 Common Metric Prefixes

Prefix	Symbol	Meaning
mega	M	1 000 000 (million)
kilo	k	1000 (thousand)
hecto	h	100 (hundred)
deca	da	10 (ten)
THE UNIT (e.g. metre)		
deci	d	0.1 (tenth)
centi	c	0.01 (hundredth)
milli	m	0.001 (thousandth)
micro	μ	0.000 001 (millionth)

Some Rules

1. Symbols are not followed by periods (except at the end of a sentence).
2. Symbols are never pluralized. For example, "m" stands for both "metre" and "metres".
3. Commas are not used in numerals. Numerals with 4 or less digits are written this way: 24   280   5285. Numerals with over 4 digits are written this way: 12 045   270 584   3 877 544.

Section Review

1. Make a summary of metric prefixes in your notes. Include their symbols and meanings.
2. State two rules for using symbols.

4.4 Length Units and Symbols

The main advantage of the metric system is that it uses a decimal system. Each unit is related to the others by factors of 10. Study Table 4-3 to see how this decimal system works.

In this table the following symbols are used:

kilometre = km
hectometre = hm
decametre = dam
metre = m
decimetre = dm
centimetre = cm
millimetre = mm

Table 4-3 Measuring Length

1 km = 10 hm	1 km = 1000 m	1 m = 0.001 km
1 hm = 10 dam	1 hm = 100 m	1 m = 0.01 km
1 dam = 10 m OR	1 dam = 10 m OR	1 m = 0.1 dam
1 m = 10 dm	1 dm = 0.1 m	1 m = 10 dm
1 dm = 10 cm	1 cm = 0.01 m	1 m = 100 cm
1 cm = 10 mm	1 mm = 0.001 m	1 m = 1000 mm

Primary and Secondary Units

Note that there is really only one unit of length, the metre. The others are just multiples or submultiples of it (by factors of 10). Thus the metre is called the primary unit of length. The others are called secondary units of length.

Only the kilometre (km), metre (m), centimetre (cm), and millimetre (mm) are commonly used.

Picking the Correct Prefix

Suppose you are 1.6 m tall. You are also 160 cm tall, 1600 mm tall, and 0.0016 km tall. Perhaps you are wondering which is the best choice. Generally speaking, it does not matter. In this case, 0.0016 km and 1600 mm seem a little awkward. Therefore it may be best to give your height in centimetres or metres.

The distance from Halifax to Winnipeg is 4 000 000 000 mm (four thousand million millimetres). It is also 4 000 000 m, 4000 km, and 4 Mm. Clearly the units kilometre and megametre are the best choices for such a long distance.

A dime is about 1 mm thick. It is also 0.001 m thick and 0.000 001 km thick. Clearly the millimetre is the best choice for such a short distance.

Here is a rule for choosing the best prefix. *Choose a prefix so that the numeral lies between 0.1 and 1000.* Thus 12 km is better than 12 000 m. And 0.2 mm is better than 0.02 cm. It is not wrong to use prefixes that give numerals outside the range in the rule. If you do, however, you may end up with awkward numerals.

Section Review

1. What is the primary unit of length?
2. Name the three commonly used secondary units of length.
3. Convert the following measurements to metres: 1000 mm, 100 mm, 10 mm, 1000 cm, 100 cm, 10 cm, 10 km.
4. Convert the following measurements to centimetres: 1000 mm, 100 mm, 10 mm.

5. Convert the following measurements to millimetres: 1.0 cm, 10 cm, 100 cm.
6. Convert the following measurements to kilometres: 1000 m, 100 m.

4.5 ACTIVITY Relationships among Length Units

This activity has two main purposes. First, it will give you practice in metric conversions. Second, it will give you an idea of the sizes of the various units of length.

This investigation is for those of you who have not had much experience with the metric system. It is a good review for the rest of you.

Materials

metre stick   pencil and notebook

Procedure

a. Copy Table 4-4 into your notebook.
b. Complete the table. Refer to Table 4-3 for help if you need it.
c. Keep the metre stick in front of you at all times. Look at it as you do the calculations. Make sure you have a good idea of the size of a metre, centimetre, and millimetre. You need to know their sizes for the next activity.

Table 4-4 Converting Length Units

10 mm = ___ cm	6000 m = ___ km	50 mm = ___ cm
20 mm = ___ cm	8500 m = ___ km	1500 m = ___ km
55 mm = ___ cm	100 000 m = ___ km	150 m = ___ km
100 mm = ___ cm	1 mm = ___ cm	5000 mm = ___ cm
10 cm = ___ dm	10 cm = ___ m	5000 mm = ___ m
75 cm = ___ dm	100 m = ___ km	1.03 m = ___ mm
10 dm = ___ m	10 m = ___ km	0.52 m = ___ cm
80 dm = ___ m	1 km = ___ m	0.52 cm = ___ m
250 dm = ___ m	5 km = ___ m	358 m = ___ km
100 cm = ___ m	1 km = ___ cm	0.35 km = ___ m
1000 cm = ___ m	1 cm = ___ mm	1.5 km = ___ m
1500 cm = ___ m	10 cm = ___ mm	5800 mm = ___ m
10 000 cm = ___ m	500 cm = ___ mm	25 dm = ___ m
1000 m = ___ km	50 cm = ___ m	6.2 dm = ___ cm

4.6 ACTIVITY Estimating and Measuring the Length of Lines

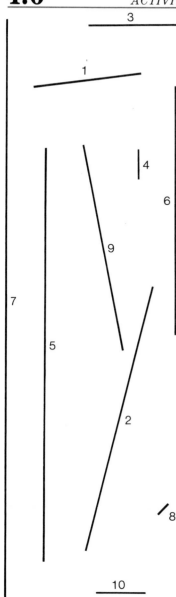

Fig. 4-4 How long are these lines?

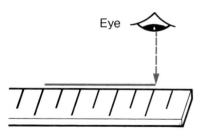

Fig. 4-5 How to avoid errors due to parallax

In this activity you are given some lines. You are to estimate how long they are. Then you are to measure them to see how good your estimate was.

Materials

ruler   pencil and notebook

Procedure

a. Copy Table 4-5 into your notebook.
b. Estimate in centimetres the length of each line in Figure 4-4. Record your estimate in your table.

c. Now measure each line in centimetres. Record your measurement in the table.
   NOTE: Put your eye *directly* above the end of the line and the mark on the ruler (Fig. 4-5). This will avoid error due to **parallax** [PAIR-eh-lax]. (Try moving your eye back and forth as you take the reading. See how the reading changes? This change is called parallax.)
d. Record the difference between each estimate and the measured length. Your teacher will look at your differences to see if you need more practice.

Table 4-5   Lengths of Some Lines

Line	Estimate (cm)	Length (cm)	Difference (cm)	Was estimate high or low?
1				
2				
3				
4				
5				
6				
7				
8				
9				
10				

Section 4.6   39

4.7 ACTIVITY Drawing Lines of Given Length

Try this activity to see how good you are at estimating length.

Materials

straight edge (not a ruler)   pencil and notebook   ruler

Procedure

a. Use the straight edge to draw lines you think are the following lengths: 1.0 cm, 2.5 cm, 15 mm, 130 mm, 7.0 cm, 25 mm, 0.8 cm.
b. Measure each line with a ruler. Put your results in a table like Table 4-6.
c. Record the difference between the given and the measured lengths. Again, your teacher will tell you if you need more practice.

Table 4-6 Drawing and Measuring Lines

Given length	Measured length (mm or cm)	Difference (mm or cm)	Was estimate high or low?
1.0 cm			
2.5 cm			
15 mm			
130 mm			
7.0 cm			
25 mm			
0.8 cm			

4.8 ACTIVITY Your Metric Measurements

It helps to know your measurements when you go shopping for clothes or shoes. By now you should be able to estimate them closely. Can you?

Materials

measuring tape   pencil and notebook   metre stick

Procedure

a. Copy Table 4-7 into your notebook.
b. Estimate each of the measurements in the table. Enter your estimate in your table.

c. Measure each factor with the tape or metre stick. Work with a partner when necessary (Fig. 4-6).

Table 4-7 My Metric Measurements

Factor	Estimate	Measurement
Height		
Hips to floor		
Vertical reach		
Neck size		
Sleeve length		
Wrist size		
Chest size		
Waist size		
Length of foot		
Length of pace		

Main Ideas

1. Quantitative observations involve measurements.
2. Quantitative observations are often more useful than qualitative observations.
3. A physical quantity has a numeral and a unit. A unit can be represented by a symbol.
4. The metric system uses standard units and is based on multiples of 10.
5. The primary unit of length is the metre.
6. The common secondary units of length are the kilometre, centimetre, and millimetre.
7. Most of the world uses the modernized metric system, SI.

Glossary

centimetre	SEN-tih-me-ter	a hundredth of a metre
kilometre	KILL-o-me-ter	one thousand metres
metre		primary unit of length
millimetre	MILL-i-me-ter	a thousandth of a metre
numeral		the number part of a physical quantity

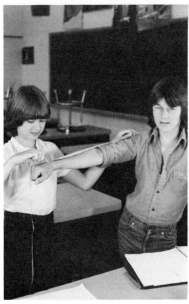

Fig. 4-6 Measuring height and sleeve length

parallax	PAIR-eh-lax	apparent change in the position of an object when the position of the eye is changed
physical quantity		a characteristic that can be measured; made up of a numeral and a unit
primary unit		unit on which others are based
secondary unit		a multiple or submultiple of a primary unit
SI		the modernized metric system
symbol		a letter or group of letters; represents a unit
unit		the word part of a physical quantity

Study Questions

A. True or False

Decide whether each of the following sentences is true or false. If the sentence is false, rewrite it to make it true. (Do not write in this book.)

1. A physical quantity is a characteristic that can be measured.
2. Qualitative observations are called measurements.
3. The kilometre is the primary unit for measuring length.
4. Kilo means one thousand.
5. Mega means one million.

B. Completion

Complete each of the following sentences with a word or phrase which will make the sentence correct. (Do not write in this book.)

1. A physical quantity is made·up of two parts, a ▬▬▬ and a ▬▬▬.
2. Secondary units are ▬▬▬ or ▬▬▬ of primary units.
3. The modernized metric system is called ▬▬▬ for short.
4. Metric prefixes are related to each other by factors of ▬▬▬.
5. The symbol for metre and metres is ▬▬▬.

C. Multiple Choice

Each of the following statements or questions is followed by four responses. Choose the correct response in each case. (Do not write in this book.)

1. Which one of the following has only qualitative observations about a dog?
   a) loud bark, brown hair, long legs, large size
   b) mass = 25 kg, height = 0.3 m, runs 30 km/h
   c) brown hair, mass = 25 kg, height = 0.3 m, green eyes
   d) brown hair, large size, runs 30 km/h, long legs
2. 358 m is equal to
   a) 0.358 km
   b) 3580 km
   c) 3.58 km
   d) 0.0358 km
3. 0.35 km is equal to
   a) 35 m
   b) 3.5 m
   c) 350 m
   d) 3500 m
4. The best estimate for the length of this line is
   a) 5 cm
   b) 10 cm
   c) 120 mm
   d) 10 mm
5. A dime is 0.001 m thick. How thick is it in mm?
   a) 0.000 001 mm
   b) 1 mm
   c) 0.1 mm
   d) 0.01 mm

D. Using Your Knowledge

1. List 5 qualitative observations about yourself.
2. List 5 quantitative observations about yourself.
3. Explain why the police like to have quantitative observations about criminals.
4. Convert the following measurements to metres: 570 cm, 650 mm, 1.5 km, 1.5 Mm.
5. Convert the following measurements to centimetres: 8.0 m, 23.5 m, 25 dm, 6.5 mm, 720 mm.
6. Convert the following measurements to millimetres: 30 cm, 8.5 cm, 175 cm, 3.0 m, 0.7 cm, 1500 μm.
7. Convert the following measurements to kilometres: 8000 m, 3500 m, 15 200 m, 2 Mm.
8. What unit is best for each of the following lengths?
   a) Distance from Regina to Montreal
   b) Thickness of a dime
   c) Distance from the earth to the moon
   d) Height of a table
   e) Length of a book
   f) Height of a horse
   g) Diameter of an L-P record
   h) Height of a door
   i) Height of a person
   j) Radius of a penny
   k) Length of a pencil
   l) Dimensions of a room
   m) Height of Niagara Falls
   n) Diameter of a human hair

E. Investigations

1. Can you estimate and measure long distances? To find out, estimate the following distances:
   a) Length of your classroom;

b) Length of the hall outside;
   c) Length of the gym;
   d) Length of a certain car.

   Now measure these lengths using a metre stick and tape measure (or trundle wheel, if available).
2. People have made two instruments for measuring very small distances. They are calipers and the micrometer. Select one of these. See if your teacher has one. Perhaps you can borrow one from the shop. Then find out how it works. To whom might it be useful?
3. Find the thickness of a page in this book. You could use a micrometer, but try to make up a method that uses only your ruler.

5 Measuring Area

5.1 Area Units
5.2 Activity: Measuring Area

In the last chapter you gained measuring skills as you measured length. In this chapter you use those skills to measure areas.

5.1 Area Units

What is Area?

Area is defined as the amount of surface. It is two-dimensional. That is, it is flat. It does not extend into the space above. A rectangle, square, triangle, and circle are two-dimensional; all have an area.

Examine Figure 5-1. The area of a rectangle is calculated using $A = l \times w$, where l = length and w = width. Thus the area of this rectangle is:

$$A = l \times w$$
$$= 4.0 \text{ cm} \times 3.0 \text{ cm}$$
$$= 12.0 \text{ cm}^2$$

The symbol cm² is read "square centimetre".

The area of a circle is calculated using $A = \pi r^2$, where π ("pi") = 3.14 and r = radius. Thus the area of this circle is:

$$A = \pi r^2$$
$$= 3.14 \times (2.0 \text{ cm})^2$$
$$= 3.14 \times 2.0 \text{ cm} \times 2.0 \text{ cm}$$
$$= 3.14 \times 4.0 \text{ cm}^2$$
$$= 12.6 \text{ cm}^2$$

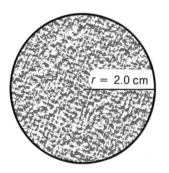

Fig. 5-1 Area of a rectangle and a circle

Area Units

The primary unit of area is the **square metre (m²)**. Several secondary units are derived from it. Commonly used multiple units are the square kilometre (km²) and the square hectometre (hm²). The square hectometre is usually called a **hectare** [HECK-tair]. It has the symbol **ha**. A hectare is equal to the area of about two football fields. The square centimetre (cm²) is the only submultiple unit that is commonly used.

Think for a moment about the sizes of the area units. Try to imagine a square metre as being a square that is 1 m long and 1 m wide. Next, imagine a square kilometre as being a square that is 1 km long and 1 km wide. Now imagine a hectare as being a square that is 1 hm (100 m) long and 1 hm (100 m) wide. Finally imagine a square centimetre as being a square that is 1 cm long and 1 cm wide.

Fig. 5-2 Each of these rectangles has an area of 1 cm².

Of course, areas are not always square. An area may be any shape. Figure 5-2 shows three rectangles with the same area of 1 cm² but different shapes. Can you imagine a circle and a triangle with an area of 1 cm²?

The commonly used units of area are listed in Table 5-1. Study this table carefully before you do the section review and the activity which follows.

Table 5-1 Common Area Units and Uses

Unit	Symbol	Uses
Square kilometre	km²	to measure areas of large surfaces such as land masses, oceans, lakes, and forests
Square hectometre or Hectare	hm² ha	to measure areas of smaller surfaces such as farms, parks, and playing fields
Square metre	m²	to measure areas of such things as floors, lawns, small gardens, and offices
Square centimetre	cm²	to measure areas of small surfaces like the sole of a shoe, a card, a leaf, an animal's footprint

Section Review

1. Define area.
2. Calculate the area of a rectangle with a length of 4.3 m and a width of 2.1 m.
3. Calculate the area of a circle with a radius of 3 cm.
4. Write the unit for each of these symbols: cm², mm², m², km².
5. What is the primary unit of area? Give its symbol and meaning.
6. Name the 3 common secondary units of area. Give their symbols and meanings.
7. State 3 uses for each of the 4 commonly used area units.

5.2 ACTIVITY Measuring Area

In this activity you will measure the areas of some surfaces provided by your teacher.

Materials

pencil and notebook
metre stick
tape measure (preferably 30 m)

Procedure

a. Copy Table 5-2 into your notebook. Your teacher may add or delete items. He/she will tell you what part of the lawn and playing field to measure.
b. Estimate the area of each surface listed in the table. Be sure to use suitable units. (See Table 5-1.)
c. Measure each of the surfaces listed in the table.
d. Calculate the area of each surface.

Table 5-2   Measuring Area

Area to be measured	Estimate of area	Linear measurements	Calculated area
Playing card	40 cm²	8.8 cm x 5.7 cm	50 cm²
Dollar bill			
Postage stamp			
Floor tile			
Desk top			
Section of chalkboard			
Classroom floor			
Part of school lawn			
Part of playing field			
Quarter			
45-r/min record			
L-P record			
Circular area in schoolyard			

Discussion

1. Compare your estimated and calculated areas. Are you better at estimating small or large areas? Why?

Main Ideas

1. The primary unit of area is the square metre (m²).
2. Common secondary units of area are the square kilometre (km²), hectare (ha), and square centimetre (cm²).

Glossary

area		the amount of surface (two-dimensional)
hectare	HECK-tair	a unit of area; one square hectometre
square metre (m²)		the primary unit of area

Study Questions

A. True or False

Decide whether each of the following sentences is true or false. If the sentence is false, rewrite it to make it true. (Do not write in this book.)
1. A rectangle, triangle, and circle all have area.
2. The primary unit of area is the square metre.
3. The area of a lawn would best be measured in hectares.

B. Completion

Complete each of the following sentences with a word or phrase which will make the sentence correct. (Do not write in this book.)
1. A rectangle with a length of 20 cm and a width of 25 cm has an area of ▒▒▒▒ .
2. A hectare is equal to one ▒▒▒▒ .

C. Multiple Choice

1. A field is 100 m long and 50 m wide. Its area in m^2 is
   a) $500 \ m^2$ b) $5000 \ m^2$ c) $50 \ m^2$ d) $2 \ m^2$
2. A rectangle has an area of $100 \ cm^2$ and a length of 25 cm. How wide is the rectangle?
   a) 4 cm   b) 2500 cm   c) 40 cm   d) 0.4 cm

D. Using Your Knowledge

1. Wall-to-wall broadloom costs $30.00/$m^2$ installed. What would it cost to carpet a room that is 6.0 m long and 4.5 m wide?
2. Sod costs $2.00/$m^2$. What would it cost to sod a lawn that is 9.5 m long and 7.0 m wide?

E. Investigations

1. Measure the football field or some other sports area of your schoolyard. Then calculate the area in hectares.
2. Estimate the total floor area of your home. Check your estimate by making the necessary measurements.

6 Measuring Volume

6.1 Volume Units
6.2 Activity: Finding the Volume of a Block by Measurement
6.3 Activity: Finding the Volume of a Liquid with a Graduated Cylinder
6.4 Activity: Finding the Volume of a Block by Displacement of Water
6.5 Activity: Finding the Volume of Irregular Solids by Displacement

In Chapter 4 you measured length. In Chapter 5 you went a step further and measured area. In this chapter you go still one step further and measure volume.

6.1 Volume Units

What is Volume?

Volume [VOL-yoom] *is defined as the amount of space an object occupies.* It is three-dimensional. Like area it covers an area on a surface. But it also extends into the space above. A brick, a baseball, and a glass of water all have volume. They take up space. You, too, have volume. You can change your volume by breathing in or out. When you breathe in your chest cavity expands. Therefore you take up more space or, you have a larger volume. When you breathe out your chest cavity shrinks. Therefore you take up less space or, you have a smaller volume.

Volume Units

Volume units are of two types. Some are **cubic units**. Others are **capacity units**. The common cubic units are the cubic metre (m^3), cubic decimetre (dm^3), and cubic centimetre (cm^3). The common capacity units are the kilolitre (kL), litre (L), and millilitre (mL). Table 6-1 shows how these units are related. The **cubic metre** is the primary unit of volume.

Table 6-1 Volume Units

A. Relationships Among Cubic Units
1 m^3 = 1000 dm^3
1 dm^3 = 1000 cm^3
B. Relationships Among Capacity Units
1 kL = 1000 L
1 L = 1000 mL

Don't memorize the relationships in Table 6-1. You can reason them out. For example, 1 dm = 10 cm. Therefore 1 dm^3 = 10 cm x 10 cm x 10 cm, or 1000 cm^3. Also, kilo means 1000. Thus 1 kL = 1000 L. Can you reason out the other two relationships in Table 6-1?

Perhaps you wonder why there are two systems for measuring volume. Table 6-2 shows that they are actually much alike. Generally cubic units are used for solids. Capacity units are used for things that pour. However, it is not wrong to speak of 50 mL or 50 cm^3 of water. Nor is it wrong to speak of 50 cm^3 or 50 mL of wood.

Table 6-2 Comparison of Cubic and Capacity Units

Cubic unit		Capacity unit
1 m^3	=	1 kL
1 dm^3	=	1 L
1 cm^3	=	1 mL

Study Figure 6-1 carefully. It will give you an idea of the sizes of the volume units. It will also make clear the relationships among them. Your teacher may have an actual cubic decimetre to show you.

Now, look at Table 6-3 to see when you would use each unit. Think of other uses for each unit.

Table 6-3 Common Volume Units and Uses

Unit	Symbol	Uses
Cubic metre	m^3	to measure large volumes of earth and gravel, the volume of a building, the volume of the hold of a ship
Kilolitre	kL	to measure the volume of water in a reservoir, the volume of gasoline in a tanker truck
Litre	L	to measure volumes of milk, gasoline, paint, ice cream; to measure capacities of pails, kettles, auto gas tanks, refrigerators, freezers
Cubic centimetre	cm^3	to measure volumes of small boxes and other small objects of regular shapes (cuboid, spherical, cylindrical, etc.)
Millilitre	mL	to measure volume of materials (usually fluid) that come in containers smaller than 1 L; for example, toothpaste, a glass of milk, soft drinks, hair shampoo, shaving lotion

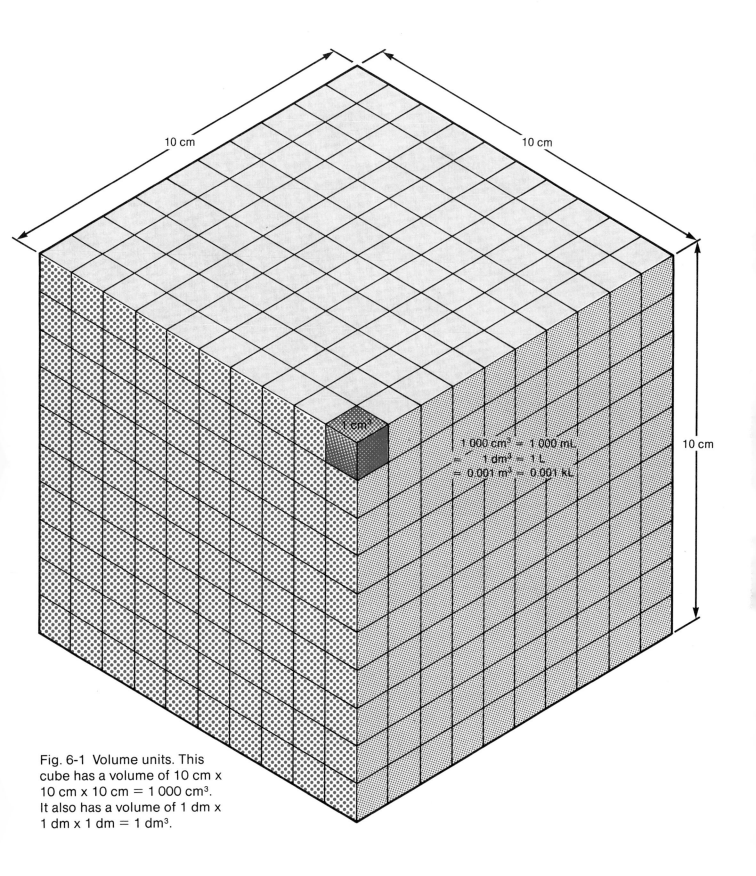

Fig. 6-1 Volume units. This cube has a volume of 10 cm x 10 cm x 10 cm = 1 000 cm³. It also has a volume of 1 dm x 1 dm x 1 dm = 1 dm³.

Section Review

1. Define volume.
2. Name the common cubic units of volume. Also, give their symbols.
3. Name the common capacity units of volume. Also, give their symbols.
4. What is the primary unit of volume?
5. What volume units would you use to measure each of the following: the volume of a milk truck; the volume of gasoline purchased for a car; the volume of milk purchased for your home; the volume of a match box; the volume of toothpaste in a tube?

6.2 ACTIVITY Finding the Volume of a Block by Measurement

The block you will use in this activity has a regular shape. All its sides are rectangles. Therefore you can find its volume by making some measurements and calculations. For example, the block in Figure 6-2 is a regular solid. Its volume is calculated using the formula $V = l \times w \times h$, where l = length, w = width, and h = height. Thus $V = 5.0$ cm x 3.0 cm x 4.0 cm = 60 cm^3.

In this activity you will estimate the volumes of four blocks. Then you will measure the blocks and calculate their volumes.

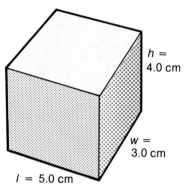

Fig. 6-2 A regular block. Note that each side is a rectangle.

Materials

4 blocks with different volumes   ruler

Procedure

a. Estimate the volume of each block in cubic centimetres (cm^3). Record your estimate in a table like Table 6-4.
b. Measure the length, width, and height of each block. Record the results in your table.
c. Calculate the volume of each block using the formula $V = l \times w \times h$.

Table 6-4 Volume of a Cuboid

Block	Estimated volume (cm^3)	Length (cm)	Width (cm)	Height (cm)	Calculated volume (cm^3)
1					
2					
3					
4					

Discussion

1. Compare your estimated and calculated volumes.
2. Calculate the volume, in cubic units, of a swimming pool that is 12 m long, 6 m wide, and 3 m deep. What is the capacity of this pool in kilolitres?

6.3 ACTIVITY Finding the Volume of a Liquid with a Graduated Cylinder

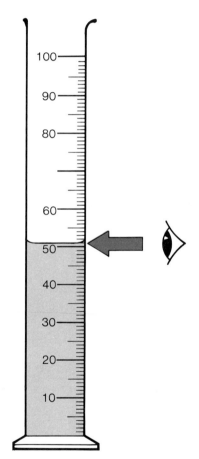

Fig. 6-3 A graduated cylinder containing a liquid. Note that the surface of the liquid is curved. The curved surface is called a *meniscus*. Always take your reading at the bottom of the curve. Make sure that your eye is directly opposite the curve. The correct reading here is 51.0 mL.

The volume of a liquid is usually measured with a **graduated cylinder**. Graduated cylinders are made in many sizes. The one shown in Figure 6-3 is a 100 mL graduated cylinder. A graduated cylinder is *graduated* or marked in regular intervals along its length. The one in Figure 6-3 is graduated in millilitres.

Most liquids have a curved surface when they are in a narrow container. This curved surface is called a **meniscus** [men-ISS-cuss]. For water and most other liquids the meniscus is lowest in the centre. *Always read the volume at the LOWEST level of the meniscus.* Make sure that your eye is directly opposite the curve. Then you will avoid errors due to parallax.

In this activity you will estimate the volume of water in several containers. Then you will pour the water from each container into a graduated cylinder and measure its volume.

Materials

graduated cylinders of several sizes
4 containers, each of a different size
funnel
water

Procedure

a. Estimate the capacity of container 1 in millilitres. Record your estimate in a table like Table 6-5.
b. Fill container 1 with water. Do not fill it so full that the water will spill when you move the container.
c. Pick a graduated cylinder large enough to hold the water from container 1.
d. Pour the water into the graduated cylinder. Use the funnel, if necessary, to prevent spillage.
e. Read the volume. Record your answer in the table.
f. Repeat steps (b) to (e) two times. Record your answers in the table. Then average the three values.
g. Repeat steps (a) to (f) for each of the other three containers.

Table 6-5 Volume of a Liquid

Container	Estimated capacity	Measured Capacity			Average
		Trial 1	Trial 2	Trial 3	
1					
2					
3					
4					

Discussion

1. a) What is a meniscus?
   b) Describe the meniscus of water in a graduated cylinder.
2. a) What was the purpose of measuring each volume three times?
   b) How accurately do you feel you can measure the volume of a liquid with each of the graduated cylinders? Express your answer this way: to the nearest 1 millilitre; to the nearest 0.5 millilitre; to the nearest 10 millilitre.
3. How well can you estimate the volume of a large container? a small container?

6.4   ACTIVITY   Finding the Volume of a Block by Displacement of Water

In Section 6.2 you found the volume of a block by measurement. You can also find the volume of a block by an indirect method. You can submerge the block in water. It will **displace** [dis-PLACE] or push away a volume of water equal to its own volume. The volume of the displaced water can be measured with a graduated cylinder.

In this activity you will find some volumes by displacement. You will use the same four blocks that you measured in Section 6.2.

Step 1

Cut top off

Materials

4 blocks of different volumes (from Section 6.2)
overflow container (overflow can or milk carton)
graduated cylinder (100 mL)
fine thread
pin
beaker or jar (150 mL or larger)
scissors, knife, or razor blade

Procedure A Making an Overflow Container

You can make an overflow container by following this procedure and

Cut 2 vertical slits 5 cm long, each 1.5 cm from the corner

Fold corner down to make a spout

Fig. 6-4 Procedure for making an overflow container

Fig. 6-5 An overflow container ready for use in finding the volume of a solid by displacement

Figure 6-4. (To be done if your school does not have overflow cans.)
a. Cut the top off a milk carton with scissors, a knife, or a razor blade.
b. In one corner cut two vertical slits. Each slit should be 5 cm long and 1.5 cm from the corner.
c. Fold the cut corner downward to form a spout.

Procedure B Using the Overflow Container

a. Fill the overflow container with water. Allow any excess water to overflow from the spout into the beaker. Wait until the water has stopped dripping. Discard the water that overflows. Return the beaker to its position under the spout (Fig. 6-5).
b. Tie a 50 cm length of thread to block 1.
c. Slowly lower block 1 into the water in the overflow container. If necessary, slowly push the block under the water with a pin. Keep it there until no more water runs out the spout. Do not drop the block into the water. Also, do not use your fingers to push it under. Why?
d. Pour the water from the beaker into a graduated cylinder. Record the volume of the water in a table like Table 6-6. This is the volume of the block.
e. Repeat steps (a) to (d) two times, still using block 1. Record your answers in the table. Then average your results.
f. Repeat steps (a) to (e) for each of the other three blocks.

Table 6-6 Volume of a Block by Displacement

Block	Volume of displaced water			Average	Volume of block
	Trial 1	Trial 2	Trial 3		
1					
2					
3					
4					

Discussion

1. What is meant by the term "displacement"?
2. Why can the displacement method be used to find the volume of a solid?
3. Compare the volumes you obtained for the four blocks in Section 6.2 with the volumes you obtained here.
4. You have measured the volume of a block by two methods. Which method do you feel is better? Why?
5. Are there any solids with which the displacement method cannot be used? Explain.
6. Are there any solids with which the measurement method cannot be used? Explain.

6.5 ACTIVITY Determining the Volumes of Irregular Solids by Displacement

Solids such as stones, nails, and sand are **irregular solids**. Their volumes cannot be found by measurement. They can, however, be found by the displacement method. In this activity, you will find the volumes of some irregular solids using the displacement method that you learned in Section 6.4. You will also measure the volumes of some very small irregular solids using a different form of the displacement method.

Materials

stone	graduated cylinder (100 mL)
bolt	overflow container (overflow can or milk carton)
nails	fine thread
pin	metal pellets (shot)
	beaker or jar (150 mL or larger)

Procedure A Finding the Volume of a Large Irregular Solid

Your teacher will give you a stone and, perhaps, some other large irregular solids. Find the volume of each of these solids with the displacement method. Make up your own procedure. Design your own data table. When you have finished the experiment, write a full report. Include purpose, procedure, results, and conclusions.

Procedure B Finding the Volume of a Small Irregular Solid

Sometimes the solid is small enough to fit inside a graduated cylinder. If so, you can find its volume by displacement very easily as follows. Try it using the bolt.

a. Estimate the volume of the bolt in cm^3. Record your estimate in a table like Table 6-7.
b. Fill the graduated cylinder about half full of water. Read the volume. Record your answer in the table.
c. Tie a 50 cm length of thread to the bolt.
d. Slowly lower the bolt into the water until it reaches the bottom of the graduated cylinder (Fig. 6-6). Do NOT drop the bolt into the cylinder.
e. Read and record the new level of the water.
f. Calculate the volume of water displaced by the bolt. This is the volume of the bolt.
g. Repeat steps (a) to (f) for any other small irregular solids that

your teacher gives to you. Some of these solids may float. If they do, push them under the water with a pin. Hold them there until you make the volume reading.

Table 6-7 Volume of an Irregular Solid

Solid	Estimated volume (cm³)	Original volume of water (mL)	Volume of water + solid (mL)	Volume of solid (cm³)
Bolt				

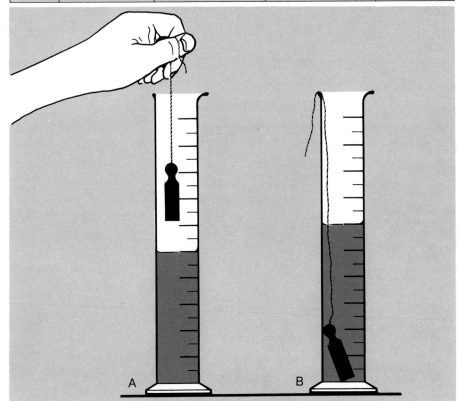

Fig. 6-6 Finding the volume of a small irregular solid by displacement

Procedure C Finding the Volume of Loosely Packed Solid Material

Your teacher will give you samples of metal pellets (shot) and nails. Use the graduated cylinder and the displacement method to find the total volume of each sample (not the volume of one pellet or one nail).

Design your own procedure for this activity. Before you begin, however, consider the following questions:

a. Can you find the volumes of the pellets and nails simply by pouring each into the graduated cylinder? (Try it if you are not sure.)

b. How can you find the volumes of the pellets and nails using the displacement method and *only* the graduated cylinder and water?

Section 6.5

Discussion A

Your report is all that is required as a writeup for Procedure A.

Discussion B

1. What is the relationship between a cubic centimetre (cm³) and a millilitre (mL)?
2. Why was the displacement method used to find the volumes of these irregular solids?
3. With small solids such as these, would the method in Procedure A (overflow container) or the method in Procedure B (graduated cylinder) give the better answer? Why?

Discussion C

1. Would the displacement method work for finding the volume of table sugar or table salt? Why?
2. Calculate the volume of one nail. Could you find its volume using only one nail and the displacement method (just as you did for the bolt)? Why?

Main Ideas

1. The primary unit of volume is the cubic metre.
2. Common cubic units of volume are the cubic metre (m³), cubic decimetre (dm³), and cubic centimetre (cm³).
3. Common capacity units of volume are the kilolitre (kL), litre (L), and millilitre (mL).
4. The cubic unit cm³ is equal to the capacity unit mL.
5. The volume of a regular solid can be found by measurement and displacement.
6. The volume of an irregular solid can be found by displacement only.

Glossary

displacement	dis-PLACE-ment	the pushing away of a liquid by a solid
irregular solid		a solid with no definite shape
litre (L)		a commonly used capacity volume unit; 1 L = 1 dm³
meniscus	men-ISS-cuss	curved surface of a liquid in a container
regular solid		a solid with a definite shape
volume	VOL-yoom	the amount of space an object takes up

Study Questions

A. True or False

Decide whether each of the following sentences is true or false. If the sentence is false, rewrite it to make it true. (Do not write in this book.)
1. The primary unit of volume is the litre.
2. Always read the volume of a liquid in a graduated cylinder at the highest part of the meniscus.
3. A solid can displace a volume of water equal to its own volume.
4. $1 \text{ cm}^3 = 1 \text{ mL}$.

B. Completion

Complete each of the following sentences with a word or phrase which will make the sentence correct. (Do not write in this book.)
1. The two types of volume units are ▒▒▒ and ▒▒▒.
2. A solid with a volume of 65 cm^3 can displace ▒▒▒ of water.
3. One litre (1 L) is equal to ▒▒▒ millilitres (mL).

C. Multiple Choice

Each of the following statements or questions is followed by four responses. Choose the correct response in each case. (Do not write in this book.)
1. A volume of 1 L is equal to how many mL?
    a) 10 mL b) 100 mL c) 10 000 mL d) 1000 mL
2. 150 cm^3 is equal to
    a) 0.15 L b) 1500 mL c) 150 dm^3 d) 0.15 m^3
3. The volume of a container is 575 mL. Its volume in cm^3 is
    a) 5750 cm^3 b) 57.5 cm^3 c) 575 cm^3 d) 5.75 cm^3
4. The gas tank of a truck has a capacity of 0.2 kL. How many litres of gasoline will it hold?
    a) 200 L b) 20 L c) 2000 L d) 0.002 L

D. Using Your Knowledge

1. Complete the following:
    1 m^3 = ▒▒▒ dm^3 1 dm^3 = ▒▒▒ L   775 mL = ▒▒▒ L
    1 dm^3 = ▒▒▒ cm^3 1 cm^3 = ▒▒▒ mL 2260 cm^3 = ▒▒▒ mL
    1 kL = ▒▒▒ L   500 mL = ▒▒▒ cm^3 2260 cm^3 = ▒▒▒ L
    1 L = ▒▒▒ mL   500 mL = ▒▒▒ L   35 L = ▒▒▒ mL
    1 m^3 = ▒▒▒ kL   775 mL = ▒▒▒ cm^3 35 L = ▒▒▒ cm^3
2. What unit of volume would you use to measure each of the following:
    a) Size of dice
    b) Size of a golf ball
    c) Capacity of a coffee mug
    d) Size of a loaf of bread
    e) Capacity of a pail
    f) Capacity of a refrigerator
    g) Size of a 4-drawer filing cabinet
    h) Capacity of a swimming pool
    i) Storage space in a farm silo
    j) Volume of water in a lake

3. Gasoline costs 50¢/L. What will it cost to fill a 55 L gas tank?
4. The fuel economy rating of a car is 7.0 L/100 km.
   a) What does this mean?
   b) How many litres will the car use travelling 500 km?
   c) What will it cost to drive this car 500 km? (Gasoline costs 50¢/L.)

E. Investigations

1. Find 10 items in your home that have their volumes stated in metric units. Look at grocery items, cosmetics, medicines, sprays, chemicals, and so on. For each item, give the name and the volume.
2. Get a cylindrical solid or a spherical solid from your teacher. Find its volume both by measurement and by displacement.
3. Car engines are often compared by their displacements. For example, a small 4-cylinder engine may have a displacement of 1.3 L. A large 8-cylinder engine may have a displacement of 6.5 L. Find out what displacement means in a car engine. Find the displacements of at least 10 car models. Find out why fuel consumption usually increases when the displacement increases.

7 Measuring Mass

7.1 Mass, Volume, and Weight
7.2 Common Mass Units
7.3 Activity: Measuring Mass

This chapter first explains the difference between mass, volume, and weight. Next it explains the common mass units. Finally, it gives you a chance to use your classroom balance to measure masses.

7.1 Mass, Volume, and Weight

Many people get the terms mass, volume, and weight confused. Let's make sure you know the meanings of these terms before you go on.

Mass and Volume

*The **mass** of an object is the amount of material in it. The **volume** of an object is the space it takes up.* An object's volume can change, but its mass stays constant. For example, when a thermometer is heated, the liquid in it rises. The *volume* of the liquid (the space it takes up) becomes larger. The *mass* (amount of material in the liquid) stays the same. No liquid was added or removed.

Mass and Weight

*The **weight** of an object is the pull of gravity on that object.* Unlike mass, weight changes with location. For example, suppose you "weighed" yourself on a bathroom scale on the earth and on the moon. You would weigh only one sixth as much on the moon (Fig. 7-1). The moon pulls you down on the scales only one sixth as hard as the earth does. However, your mass (the amount of material in you) would be the same in both places.

The **pull of gravity** between a planet and an object like yourself depends on three things. They are the **mass** of the planet, your mass, and the **distance** you are from the centre of the planet. Jupiter is much more massive than Earth. Thus you would weigh much more on Jupiter than on Earth. Also, you would weigh less on a mountain top than in a valley. You are further from the centre of the Earth. But, no matter where you are, your mass is the same.

Many people use "weight" to mean both "mass" and "force of gravity". Strictly speaking, this is wrong. To prevent confusion you should use *"mass"* only when you mean the quantity of material in an object. Use *"weight"* when you mean the pull of gravity on an object.

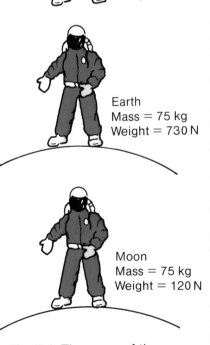

Space
Mass = 75 kg
Weight = 0 N

Earth
Mass = 75 kg
Weight = 730 N

Moon
Mass = 75 kg
Weight = 120 N

Fig. 7-1 The mass of the astronaut stays the same. Note how the weight changes. (N means newtons. You will study this weight unit later.) Why is the weight 0 N in space?

Section 7.1   61

Section Review

1. Define mass.
2. Explain the difference between mass and volume.
3. Explain the difference between mass and weight (force of gravity).

7.2 Common Mass Units

The Standard Kilogram

The primary unit of mass is the **kilogram**. A standard kilogram mass is kept at Sèvres, France. It is a platinum-iridium cylinder. All countries using the metric system standardize their kilogram against this one.

Common Mass Units

The common mass units in the metric system are the **tonne (t)**, **kilogram (kg)**, **gram (g)**, and **milligram (mg)**. Table 7-1 shows how these are related. The tonne is a multiple unit of the kilogram. The gram and milligram are submultiple units of the kilogram.

"Tonne" sounds like "ton" (2000 pounds). Therefore, in spoken language we commonly call a tonne a "metric ton". You could also call a tonne a **megagram (Mg)**, since it is 1 000 000 g.

Table 7-1 Common Mass Units

1 t	= 1000 kg
1 kg	= 1000 g
1 g	= 1000 mg

Look at the set of masses given to you by your teacher. Try to get an idea of the sizes of the mass units. Then study Table 7-2 to see when you would use each unit. Try to think of other uses for each unit.

Table 7-2 Common Mass Units and Uses

Unit	Symbol	Uses
Tonne or Megagram	t Mg	Masses of large objects like trucks, tractors, airplanes, and ships; masses of loads of earth, grain, ore, etc.
Kilogram	kg	Masses of sugar, flour, meat, and other grocery items; masses of people, horses, and other animals
Gram	g	Masses of smaller grocery items like butter, powdered milk, yogurt, cheese, and meat slices
Milligram	mg	Masses of vitamins and minerals in pills, cereal, or bread; masses of ingredients in medical products

Section Review

1. What is the primary unit of mass?
2. State three secondary mass units and give their symbols.
3. What is a megagram?
4. Complete the following:
   1 t = ▩ kg   1 t = ▩ g
   1 kg = ▩ g   1 t = ▩ Mg
   1 g = ▩ mg
5. What mass unit would you use to measure each of the following: a farm tractor, a roast of beef, yourself, a container of margarine, a package of vitamin pills?

7.3 ACTIVITY Measuring Mass

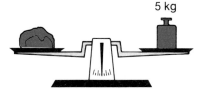

Fig. 7-2 An equal-arm balance, also called a double-pan balance. The mass of the stone is 5 kg — since it balances with a known 5 kg mass.

The mass of an object can be measured with a double-arm balance. The double-arm balance in Figure 7-2 is an equal arm balance (both arms are the same length). With it you measure the mass of an object by balancing the object with known masses.

You may use some other type of balance. Figure 7-3,B, shows a four-beam balance. Look at it closely. It is still a double-arm balance. However, one arm is shorter than the other.

In this activity you will use a balance to find the masses of several objects. You will also use the balance to prepare given masses of a solid and a liquid. Follow your teacher's instructions carefully regarding how to use the balance. A balance is a delicate, expensive instrument. It can be easily damaged.

Fig. 7-3 Three common balances: A triple beam (A), a four-beam (B), and a Dial-o-gram (C).

Materials

balance
set of masses
a bottle of salt
graduated cylinder 100 mL
variety of solid objects (stone, piece of metal, piece of wood, rubber stopper, etc.)
beaker (150 mL or larger)
water
spatula or small spoon

Procedure A Measuring the Mass of a Solid Object

a. Make sure that the pan of the balance is clean and dry.
b. Zero the balance (set the pointer to zero) by carefully following your teacher's instructions.
c. Estimate the mass of each object. Enter your estimate in a table like Table 7-3.
d. Find the mass of the object using the balance.
e. Record the mass in the table. Compare your results with those of your classmates and with your estimates. Repeat any measurements that seem wrong.

Table 7-3 Measuring Mass

Object	Estimate (g)	Measured mass (g)

Procedure B Preparing a Certain Mass of Solid

In some activities that follow in this text, you will require a certain mass of solid. For example, you may need 5 g of salt for an experiment. Practice the following procedure so you can do it without error.

a. Place a piece of smooth clean paper on the balance pan.
b. Move the front sliding mass along until the pointer rests at zero.
c. Add a 5 g mass to the balance.
d. Using the spatula (or small spoon), carefully add salt to the paper on the balance pan until the balance is zeroed again. You should now have 5 g of salt on the paper.

Procedure C Measuring the Mass of a Liquid

a. Place a 150 mL beaker on the balance pan. Find its mass as directed in Procedure A. Record your answer in your notebook.
b. Measure out 50 mL of water in the graduated cylinder. Pour this water into the beaker.
c. Find the mass of the beaker plus the water in it. Record your answer.
d. Calculate the mass of water in the beaker. (Subtract the mass of the beaker from the mass of the beaker plus water.)

Discussion A

1. Will objects with equal masses have equal volumes? Why?
2. Will objects with equal volumes have equal masses? Why?
3. Does the object with the largest volume always have the largest mass? Explain.

Discussion B

1. Why was a piece of paper placed on the balance pan?
2. Why was the balance zeroed after the paper was placed on the pan?

Discussion C

1. a) What would be a faster way to measure out 50 g of water?
   b) Do you think your suggested method would work for liquids other than water? Why?
2. a) How will temperature affect the volume of a liquid?
   b) How will temperature affect the mass of a liquid?
   Hint: Think about the liquid in a thermometer.

Main Ideas

1. The mass of an object is the amount of material in it.
2. An object's volume can change, but its mass stays constant.
3. The weight of an object is the pull of gravity on the object.
4. Mass and distance affect weight.
5. The primary unit of mass is the kilogram (kg).
6. Secondary units of mass are the tonne (t), gram (g), and milligram (mg).
7. Mass can be measured with a double-arm balance.

Glossary

gram (g)	one thousandth of a kilogram
kilogram (kg)	primary unit of mass
mass	amount of material in an object
megagram (Mg)	one tonne (1000 kg)
tonne (t)	one thousand kilograms
volume	amount of space an object takes up
weight	pull of gravity on an object

Study Questions

A. True or False

Decide whether each of the following sentences is true or false. If the sentence is false, rewrite it to make it true. (Do not write in this book.)

1. The mass of an object is the space it takes up.

2. An astronaut has the same weight on the earth as on the moon.
3. An astronaut would have the same mass on Jupiter as on earth.
4. The primary unit of mass is the gram.
5. The mass of a cow is best measured in kilograms.

B. Completion

Complete each of the following sentences with a word or phrase which will make the sentence correct. (Do not write in this book.)
1. When a thermometer is placed in hot water, the ▓▓▓▓ of the fluid in it increases but the ▓▓▓▓ does not change.
2. Weight is the ▓▓▓▓ on an object.
3. One tonne (1 t) is equal to ▓▓▓▓ megagrams (Mg).
4. 5 kg = ▓▓▓▓ g.
5. A small car could weigh about 1100 ▓▓▓▓ .

C. Multiple Choice

Each of the following statements or questions is followed by four responses. Choose the correct response in each case. (Do not write in this book.)
1. An object has a mass of 12 kg. What is its mass in grams?
    a) 12 000 g   b) 1200 g   c) 0.12 g   d) 120 g
2. 250 mg is equal to how many grams?
    a) 2.5 g   b) 2500 g   c) 0.25 g   d) 25 000 g
3. A kilogram is the best unit for measuring the masses of
    a) airplanes, ships, and tractors
    b) sugar, flour, and meats in a grocery store
    c) butter, meat slices, and cheese
    d) vitamin pills and pepper in a shaker
4. 535 kg is equal to
    a) 5.35 t   b) 53.5 t   c) 5350 t   d) 0.535 t
5. The label on a bottle of vitamins says: "Each tablet contains 2.5 mg of ascorbic acid, 3.0 mg of thiamine, and 2.5 mg of riboflavin. Each tablet has a total mass of 60 mg." How much of the tablet, in grams, is *not* ascorbic acid, thiamine, and riboflavin?
    a) 52 g   b) 8 g   c) 0.008 g   d) 0.052 g

D. Using Your Knowledge

1. a) What could you do to change your volume in just a few seconds?
   b) What must you do to change your mass?
   c) State three quite different things you could do to change your weight.
2. Dry cereal such as cornflakes is sold by mass and not by volume. Why?
3. Complete the following:
   50 t = ▓▓▓▓ kg   2500 g = ▓▓▓▓ kg
   22 kg = ▓▓▓▓ g   1370 kg = ▓▓▓▓ t
   26 g = ▓▓▓▓ mg   530 g = ▓▓▓▓ kg

250 mg = ▨ g   6530 mg = ▨ g

4. What unit of mass would you use to measure each of the following: a jet plane, an elephant, a chicken, a jar of peanut butter, a box of cereal, an aspirin tablet?

E. Investigations

1. Find 10 items in your home that have their masses stated in metric units. Look at grocery items, drugstore items, hardware items, and so on. For each item, give the name and the mass.
2. Make an equal arm balance using common things you have at home.
3. Examine a type of balance other than the one you have been using. Write a report on how it works.
4. Prepare a chart or poster to show the difference between mass, weight, and volume.

The Physical Properties of Matter

CHAPTER 8
Solids, Liquids, and Gases

CHAPTER 9
Changes of State and the Particle Theory

CHAPTER 10
Using the Particle Theory

CHAPTER 11
Everyday Uses of the Particle Theory

Every object or thing is matter. This includes solids such as rocks, wood, and ice. It also includes liquids such as water, alcohol, and gasoline. And it includes gases such as oxygen and carbon dioxide. Do you know of anything that is not matter?

This unit deals with the physical properties of matter. You will investigate many physical properties. Then you will study a theory that explains those properties. Finally you will learn how to use the theory. This theory will help you answer many interesting questions: Why does tire pressure go up on a hot day? Why does motor oil get thick on a cold day? How does frost form? Why does salt melt ice?

Fig. 8-0 Matter comes in three forms: solid, liquid, and gas. Can you see all three forms in this picture?

8 Solids, Liquids, and Gases

8.1 Matter and Energy
8.2 The Three States of Matter
8.3 The Solid State of Matter
8.4 The Liquid State of Matter
8.5 The Gaseous State of Matter

This chapter explains what matter is. Then it describes the properties of solids, liquids, and gases. You likely know much of this already. However, read the chapter carefully. Make sure you understand everything in it. Then the chapters that follow will be easier for you.

8.1 Matter and Energy

Our world consists of matter and energy. This book is about the properties of matter and energy. Just what are matter and energy?

Matter

Matter *is anything that has mass*. Since it has mass, it takes up space. That is, it has volume, too.

You can see many examples of matter around you. Your chair is matter; so is this book. Your food is matter. The water and milk you drink are matter. All of these have mass and take up space. Even the air you breathe is matter, but you cannot see it. Can you prove it has mass and takes up space? The student in Figure 8-1 is trying to do that. Will her experiment work?

There are two kinds of matter, **living** and **nonliving**. Rocks, soil, water, ice, metals, and wood are nonliving matter. Plants and animals are living matter. Living things can grow. They can also reproduce themselves. Eventually they die. Nonliving things cannot do these things.

Fig. 8-1 Do you think she can prove air has mass? Try the experiment yourself.

Energy

Energy *is the ability to do work*. Heat, sound, light, and electricity are forms of energy. They differ greatly from matter such as rocks, plants, and water. They have no mass. Light energy makes plants grow. Heat energy can melt ice and boil water. Electrical energy runs motors and lights.

You will learn more about energy in the last half of this book. For now, just be sure you know how energy differs from matter.

The Properties of Matter

All types of matter have properties. These properties are of two types — chemical and physical.

Chemical properties *are those properties that involve the formation of a new substance.* The fact that iron rusts is a chemical property of iron. A new substance, rust, is formed. The fact that wood burns is a chemical property of wood. New substances such as water and carbon dioxide are formed. You will study chemical properties in Unit 5.

Physical properties *are properties that do not involve the formation of a new substance.* Colour, odour, and taste are physical properties. Melting point and boiling point of a substance are also physical properties.

In this unit you will study the physical properties of matter. Some physical properties that you will study are called characteristic physical properties. **Characteristic physical properties** *are physical properties that are different from those of most other substances.* Freezing point is a characteristic physical property. Few liquids except water freeze at 0°C. Boiling point is another characteristic physical property.

Section Review

1. Define matter.
2. Name two kinds of matter.
3. Define energy.
4. How can you tell the difference between matter and energy?
5. How do chemical and physical properties differ?
6. What is a characteristic physical property?
7. Name two characteristic physical properties.

8.2 The Three States of Matter

All matter occurs in one of three forms, or **states**. The three states of matter are solid, liquid, and gaseous. Rocks, metals, ice, and wood are in the solid state. Gasoline, sulfuric acid, and water are in the liquid state. Air, oxygen, and water vapour are in the gaseous state.

Water Occurs in Three States

Many substances can occur in all three states. Water is an example (Fig. 8-2). You have seen solid water, or ice. You have seen liquid water. However, you cannot see gaseous water, or water vapour. It is invisible. The cloud you see coming out of a kettle is not water vapour. It is tiny droplets of liquid water. Look at Figure 8-3. The arrow points to a place where there is only water vapour. Look there next time you see a kettle boiling.

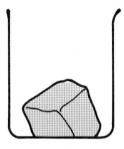
Solid state (Ice)

Liquid state (Water)

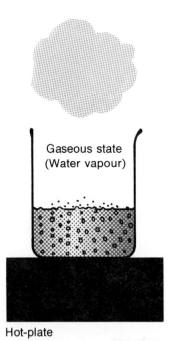
Gaseous state (Water vapour)
Hot-plate

Fig. 8-2 Water exists in all three states — solid, liquid, and gaseous.

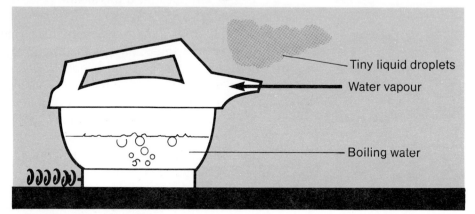
Fig. 8-3 Water vapour is invisible. When it cools, it forms a cloud of tiny water droplets.

Fig. 8-4 Liquid iron can be poured like any other liquid.

Iron Occurs in Three States

Iron can also occur in all three states. You have seen solid iron. Railway tracks and parts of bridges and cars are solid iron. Solid iron can be melted into liquid iron (Fig. 8-4). The hot liquid iron gives off iron vapour. It, like water vapour, cannot be seen.

Section Review

1. Name the three states of matter.
2. Name two substances that can occur in all three states.

8.3 The Solid State of Matter

Iron, rocks, wood, and ice are typical solids. Think of them as you read what follows.

Shape

A solid has a definite shape. It is rigid or stiff. It is not easily pulled or squeezed out of shape. However, under very high pressures a small change in shape can occur.

Volume

Fig. 8-5 An expansion joint in a bridge. On a hot day the bridge would buckle if it had no expansion joints.

A solid takes up space. The space it takes up is called its volume. A solid has a definite volume. It cannot be easily pressed together to make its volume smaller. Nor can it be easily pulled apart to make its volume larger.

Under very high pressures some solids become slightly smaller in volume. Most solids expand (increase in volume) when heated. The change is not great, however. You may have seen expansion joints in bridges (Fig. 8-5). What do you think would happen to a bridge on a hot day if these joints were not there?

Mass

The mass of a solid does not change. Ten grams (10 g) of a solid remains 10 g. Warming a solid will not change its mass. Nor will pressure change its mass. The amount of material in the solid (its mass) always remains the same. Pieces of the solid must be removed to change its mass.

Section Review

1. Write a short paragraph that describes the solid state of matter.
2. Why do bridges need expansion joints?

8.4 The Liquid State of Matter

Think of liquids such as water, gasoline, and oil as you read what follows.

Shape

A liquid is said to be **fluid**. This means it flows and can be poured. Its shape is not definite like that of a solid. Instead, a liquid flows to take the shape of its container. You can make a liquid have almost any shape. Just pour it into a container with the desired shape.

Volume

A liquid has a definite volume. It takes up the same space regardless of its shape (Fig. 8-6). Like a solid, a liquid cannot be compressed much. The brakes of cars depend on this fact. When you push the brake pedal, you put pressure on a liquid called brake fluid. This fluid transmits that pressure to the brakes. Suppose liquids were easily compressed. Then the brake fluid would simply shrink in volume when you pushed the pedal.

Liquids increase in volume a little when heated. Watch a thermometer when it is heated. The liquid in it (mercury or alcohol) rises. Its volume has increased.

Mass

The mass of a liquid does not change. Warming a liquid will not change its mass. When a thermometer is warmed, the volume of liquid increases. But the thermometer still contains the same amount of material. Therefore its mass does not change.

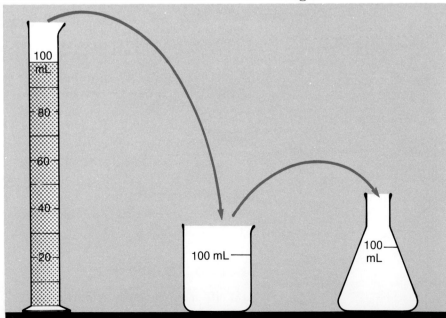

Fig. 8-6 Pour 100 mL of water from a graduated cylinder to a 100 mL beaker, then to a 100 mL flask. The shape of the liquid changes. What happens to the volume?

Section Review

1. Write a short paragraph that describes the liquid state of matter.
2. What does "fluid" mean?
3. Describe how car brakes work.
4. Why is it important that brake fluid be noncompressible?
5. How are liquids like solids?
6. How do liquids differ from solids?

8.5 The Gaseous State of Matter

Think of gases like air, oxygen, and water vapour as you read what follows.

Shape

A gas has no definite shape. It usually expands to fill its container. Suppose you spray room deodorant into one corner of a room. The deodorant does not stay there. It gradually spreads across the room. Eventually it fills the room. In other words, a gas takes the shape of its container.

Gases, like liquids, are fluid. They can be poured. If you were not aware of this fact, try this experiment. You can even do it at home. Put a spoonful of baking soda in a tall glass. Then pour a few millilitres of vinegar into the glass. The mixture will fizz. It produces carbon dioxide gas. After a minute or two the glass will be full of carbon dioxide. Pour the gas (but not the vinegar and baking soda) into a second glass that contains a short, burning candle (Fig. 8-7). Does the gas pour? (Hint: Carbon dioxide gas is used to put out fires.)

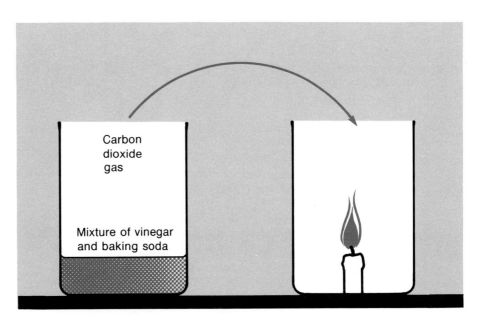

Fig. 8-7 What proof do you have that a gas will pour?

Volume

A gas does not have a definite volume. It fills any container you put it in. But, unlike solids and liquids, gases are easily compressed. A large volume of gas can be squeezed into a small volume by using pressure. You have probably seen "bottles" of compressed gases.

A gas expands a great deal when heated. A car engine makes use of this fact. Gasoline is burned in a cylinder of the engine. This heats up the air and other gases in the cylinder. These gases expand, pushing the piston in the cylinder down.

Mass

The mass of a gas does not change. Warming a gas expands it. But the amount of material remains the same. Also, compressing a gas reduces its volume. But the mass remains unchanged.

Gas and Vapour

The gaseous state of some substances is called a gas. The gaseous state of other substances is called a vapour. What is the difference?

*A **vapour** is the gaseous state of a substance that is a solid or liquid at room temperature.* Thus we say water vapour instead of water gas. Alcohol is a liquid at room temperature. Therefore its gaseous state is called alcohol vapour. Gasoline is also a liquid at room temperature. What is its gaseous state called?

The smelly substance in some cold remedies is called camphor. It is a solid at room temperature. Thus the gaseous material you inhale is called camphor vapour.

*A **gas** is a substance that is in the gaseous state at room temperature.* Oxygen, nitrogen, and carbon dioxide are called gases instead of vapours. They are in air. Another common gas is methane. It is used for cooking, heating, and running Bunsen burners. Most Canadian gas comes from wells in western Canada and the Northwest Territories. It is pumped through pipes to cities and towns throughout most of Canada.

Section Review

1. Write a short paragraph that describes the gaseous state of matter.
2. Explain the difference between a gas and a vapour.
3. Name 5 vapours.
4. Name 5 gases.
5. Copy Table 8-1 into your notebook. Then complete it to compare solids, liquids, and gases.

Table 8-1 Comparing Solids, Liquids, and Gases

Property	Solid	Liquid	Gas
Shape			
Volume			
Effect of compression on volume			
Effect of adding heat on volume			
Mass			

Main Ideas

1. Matter is anything that has mass. Energy is the ability to do work.
2. Physical properties do not involve the formation of a new substance.
3. Characteristic physical properties are physical properties that are different from those of most other substances.
4. There are three states of matter — solid, liquid, and gaseous.
5. A solid has a definite shape, volume and mass.
6. A liquid takes the shape of its container. It has a definite volume and mass. When the temperature changes, the volume changes.
7. A gas has a definite mass but no definite shape or volume.

Glossary

fluid	a substance that can be poured
gas	a state of matter with no definite shape or volume
matter	anything that has mass
liquid	the state of matter with a definite volume but no definite shape
physical property	a property that does not involve formation of a new substance
solid	the state of matter with definite shape and volume
vapour	the gaseous state of a substance that is a liquid or solid at room temperature

Study Questions

A. True or False

Decide whether each of the following sentences is true or false. If the sentence is false, rewrite it to make it true. (Do not write in this book.)
1. Energy has mass.
2. Physical properties involve formation of a new substance.
3. Iron can occur in all three states.
4. A solid has a definite shape and volume.

B. Completion

Complete each of the following sentences with a word or phase which will make the sentence correct. (Do not write in this book.)
1. Matter is anything that has ▓▓▓ .
2. Anything that can be poured is called a ▓▓▓ .
3. If the temperature increases, the volume of a liquid will ▓▓▓ .
4. If the pressure increases, the volume of a gas will ▓▓▓ .
5. The gaseous state of a solid (at room temperature) is called a ▓▓▓ .

C. Multiple Choice

Each of the following statements or questions is followed by four responses. Choose the correct response in each case. (Do not write in this book.)
1. The two kinds of matter are
    a) mass and energy
    b) living and nonliving
    c) solid and fluid
    d) physical and chemical
2. Which one of the following statements is true?
    a) Both liquids and gases have a definite volume.
    b) Liquids decrease in volume when heated.
    c) Both liquids and solids have a definite shape.
    d) Liquids and solids cannot be compressed much.

3. A vapour is defined as
    a) the gaseous state of matter
    b) any heated gas
    c) the gaseous state of any substance that is a solid or liquid at room temperature
    d) a cloud of liquid droplets
4. Which one of the following statements is true?
    a) A gas has no definite shape but does have a definite volume.
    b) A gas will not expand much when heated.
    c) A gas cannot be compressed.
    d) A gas can be poured.

5. When a gas is compressed
   a) it loses mass
   b) it gains mass
   c) its volume gets larger
   d) its volume gets smaller

D. Using Your Knowledge

1. Classify the gaseous state of each of the following as either a gas or vapour: water, nitrogen, oxygen, rubbing alcohol, moth crystals, helium, turpentine, fuel oil, dry ice (frozen carbon dioxide).
2. Look back to Figure 8-7. Carbon dioxide gas can be poured through air. This is because carbon dioxide is more dense than air. Helium gas is less dense than air. That is why it is used in weather balloons. Can it be poured through air? Explain.
3. Are butter, margarine, lard, and shortening solids? Explain your answer. (Hint: Review the properties of solids.)
4. Sand can be poured. Is it a fluid? Explain.

E. Investigations

1. For this investigation you will need: two volleyballs, a metre stick, a tire pump, an air pin (for blowing up the volleyballs), masking tape, and string. Use this material to show that air has mass.
2. For this investigation you will need: modeling clay, string, a graduated cylinder, a balance, and water. Use this material to prove that the shape of the clay can be changed without the mass and volume changing.
3. Find a diagram of the inside of a car engine. Make a simple copy of it. Then use your diagram to help explain how the engine works.
4. Find the names of at least 10 gases in air. What percent of the air is each of these?

9 Changes of State and the Particle Theory

9.1 Changes of State: What Are They?
9.2 Activity: Melting of Ice
9.3 Activity: Effect of Heat on the Temperature of Melting Ice
9.4 Activity: Effect of Heat on the Temperature of Boiling Water
9.5 Activity: Effect of Heat on Iodine Crystals
9.6 The Particle Theory of Matter
9.7 Changes of State: An Explanation

There are three states of matter — solid, liquid, and gaseous. Some substances occur in all three states. Water is an example. It can occur as ice, liquid water, and water vapour. In this chapter you will study changes from one state to another. Then you will use the particle theory to explain the changes.

9.1 Changes of State: What are they?

There are six possible changes of state. You need to know what they are before you proceed with this chapter. Refer to Figure 9-1 as you read the following descriptions of changes of state. Remember that water is not the only substance that exists in three states. At high temperatures, even metals like iron and nickel can be liquids and vapours.

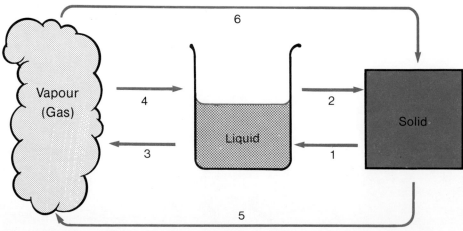

Fig. 9-1 Changes of state. Can you name them?

Like Chapter 8, much of this will be review for you. However, study it carefully. You need the knowledge for the activities that follow.

1. Melting

Melting *is the change of state from a solid to a liquid.* It is also called **liquefaction** [LIK-wi-FAK-shun]. Liquefaction means "making a liquid". Heat is needed to melt ice. In fact, heat is needed to melt all solids. Usually the liquid formed has a greater volume than the solid. Water is an exception. The volume decreases when ice melts.

2. Freezing

Freezing *is the change of state from a liquid to a solid.* It is also called **solidification** [so-LID-if-fi-KAY-shun]. Solidification means "forming a solid". You have to cool water to make it into ice. That is, heat must be removed. Heat must be removed to freeze all liquids. The volume decreases when most liquids freeze. Water is an exception. The volume increases when water freezes.

3. Evaporation

Evaporation [ee-VAP-or-AY-shun] *is the change of state from a liquid to a vapour or gas.* It is also called **vapourization** [VAY-pur-iz-AY-shun]. Vapourization means "forming a vapour". Heat is needed to evaporate a liquid. If gentle heat is provided, the rate of evaporation is slow. You have seen how long it takes for spilled water to evaporate. The heat has to come from the air. If the heat is provided quickly, the rate of evaporation is rapid. Bubbles of vapour form throughout the liquid. This is called **boiling**. The volume increases when evaporation occurs. That is, the volume of vapour is far greater than the volume of the original liquid.

4. Condensation

Condensation [kon-den-SAY-shun] *is the change of state from a vapour or gas to a liquid.* It is also called **liquefaction**. (A liquid is also formed here.) Be careful when you use the word liquefaction. It can mean either melting or condensation.

Heat must be removed for condensation to occur. You have seen condensed water on a glass of cold pop. The cold pop removes heat from water vapour in the air. Then the vapour condenses. The volume decreases during condensation. That is, the volume of liquid is less than the volume of the vapour.

5. and 6. Sublimation

Sublimation [sub-li-MAY-shun] *is the change of state directly from a solid to a vapour and/or from a vapour to a solid.* No liquid is formed. Solid bathroom deodorizers sublime from solid to vapour. Frost is formed when water sublimes from vapour to solid. Sublimation from solid to vapour requires heat. Also, the volume increases. Sublimation from vapour to solid requires removal of heat. Also, the volume decreases.

Chapter Review

1. Define melting, freezing, evaporation, condensation, and sublimation.
2. Copy Table 9-1 into your notebook. Then complete it. It summarizes the changes in volume and heat that occur during changes of state.

Table 9-1 Heat and Volume Changes During Changes of State

Change of State	Heat		Volume	
	Add	Remove	Increases	Decreases
Melting				
Freezing				
Evaporation				
Condensation				
Sublimation (#5)				
Sublimation (#6)				

9.2 ACTIVITY Melting of Ice

What happens to the mass when ice melts? What happens to the volume when ice melts? Write answers to these questions in your notebook. These are your predictions. Then do the activity to test your predictions.

Problem

Do the mass and volume of ice change when it melts?

Materials

small ice cubes
graduated cylinder (100 mL)
balance
dissecting needle

Procedure

a. Copy Table 9-2 into your notebook. Record all your results in it.
b. Place 50 mL of warm water (60-70°C) in the graduated cylinder.
c. Add 3 or 4 ice cubes to the water. Quickly push them under the water. Use the dissecting needle (Fig. 9-2).

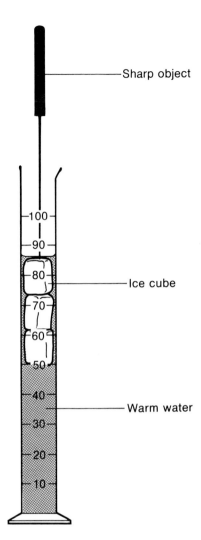

Fig. 9-2 Push all the ice cubes under the water. Then read the volume.

d. Quickly read the volume of water + ice cubes.
e. Quickly place the graduated cylinder on the balance. Measure the mass of the cylinder + water + ice. Leave the cylinder on the balance for the rest of the activity.
f. After the ice has melted, measure the mass again.
g. Measure the volume of water after the ice has melted.
h. Calculate the volume of ice.
i. Calculate the volume of melted ice.
j. If time permits, repeat the activity twice.

Table 9-2   Melting of Ice

Measurement or calculation	Trial 1	Trial 2	Trial 3
Volume of warm water			
Volume of water + ice			
Volume of ice			
Volume after melting			
Volume of melted ice			
Mass before melting			
Mass after melting			

Discussion

1. What change of state occurred in this activity?
2. What happens to the mass when ice melts?
3. What happens to the volume when ice melts?
4. Is ice more dense or less dense than water? Use your results to prove your answer.
5. Give further proof that ice is less dense than water.

9.3 ACTIVITY Effect of Heat on the Temperature of Melting Ice

In this activity you will measure the temperature of ice as it melts. Remember that heat is needed to melt the ice. Now read the problem and make your prediction. Write it in your notebook.

Problem

What happens to the temperature as ice melts?

Materials

finely crushed ice (or snow)   ring stand

Section 9.3   83

beaker (250 mL)
thermometer ($-10°C$ to $110°C$)
glass or plastic stirring rod
Bunsen burner
string
iron ring
wire gauze
adjustable clamp
watch or clock with second hand
graph paper

CAUTION: Wear safety goggles during this activity.

Procedure

a. Prepare a data table with the headings "Time" and "Temperature".
b. Half fill a 250 mL beaker with crushed ice.
c. Set up the apparatus as shown in Figure 9-3. Be sure the flame is **very low**.
d. Stir the ice continually. Use the stirring rod. Do not stir with the thermometer — it may break.
e. Keep the thermometer near the centre of the beaker. Do not let it touch the bottom.
f. Record the temperature every 30 s. Keep doing this until the ice is all gone.
g. Now keep recording the temperature every 30 s for 5-10 min after the ice is gone.

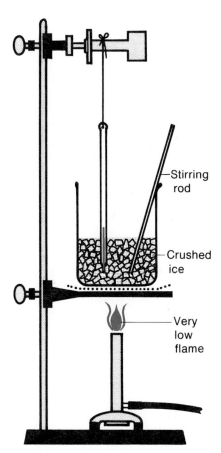

Fig. 9-3 Finding the effect of heat on the temperature of melting ice

Discussion

1. What happens to the temperature when the ice is melting?
2. What happens to the heat that the Bunsen burner put into the ice-water mixture?
3. Does the melting of ice require heat? Or, does it require removal of heat?
4. What is the melting point of ice?
5. What is the freezing point of water?
6. How are the melting point and freezing point of a substance related?
7. Why is melting point called a characteristic physical property?

9.4 ACTIVITY Effect of Heat on the Temperature of Boiling Water

In this activity you will measure the temperature of water as it boils. Remember that heat is needed to boil water. Now read the problem and make your prediction. Write it in your notebook.

Problem

What happens to the temperature as water boils?

Materials

round-bottomed flask (250 mL)
thermometer (−10°C to 110°C)
distilled water
boiling chips
Bunsen burner

ring stand
iron ring
wire gauze
2 adjustable clamps

CAUTION: Wear safety goggles during this activity.

Procedure

a. Prepare a data table with column headings "Time" and "Temperature".
b. Set up the apparatus as shown in Figure 9-4. Fill the flask about half full of water. Put the thermometer bulb *just below* the surface of the water. The boiling chips are used to make the water boil smoothly.
c. Record the temperature of the water.
d. Heat the water strongly. Record the temperature every 30 s. Keep doing this until the water is boiling rapidly. Note the temperature at which boiling began.
e. Keep the water boiling for about 10 min. As it boils, record the temperature every 30 s. (Make sure the bulb is always just below the surface of the water.)

Discussion

1. What happens to the temperature while water is boiling?
2. What happens to the heat that the Bunsen burner puts into the water?
3. What is the boiling point of water?
4. How does the mass of water affect the boiling point? (Hint: Compare your results with those of some classmates. You all likely began with different masses of water.)
5. Suppose you repeated this activity with another liquid. In what ways would the results be similar to those for water? In what ways would they differ from those for water?
6. Why is boiling point called a characteristic physical property?

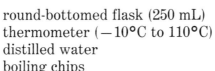

Fig. 9-4 Studying boiling water

9.5 ACTIVITY Effect of Heat on Iodine Crystals

In this activity you will heat iodine crystals gently. Watch for evidence of sublimation.

Problem

Does iodine sublime when heated?

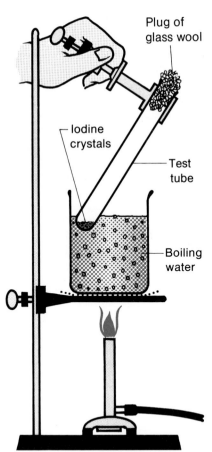

Fig. 9-5 Will gentle heating cause sublimation of iodine crystals?

Materials

test tube (20 x 150 mm) containing iodine crystals and plugged with glass wool
Bunsen burner
beaker (250 mL)
ring stand
iron ring
wire gauze
adjustable clamp

CAUTION: Wear safety goggles during this activity. Iodine is very corrosive. It will damage skin and clothing. Do not remove the glass wool. Do not touch or smell the iodine. If you spill any, tell your teacher.

Procedure

a. Set up a water bath as shown in Figure 9-5.
b. Heat the water until it is boiling.
c. Get a test tube with iodine crystals from your teacher. Describe the crystals.
d. Hold the bottom of the test tube in the boiling water as shown in Figure 9-5. Keep the top part as cool as possible.
e. Continue heating the crystals in this way for at least 10 min (or until the crystals are all gone). Let the test tube cool.
f. Make a careful record of all observations.
g. Give the test tube and contents to your teacher. Do not try to wash it.

Discussion

1. What changes of state took place in this activity? How do you know?
2. The melting point of iodine is 113.5°C. Why was a water bath used? Your teacher will show you what happens if iodine is heated without the water bath. CAUTION: Use a fume-hood.
3. Compare the volume of the iodine vapour formed to the volume of the crystals.

9.6 The Particle Theory of Matter

You have likely heard of **atoms** and **molecules** [MOL-i-kuls]. They are both **particles** of matter. Don't worry about the difference between them now. You will learn about them in Unit 5.

No one has seen these particles. But, if we assume matter is made of particles, we can explain changes of state and other things. The particle theory is a model for matter. We do not know for sure that it is right. But it has stood the test of time. And that is a good reason for believing in it. Here is what it says:

1. *All matter is made of small particles.* How small is "small"? Suppose everyone on earth worked together to count the atoms (particles) in the head of a pin. Even working full time, it would take their entire lives to do so!

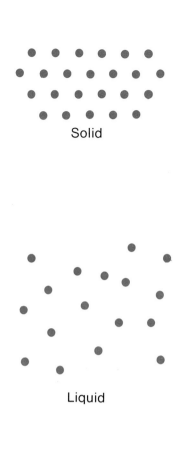

Fig. 9-6 The particle nature of solids, liquids, and gases

2. *All particles of the same substance are identical.* Ice, liquid water, and water vapour are the same substance. Thus a particle of ice, a particle of liquid water, and a particle of water vapour are all the same. They have the same mass and volume.
3. *The particles in matter attract one another. These "attractive forces" get stronger as the particles get closer together.* This explains why solids are more rigid than liquids or gases. The particles in a solid are closer together (Fig. 9-6). This makes the attractive forces stronger. Thus the solid is more rigid.
4. *The spaces between the particles are large compared to the sizes of the particles themselves.* In fact, the spaces are much larger than Figure 9-6 suggests. The spaces are smallest in a solid and largest in a gas. This explains why gases can be compressed (reduced in volume). Solids and liquids cannot be easily compressed. The particles are already closer together than in a gas. This point also explains why solids are usually the most dense. Can you see how?
5. *The particles in matter are always in motion.* Their speed depends on the temperature. As the temperature goes up, the particles have more energy. Therefore they move more quickly.

The motion of the particles causes them to bump into the walls of their container causing pressure. The pressure in a bicycle tire is caused this way. The moving air particles push out on the tire.

You cannot see the particles of matter, even with a microscope. They are far too small. However, these particles often bump into larger particles that we can see. And they make the larger particles move. This motion is called **Brownian Motion**. It was first seen by Robert Brown in 1827. The large particles move much like the smaller ones. Therefore, by studying Brownian Motion we can get an idea of how small particles move. Your teacher will show you Brownian Motion.

Section Review

1. List the five points in the particle theory.
2. Why is it easier to compress a gas than a liquid?
3. Why is solid iron more dense than liquid iron?
4. What causes pressure in a car tire?
5. Why does the air pressure in a car tire go up on a hot day?

9.7   Changes of State: An Explanation

You did several activities involving changes of state. We will first summarize your findings. Then we will see if the particle theory explains them.

Summary of Findings

1. Melting, evaporation, and sublimation (solid to gas) require heat.

2. Freezing, condensation, and sublimation (gas to solid) require removal of heat.
3. The volume of matter increases during melting, evaporation, and sublimation (solid to gas). Note: The behaviour of water when it melts is **abnormal**, or unusual. You found that the volume decreased. Most substances do the opposite. For example, when hot liquid gold freezes to a solid, its volume decreases.
4. The volume decreases during freezing, condensation, and sublimation (gas to solid). Again, water is abnormal.

Now, put 1 and 3 together and 2 and 4 together. Here are the final conclusions:

1. *Heat is required for a volume increase during a change of state.*
2. *Heat must be removed for a volume decrease during a change of state.*

Explanation Using the Particle Theory

Such conclusions don't happen by accident. They are true because of the basic nature of matter. The particle theory describes the basic nature of matter. Therefore this theory should explain our results. Let's see if it does.

For the volume to increase, the particles must move further apart. But forces hold the particles together. Suppose we heat the particles. They now have more energy. When they have enough energy, they can overcome the forces and move further apart.

For the volume to decrease, the particles must move closer together. To do so, they must lose energy. Otherwise the forces will not be able to hold them closer together.

Summary

The particle theory explains changes of state. It also explains many other physical properties. But it is still a theory and must constantly be updated as new findings are made.

Section Review

1. Which changes of state require heat?
2. Which changes of state have a volume increase?
3. Which changes of state require removal of heat?
4. Which changes of state have a volume decrease?
5. What is meant by the abnormal behaviour of water?
6. Why does a volume increase during a change of state require heat?
7. Why does a volume decrease during a change of state require removal of heat?

Main Ideas

1. There are six changes of state — melting, freezing, evaporation, condensation, sublimation (solid to gas), and sublimation (gas to solid).
2. Melting, evaporation, and sublimation (solid to gas) require heat. The volume increases during these changes of state.
3. Freezing, condensation, and sublimation (gas to solid) require removal of heat. The volume decreases during these changes of state.
4. Water is abnormal. Its volume decreases during melting and increases during freezing.
5. All matter is made of small particles.
6. The particles of matter are in constant motion.
7. The particles of matter have attractive forces and spaces between them.
8. The Particle Theory explains changes of state. It also explains many other physical properties.

Glossary

condensation	kon-den-SAY-shun	the change of state from gas to liquid
evaporation	ee-VAP-or-AY-shun	the change of state from liquid to gas (vapour)
freezing		the change of state from liquid to solid
liquefaction	LIK-wi-FAK-shun	the change of state from solid to liquid and/or from gas to liquid
melting		the change of state from solid to liquid
solidification	so-LID-if-fi-KAY-shun	the change of state from liquid to solid
sublimation	sub-li-MAY-shun	the change of state directly from solid to gas and/or from gas to solid
vapourization	VAY-pur-iz-ay-shun	the change of state from liquid to gas (vapour)

Study Questions

A. True or False

Decide whether each of the following sentences is true or false. If the

sentence is false, rewrite it to make it true. (Do not write in this book.)
1. Evaporation and vapourization both refer to the change of state from liquid to vapour (or gas).
2. If 10 g of ice melts to form 10 g of water, the volume will decrease.
3. The temperature of melting ice gradually goes up as the ice is heated.
4. Iodine will sublime from solid to vapour and also from vapour to solid.
5. The attractive forces in a gas are greater than those in a liquid.

B. Completion

Complete each of the following sentences with a word or phrase which will make the sentence correct. (Do not write in this book.)
1. Condensation is the change of state from ▬▬▬ to ▬▬▬ .
2. During sublimation no ▬▬▬ state is formed.
3. Boiling is just a rapid form of ▬▬▬ .
4. The spaces between particles are ▬▬▬ compared to the sizes of the particles themselves.
5. The speed of particles increases as the ▬▬▬ increases.

C. Multiple Choice

Each of the following statements or questions is followed by four responses. Choose the correct response in each case. (Do not write in this book.)

Table 9-3 Melting of a Solid

Time (min)	Temperature (°C)
0	10°C
5	10°C
10	10°C
15	20°C
20	35°C
25	50°C

Table 9-4 Heating of a Liquid

Time (min)	Temperature (°C)
0	0°C
5	15°C
10	30°C
15	30°C
20	30°C
25	30°C
30	45°C
35	60°C

1. Table 9-3 shows the results of an experiment to study the melting of a solid. The experiment was done in the same way you studied the melting of ice. The solid had just finished melting completely
   a) at a time of 10 min
   b) somewhere between 10 and 15 min
   c) at a time of 25 min

      d) at a time between 0 and 10 min
2. Look at Table 9-3 again. The freezing point (melting point) of the substance is
   a) 20°C   **b)** 50°C   **c)** 10°C   **d)** not known from these data
3. Look at Table 9-3 again. After 15 min, the substance is mainly
   a) a liquid   **b)** a solid   **c)** a vapour   **d)** a gas
4. Table 9-4 shows the results of an experiment in which a liquid was heated. The experiment was done in the same way you studied the heating of water. The boiling point of this liquid is
   a) 0°C   **b)** 30°C   **c)** 45°C   **d)** 100°C
5. Look at Table 9-4 again. After 30 min the substance is
   a) a mixture of all three states   **c)** liquid
   b) solid   **d)** gaseous

D. Using Your Knowledge

1. Imagine a lake with large chunks of ice floating at the surface. What will the water temperature be among the ice? Will it be the same well below the ice? Why?
2. How would you make ice colder than 0°C?
3. Describe two examples of sublimation other than iodine.
4. Many vegetable crops are damaged by temperatures below 0°C. Farmers often turn on their sprinkling systems when the temperature drops to 0°C. Ice may form on the vegetables. Yet they do not freeze. Why?
5. The freezing of a lake in the fall warms the land nearby. Why?
6. The thawing of a lake in the spring cools the land nearby. Why?
7. A carton of milk will often split if it is frozen. Why?
8. If water did not have an abnormal behaviour, fish would be killed when the water froze. Why?
9. Use the particle theory to answer these questions:
   a) Why is it easier to move your hand through air than through water?
   b) Why is solid iron more dense than liquid iron?
   c) Why is a gas easily compressed?
   d) What causes pressure inside a balloon?
   e) Why might an over-inflated tire burst at high speeds?

E. Investigations

1. Look up the abnormal behaviour of water in a book. What does the term abnormal mean? What causes the abnormal behaviour?
2. Design and do an experiment to show that camphor sublimes.
3. Find out what dry ice is. Why do people often say it will "burn" your fingers if you touch it?
4. Find out how liquid air is made. What is it used for?

10 Using the Particle Theory

10.1 Activity: Effect of Temperature on the Volume of a Gas
10.2 Activity: Effect of Temperature on the Volume of a Liquid
10.3 Effect of Temperature on the Volume of a Solid
10.4 Effect of Temperature on the Pressure of a Gas
10.5 Activity: Relationship Between Pressure and Volume of a Gas
10.6 Effect of Pressure on the Boiling Point of Water
10.7 Activity: Diffusion
10.8 Activity: Crystallization of Salol

At the end of Chapter 9 you met the particle theory. You saw how it explains changes of state. You also learned that it explains most other physical properties. It seems like a good theory. But is it?

This chapter tests the theory. A good theory should be able to predict the results of experiments before you do them. That is what this chapter is all about. For each topic you will predict the results using the particle theory. Then, in some cases, you will do an activity to see if your prediction was right. In other cases your teacher will demonstrate the activity.

10.1 ACTIVITY Effect of Temperature on the Volume of a Gas

Turn back to the particle theory on page 36. What does it say will happen to the particles if the temperature is raised? What will this do to the volume of a gas? Write the answer to the last question in your notebook. It is your prediction. Does it agree with common sense? Now do the activity to test your prediction.

Problem

How does temperature affect the volume of a gas?

Materials

250 mL flask
2-hole rubber stopper
thermometer (−10°C to 110°C)
glass tubing (50 cm)
coloured water
marking pen

CAUTION: Wear safety goggles during this activity.

Procedure

a. Draw a table in your notebook with two columns. Title them "Slug Position" and "Temperature".
b. Set up the apparatus as shown in Figure 10-1. The glass tubing and thermometer should reach halfway into the flask. The slug of coloured water should be about 5 mm long. At the start of the activity it should be just above the stopper. Your teacher will tell you how to get it there if you have trouble.
c. Warm the flask gently. Do so by wrapping your hands around it for a few minutes. Mark the new position of the slug of water. Record the temperature. Then record the position of the slug above the stopper.
d. Warm the flask still more. Do so by moving a low flame over it. Touch the flask from time to time so you won't overheat it. (CAUTION: The thermometer will break if the temperature goes above its range.) Continue marking new positions of the slug for 3 or 4 more temperatures. Do so until the temperature is in the 70°C-80°C range.
e. Observe the slug of water as the flask cools.

Discussion

1. How did the slug of water move when the air was heated?
2. How did the slug of water move when the air was cooled?
3. What is the relationship between temperature and the volume of a gas? How does this agree with your prediction?
4. Explain your conclusion using the particle theory.
5. Does the particle theory suggest that your conclusion may be true for gases other than air? Explain.

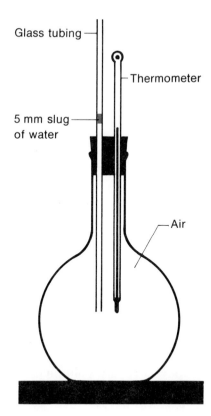

Fig. 10-1 Finding the effect of temperature on the volume of a gas

10.2 ACTIVITY Effect of Temperature on the Volume of a Liquid

Write a prediction of the effect of temperature on the volume of a liquid. Three things will help you make this prediction:
1. The results of Activity 10.1;
2. The particle theory;
3. Your everyday experiences.

Now, try the activity to test your prediction.

Problem

How does temperature affect the volume of a liquid?

Materials

1000 mL beaker   adjustable clamp

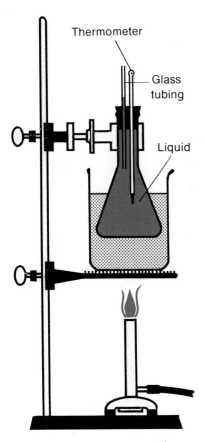

Fig. 10-2 Finding the effect of temperature on the volume of a liquid

2-hole rubber stopper
thermometer (−10°C to 110°C)
250 mL flask
ring stand, iron ring, wire gauze

water, glycerine, ethylene glycol
marking pen
glass tubing (50 cm)
Bunsen burner

CAUTION: Wear safety goggles during this activity.

Procedure

a. Prepare a data table with the column headings "Position of water" and "Temperature".
b. Set up the apparatus as shown in Figure 10-2. Make sure the liquid in the glass tube is just above the rubber stopper when you start. Use the liquid suggested by your teacher.
c. Before you begin heating, mark the position of the liquid. Then record the position (the distance above the stopper). Also, record the temperature.
d. Begin heating the liquid slowly. Record the position of the liquid at the following temperatures: 30°C, 35°C, 40°C, 45°C, 50°C, 55°C, 60°C, 65°C. Do not heat the liquid beyond 65°C.
e. If time permits, repeat the experiment using another liquid.

Discussion

1. What is the relationship between the temperature and volume of a liquid? How does this agree with your prediction?
2. Use the particle theory to explain the results of this activity. There are two things to explain:
   a. Why do the volumes change as they do?
   b. Why do different liquids give different results?

10.3 Effect of Temperature on the Volume of a Solid

In Activity 10.1 you found that gases expand when heated. In Activity 10.2 you found that liquids behave in the same way. What will solids do? Let us first review the behaviour of gases and liquids. Then we will look at the behaviour of solids.

Gases and Liquids: A Review

Both gases and liquids expand when heated. The particle theory explains this behaviour. It says that particles move faster at higher temperatures. They have more energy. For a substance to expand, the particles must move further apart. But attractive forces hold them together. Energy is needed to overcome the attractive forces. This energy is added by heating to higher temperatures. Therefore **expansion** [ex-PAN-shun] occurs at higher temperatures.

Solids

According to the particle theory, solids should also expand when heated. However, the expansion will not be as great (per degree) as for gases and liquids. The attractive forces in solids are very strong. Thus much more heat is needed to overcome them. The expansion is so little you cannot easily see it. Your teacher may show you how to "see" small expansions of solids.

The expansion of solids may be slight. But it cannot be ignored. Here are some examples that show why.

Power Transmission Lines

Fig. 10-3 Power transmission lines in the summer. What will they look like in the winter?

Transmission lines for electrical power show that solids expand when heated (Fig. 10-3). The lines hang lower in summer than in winter. The construction crews must remember this. If they build the line in the winter, they must not leave too much slack. If they do, the line will sag too much in the summer. What must they remember if they build the line in the summer?

Railway Expansion Joints

Fig. 10-4 Expansion joint in a railway track. Why is it needed?

Have you ever heard the clack...clack...clack between train wheels and the track? This noise is caused by gaps between the rails (Fig. 10-4). These gaps are important. They allow the rails to expand on warm days. Without them, the rails would buckle.

Bridge Expansion Joint

Look back to Figure 8-5 on page 73. This is a photograph of a bridge expansion joint. It allows the bridge to expand on warm days. Without it, the pavement and steel beams might buckle.

Pendulum Clocks

You learned in Activity 2.3 (page 15) that the period of a pendulum depends on its length. The longer the pendulum, the longer the period. That is, the longer the pendulum, the slower it swings. Pendulums are made of metal. Therefore they expand a little on warm days. As a result, a pendulum clock would run slow on a warm day. And it would run fast on a cold day. Clockmakers know this so they build a part into the clock that allows for changes in the pendulum. They also make pendulums using metals that expand very little.

Metal Jar Tops

How do you get the metal top off a jar? If you put it under a hot water tap it usually turns off easily. The heat causes the metal ring to expand. It also causes the glass jar to expand, but the metal expands more than the glass. This loosens the ring.

Section Review

1. Explain why solids, liquids, and gases expand when heated. Use the particle theory.
2. Construction crews build most power transmission lines in the summer. What must they do to allow for expansion and contraction?
3. Explain why expansion joints are needed in railways and bridges.
4. Why do clockmakers need to know about expansion of solids?

10.4 Effect of Temperature on the Pressure of a Gas

How will temperature affect the pressure of a gas? Let's make a prediction, or hypothesis, using the particle theory. The particles are always moving. They bump into the walls of their container. This causes pressure. If the gas is heated, the particles move faster. They strike the walls more often and also harder. Therefore, the pressure goes up. We can now make this prediction: If the temperature of a gas is increased, its pressure will increase. Now, let's check the prediction.

Figure 10-5,A shows the apparatus we used. If your school has one, your teacher will show you this experiment. We put some air in the bulb. Then we adjusted the pressure dial to 0. Next we heated the bulb. Figure 10-5,B shows the result. The pressure did increase.

Our prediction was correct. Once again, the particle theory has passed the test. We have never seen the particles of matter. But if we assume they exist, we can explain many properties of matter.

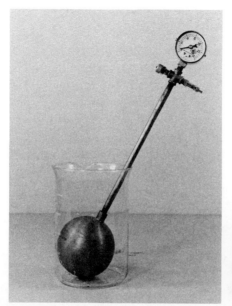

Fig. 10-5 How does temperature affect the pressure of a gas?

A.

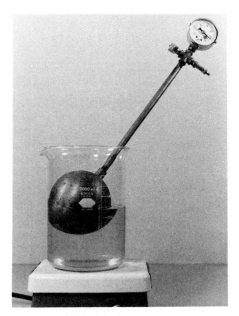

B.

Section Review

1. How does temperature affect the pressure of a gas?
2. How would you show, with this apparatus, that a decrease in temperature lowers the pressure?

10.5 ACTIVITY Relationship Between Pressure and Volume of a Gas

How will pressure affect the volume of a gas? You can guess the answer. Imagine some gas in a balloon. Suppose you squeeze (put pressure on) the balloon. What will happen to the volume?

Problem

What is the relationship between the pressure and volume of a gas?

Prediction

Let's use Figure 10-6 to make a better prediction or hypothesis.

Look at Part A. This diagram shows a cylinder with a piston in it. The piston can be moved up and down. Some gas is trapped by the piston. Its volume is V. The piston is being held down by a pressure p_d. The piston is not moving. Therefore the gas inside must be pushing up with a pressure that equals p_d. It is called p_u. The brown dots are the particles of the gas. These particles cause the pressure p_u. They move around, bumping into the piston (and the walls of the cylinder). This causes the pressure.

Now look at Part B. The volume has been cut in half. The diagram suggests that the pressure was doubled to do this. Does this agree

Section 10.5

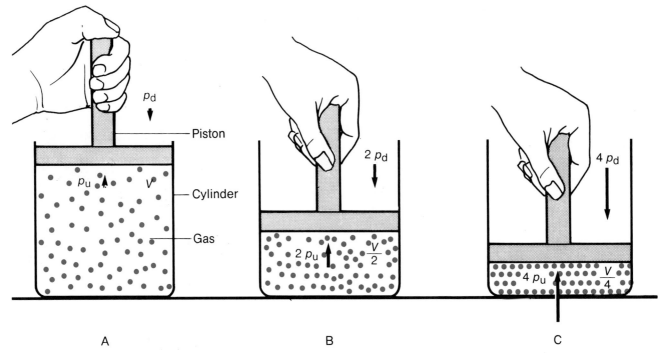

A   B   C

Fig. 10-6 Making a hypothesis about the effect of pressure on volume

with the particle theory? That is, will the pressure double if the same number of particles are in half the volume?

Now look at Part C. Think about it as you did Part B.

What is your prediction for the relationship between the pressure and the volume of a gas?

Materials

syringe	six identical textbooks
ring stand	rubber stopper with hole part way through
adjustable clamp	(The tip of the syringe must fit *tightly* into this hole.)

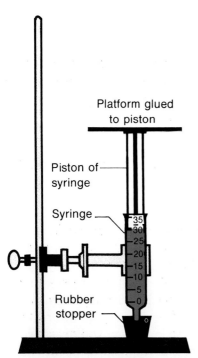

Fig. 10-7 Studying the pressure-volume relationship for a gas

Procedure

a. Set the piston of the syringe at the 30 mL mark.
b. Place the rubber stopper over the tip of the syringe.
c. Set up the apparatus as shown in Figure 10-7.
d. Carefully place one textbook on the platform. Record the new volume in Column A of a table like Table 10-1.
e. Leave the book on the platform. Now press the piston until the volume is a few millilitres less than the one you just recorded. Release the piston. Record the volume at which it stops in Column B.
f. Average the volumes in Columns A and B. Record your answer in Column C. This averaging reduces error due to friction.
g. Now put 2 books on the platform. Repeat steps (c) to (e). Repeat again using 3, 4, 5, and finally 6 books.

Table 10-1 Pressure-Volume Data

Number of books	A	B	C (Average)
0			
1			
2			
3			
4			
5			
6			

Discussion

1. What is the relationship between the pressure and the volume of a gas?
2. Do your results agree with your prediction? Explain.
3. Account for your results using the particle theory.

10.6 Effect of Pressure on the Boiling Point of Water

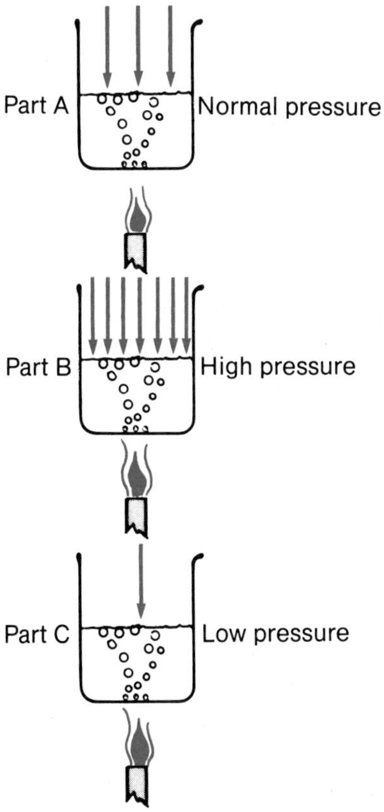

Look at Part A of Figure 10-8. It shows some water boiling. The arrows are the air pressure pushing down on the water. This pressure is caused by air particles as they bump into the surface of the water. For water to evaporate or boil, its particles must push out through the air pressure. They need energy to do so.

Increased Pressure

Suppose the pressure is increased (see Part B). The air pushes harder on the water. And the water particles need more energy to evaporate. To get more energy they need to be at a higher temperature. Therefore the water will have to be heated to a higher temperature before it will boil. *The boiling point increases when the pressure increases.*

Decreased Pressure

Suppose the pressure is decreased (see Part C). The air does not push as hard on the water. Thus the water particles need less energy to evaporate. Therefore the water will not have to be heated to as high a temperature to get it to boil. *The boiling point decreases when the pressure decreases.*

Fig. 10-8 How does pressure affect boiling point?

Are the Conclusions Right?

Our conclusions (in italics) are really just hypotheses. We made them using the particle theory. But we have not done any experiments. If time permits, your teacher will demonstrate the experiments. However, the data in Table 10-2 come from a real experiment. They support our hypotheses: **The boiling point increases when the pressure increases and decreases when the pressure decreases.**

Note that the boiling point of water is 100°C only when the pressure is 101.3 kPa (kilopascals). Suppose the pressure is lowered to 2.3 kPa. Water will now boil at 20°C, or room temperature. You could put your hand in this boiling water. Suppose the pressure is raised to 1550 kPa. Water will not boil until its temperature is 200°C. It will be twice as hot as water at its normal boiling point.

Table 10-2   Boiling Point of Water at Different Pressures

Pressure (kPa)	Boiling point (°C)
1.2	10
2.3	20
12.3	50
101.3 Standard air pressure	100 Normal boiling point
476.0	150
1550.0	200
16 528.0	300

Section Review

1. What happens to the boiling point if the pressure is raised?
2. Use the particle theory to explain your answer to question 1.
3. What happens to the boiling point if the pressure is lowered?
4. Use the particle theory to explain your answer to question 3.
5. What is meant by "normal boiling point"?

10.7   ACTIVITY Diffusion

Release some room deodorizer into one corner of a room. What happens? It spreads, or **diffuses** [di-FEW-sehs], throughout the room. This happens even if there are no air currents in the room. The odour moves by itself. How does it do this?

In this activity you will study the diffusion of a solid through a liquid. Set up the activity as described. Then use the particle theory to predict what will happen.

Problem

What will happen if potassium permanganate crystals are placed in water?

Materials

petri dish
water
2 or 3 potassium permanganate crystals (in small container)

CAUTION: Potassium permanganate is corrosive. Do not touch it. Do not spill any or get it on your clothes.

Fig. 10-9 Studying diffusion of a solid through a liquid

Procedure

a. Fill the petri dish with water.
b. Pour the potassium permanganate into the centre of the dish. Do not disturb the dish from this point on (Fig. 10-9).
c. Observe the dish for 10 min. Make notes and sketches of the changes that occur. Do this every 2 or 3 min.

Discussion

1. Describe the changes that occur as a solid diffuses through a liquid.
2. Use the particle theory to explain what happened in this activity.
3. Diffusion is usually defined this way: **Diffusion** [di-FEW-zhun] *is the process by which a substance spreads spontaneously* [spon-TAY-nee-us-lee] *in all directions from a region where there is a high concentration of the substance to a region where there is a lower concentration (perhaps none) of the substance.* That's a long definition. But think about it as you answer these questions.
    a) "Spontaneous" means "by itself". Do you agree with this part of the definition? Why?
    b) Where was the highest concentration of potassium permanganate?
    c) Where was the lowest concentration of potassium permanganate?
    d) Do you agree that a substance moves as stated in the definition. Why?

10.8 ACTIVITY The Crystallization of Salol

Crystals [KRIS-tals] are solids with regular shapes. They also have smooth flat surfaces. Many solids exist in the form of crystals. Each solid has a certain shape of crystal. For example, common salt is made of crystals that are small cubes. The surfaces of each crystal are always flat and the corners are always 90°.

What causes crystals to have regular shapes? You know that the

Section 10.8 101

particles in matter exert attractive forces on one another. Also, as a liquid is cooled, the particles are pulled together to form a solid. But the solid is always a crystal with the same shape. Perhaps the attractive forces are stronger in certain directions than in others.

This activity lets you watch crystals of a solid form. Perhaps you can verify the hypothesis just made.

Problem

How do crystals of salol form?

Materials

salol (phenyl salicylate)
glass petri dish
Bunsen burner
ice

hand lens (10 X)
test tube
beaker (250 mL)
ring stand, iron ring, wire gauze

CAUTION: Wear safety goggles during this activity.

Procedure

a. Pour a thin layer of solid salol into the petri dish. Place the top on the dish. Gently heat the dish until all of the salol has melted.
b. Turn off the Bunsen burner as soon as the salol has melted.
c. Let the petri dish cool until it is warm to the touch but not hot.
d. Drop a small piece of salol crystal into the liquid salol. Using the hand lens, observe what happens. How long does it take crystals to form? Write a description of what you observe. Draw several sketches. What shape are the crystals? Are they all alike? Note: You can remelt the salol and repeat steps (b) to (d) to check your results, if you wish.
e. Melt the salol again. The let it cool as described in step (c). This time sprinkle some powdered salol over the entire surface of the liquid salol. (Prepare the powdered salol by crushing a few small crystals of salol.) Describe what happens to the salol. Compare the time required for crystals to form with the time required for crystals to form in step (d). Compare the size and shape of the crystals with those in step (d).
f. Melt 2-3 cm^3 of salol in a test tube. Immediately plunge the lower half of the test tube into a beaker of ice water. Describe what happens. Compare the time required for crystals to form with the times required in steps (d) and (e). Compare the size and shape of the crystals with those in steps (d) and (e).

Discussion

1. Do the results of step (d) of the procedure support the hypothesis made in the introduction? Explain.
2. Account for what you observed in step (e) of the procedure.
3. Account for what you observed in step (f) of the procedure.

Main Ideas

1. An increase in temperature causes an increase in the volume of solids, liquids, and gases.
2. An increase in temperature increases the pressure of a gas.
3. As the pressure increases, the volume of a gas decreases.
4. As the pressure increases, the boiling point of a liquid increases.
5. The particle theory explains ideas 1 to 4. It also explains diffusion and crystallization.

Glossary

contraction	con-TRAK-shun	a decrease in volume
diffusion	di-FEW-zhun	the spontaneous movement of a substance from a region of high to low concentration of that substance
expansion	ex-PAN-shun	an increase in volume
spontaneous	spon-TAY-nee-us	proceeding by itself
crystallization	KRIS-tal-eye-ZAY-shun	the formation of crystals (solids of regular form)

Study Questions

A. True or False

Decide whether each of the following sentences is true or false. If the sentence is false, rewrite it to make it true. (Do not write in this book.)

1. An increase in temperature decreases the volume of a gas.
2. Most solids expand when heated.
3. The pressure of a gas increases when the temperature goes up.
4. The boiling point of water is always 100°C.
5. A mixture must be stirred before diffusion will occur.

B. Completion

Complete each of the following sentences with a word or phrase which will make the sentence correct. (Do not write in this book.)

1. When the temperature decreases, the volume of a liquid ________ .
2. Bridges and railways need ________ to protect them from damage due to temperature changes.
3. When the pressure increases, the volume of a gas ________ .
4. The boiling point of a liquid increases when the pressure ________ .
5. Solids with regular shapes are called ________ .

C. Multiple Choice

Each of the following statements or questions is followed by four responses. Choose the correct response in each case. (Do not write in this book.)

1. A rise in temperature will usually cause an increase in the volume of
   a) a gas, but not a liquid or solid
   b) a gas and a liquid, but not a solid
   c) a gas and a solid, but not a liquid
   d) a gas, a liquid, and a solid
2. When water is boiling on a kitchen stove, the temperature
   a) stays constant if the pressure does not change
   b) is always 100°C
   c) increases when the burner is turned to a higher setting
   d) increases as more and more water evaporates
3. Suppose a glass top on a steel jar is stuck. It could be more easily removed if you first
   a) put the top of the jar under a hot water tap
   b) reduced the air pressure outside the jar
   c) put the top of the jar in an ice bath for a few minutes
   d) put the whole jar in boiling water
4. A car manual tells the owner to check the tire pressure when the tires are cool. This is because
   a) tire gauges work best when cool
   b) warm tires will have a higher pressure than normal
   c) the pressure may change as you are taking it
   d) the pressure reading will be lower than normal
5. Use Table 10-2 (page 100) to help you with this question: Suppose you took a bottle of water to a planet. When you got out of the rocket ship, the temperature was 12°C. The pressure was 1.2 kPa. If you take the top off the bottle, the water will
   a) begin to evaporate slowly   c) freeze
   b) begin to boil   d) remain unchanged in the bottle

D. Using Your Knowledge

1. Aerosol cans have a warning on them. It says "DO NOT INCINERATE". Why?
2. An automobile radiator is kept under a pressure that is above normal when the car is running. How will this affect the boiling point of the liquid in the radiator? Why is the radiator kept at a pressure that is above normal?
3. Will water always boil at the same temperature in your home? Explain.
4. The instructions on a cake mix say: "Bake at 175°C. (At altitudes over 600 m, bake at 190°C.)" Why is this so?
5. Potatoes are often cooked in boiling water. It will take longer to do so at the top of a mountain than at its foot. Why?
6. A teaspoon of sugar is put in a glass of iced tea. A teaspoon of

sugar is also put in the same amount of hot tea. Neither is stirred. In which one will the sugar dissolve faster? Why?

E. Investigations

1. Get the apparatus of Section 10.4 from your teacher. Use it to show that the pressure of a gas decreases as the temperature decreases.
2. Find out why a pressure cooker is used to cook some foods. How does it work?
3. Study a dial thermometer. Draw a diagram of its working parts. Explain how it works.
4. Measure the air pressure in bicycle tires before and after a ride on a hot day. Describe and explain the results.
5. Find out how large crystals can be grown. Grow the largest crystal of alum that you can in one week.

11 Everyday Uses of the Particle Theory

11.1 Activity: Evaporation of Rubbing Alcohol
11.2 Activity: Evaporation of Water
11.3 How Does a Refrigerator Work?
11.4 Activity: Why Does Salt Melt Ice?
11.5 Activity: How Does Antifreeze Work?
11.6 Activity: Does Antifreeze Help a Car in the Summer?
11.7 Activity: Selecting Motor Oil for a Car

Many everyday things can be explained using the particle theory. This chapter lets you study some of them.

11.1 ACTIVITY Evaporation of Rubbing Alcohol

In Chapter 10 you learned that evaporation requires heat but you didn't do any activities on evaporation. Now you will do two. When you know the process well, you will understand how refrigerators work (in Section 11.3).

Problem

How does evaporation of alcohol affect temperature?

Materials

2 thermometers ($-10°C$ to $100°C$)
2 evaporating dishes (or other flat containers)
rubbing alcohol (50 mL)
droppers
graduated crylinder (50 mL)
electric fan (optional)

CAUTION: Rubbing alcohol is a deadly poison.

Procedure

a. Use the dropper to place a few drops of the alcohol on the back of your hand. Record your observations. Add more alcohol, if necessary. Then fan the wet spot with your hand. Record your observations.

b. Place 25 mL of alcohol in each of the 2 evaporating dishes. Take the air temperature. Then take the temperature of the alcohol in each dish.

c. Set one dish aside as a control. Fan the other quickly for at least 10 min. If possible, set it in front of an electric fan. Take the temperature of the alcohol in each dish every minute. Record your results in a table.

d. Pour the alcohol from each dish back into the graduated cylinder. Record the volume from each dish.

Discussion

1. Describe the results when you fanned the alcohol on your hand. Why did this happen? (Hint: What heat exchange accompanies evaporation?)
2. Copy Table 11-1 into your notebook. Complete it to compare the results from the two evaporating dishes.
3. Draw a conclusion from your data.

Table 11-1 Evaporation of Alcohol

	Control	Fanned dish
Temperature change		
Rate of evaporation		
Amount of evaporation		

4. Use the particle theory to explain the differences between the control and fanned dish.

11.2   ACTIVITY   Evaporation of Water

You have already studied the evaporation of water at its boiling point. Here you will study the evaporation of water at room temperature. Water evaporates more slowly than alcohol. Thus a wick is used to get the fastest possible rate.

Problem

How does evaporation of water affect temperature?

Materials

cotton wick
ring stand, adjustable clamp
beaker

2 thermometers ($-10°C$ to $110°C$)
water
electric fan (optional)

Fig. 11-1 Studying the evaporation of water

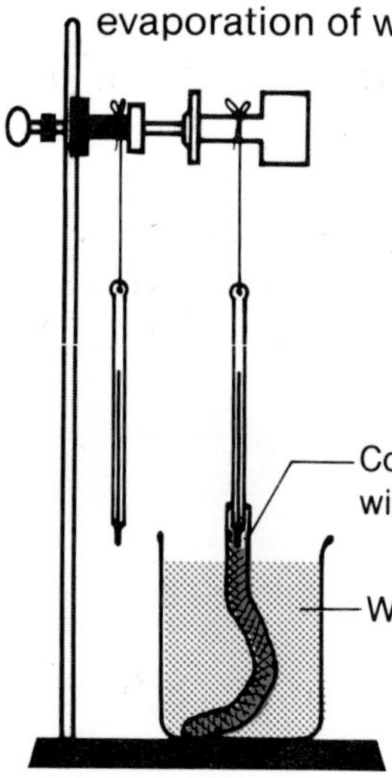

Procedure

a. Place one end of the cotton wick over the bulb of a thermometer. Hang the thermometer so that the other end of the wick is in a beaker of water (Fig. 11-1). Hang the second thermometer beside the first one. Make sure its bulb is dry.
b. Fan the thermometers with a book for at least 5 min.
c. Record the temperatures of both thermometers.

Discussion

1. Draw a conclusion from your experiment.
2. Use the particle theory to explain your results.
3. Why does water evaporate more slowly than alcohol? (Hint: Use the particle theory.)

11.3 How Does a Refrigerator Work?

Many home appliances make use of the cooling effect of evaporation. These are the refrigerator, air conditioner, dehumidifier, freezer, and heat pump. This section explains how a refrigerator works. The other appliances work on the same principle. You will study some of them at the end of this chapter.

Introduction

The refrigeration unit of a refrigerator contains a substance called a **freon** [FREE-on]. When a freon is in the gaseous state, it can be easily compressed to a liquid. Also, when it is in the liquid state, it can be easily evaporated. A condensation-evaporation cycle of the freon takes place in a refrigerator. It produces the cooling of materials in the refrigerator.

Evaporation Step

All a refrigerator does is move heat from within itself to the air in the room. Inside the refrigerator are cooling coils (Fig. 11-2). Liquid freon is forced through a small opening called a needle valve into these coils. As it passes through the valve, the freon becomes a vapour. That is, it evaporates. Evaporation requires heat. The heat comes from the cooling coils inside the refrigerator. It also comes from the air in the refrigerator. These, in turn, get heat from the contents of the refrigerator. Thus the contents are cooled.

Condensation Step

The freon vapour must now be condensed to a liquid. Then it can be forced through the needle valve again. A compressor is used to force the freon particles closer together. This forms a liquid. Condensation

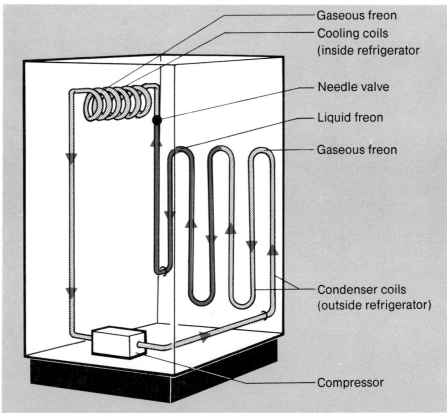

Fig. 11-2 The refrigerator uses the cooling effect of evaporation

produces heat. We do not want that in the refrigerator. Therefore the compression is done in the condenser coils outside the refrigerator box. You have probably felt the heat. It leaves the condenser coils on the back of the refrigerator when the compressor is running.

Section Review

1. Why is a freon used in a refrigerator?
2. Describe the evaporation step in the refrigeration cycle.
3. Describe the condensation step in the refrigeration cycle.

11.4   ACTIVITY   Why Does Salt Melt Ice?

You know salt melts ice. But have you ever wondered why? How can one solid make another solid melt?

Problem

Why does salt melt ice?

Materials

beaker (250 mL)   10 mL measure
thermometer   finely crushed ice (as small as coarse sand)
table salt   stirring rod

Section 11.4   109

Procedure

a. Prepare a data table with the column headings "Trial" and "Temperature".
b. Pour crushed ice into the beaker until it is half full.
c. Take the temperature of the ice while you are stirring it. Do not stir with the thermometer. Use the stirring rod. Put "1" in the "Trial" column and the temperature in the "Temperature" column.
d. When the thermometer reads close to 0°C, add 10 mL of salt to the ice. Continue stirring. Record the temperature after it stops changing. This is Trial 2.
e. Add another 10 mL of salt. Continue stirring. Record the temperature after it stops changing. This is Trial 3.
f. Repeat step (e) a few more times.

Discussion

1. What is the normal melting point of ice?
2. When ice is at its melting point, both solid and liquid states are present. Was the ice at its melting point throughout this activity?
3. What effect does salt have on the melting point of ice?

Explanation

How does salt melt ice on streets? All ice has a thin layer of water in it. Some of the salt dissolves in this water. As you discovered, the salt lowers the freezing (melting) point of the water-ice mixture. Suppose the air temperature is −10°C. Water would normally be in the solid state at this temperature. But, if the salt lowers the freezing point to −15°C, the water would be in the liquid state at −10°C. Therefore the ice melts.

11.5   ACTIVITY   How Does Antifreeze Work?

Antifreeze is used to protect car radiators from freezing. Its chemical name is **ethylene glycol** [ETH-i-leen GLY-kol]. It is never used alone; it is mixed with water. A 70:30 mixture gives the best protection. But a 50:50 mixture is good enough in most climates.

Problem

How does antifreeze keep the water in the radiator from freezing?

Materials

beaker
thermometer (−10°C to 110°C)
finely crushed ice (as small as coarse sand)

antifreeze (ethylene glycol)
graduated cylinder (25 mL)
stirring rod

CAUTION: Ethylene glycol is poisonous.

Procedure

a. Prepare a data table with the column headings "Trial" and "Temperature".
b. Pour crushed ice into the beaker until it is half full.
c. Take the temperature of the ice while you are stirring it. Do not stir with the thermometer. Use the stirring rod. Put "1" in the "Trial" column and the temperature in the "Temperature" column.
d. When the thermometer reads close to 0°C, add 15 mL of antifreeze to the ice. Continue stirring. Record the temperature after it stops changing (Fig. 11-3). This is Trial 2.
e. Add another 15 mL of antifreeze. Continue stirring. Record the temperature after it stops changing. This is Trial 3.
f. Repeat step (e) a few more times.

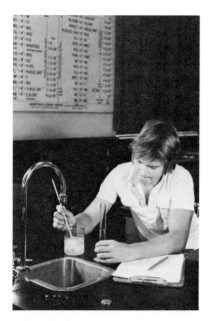

Fig. 11-3 Studying the effect of antifreeze on melting ice

Discussion

1. What effect does antifreeze have on the melting point of ice?
2. What effect does antifreeze have on the freezing point of water?
3. Why is antifreeze added to the water in the radiator of cars for winter driving?

Explanation

Water alone freezes at 0°C. Ethylene glycol alone freezes at −13°C. But a mixture of the two freezes at a much lower temperature. For example, a 70:30 mixture of ethylene glycol and water freezes at −64°C. And a 50:50 mixture freezes at −35°C. Like salt, ethylene glycol lowers the freezing point of water.

11.6 ACTIVITY Does Antifreeze Help a Car in the Summer?

Most car makers tell us to leave the antifreeze in the car radiator during the summer. Why?

Problem

Why do we leave antifreeze in a car radiator in the summer?

Materials

round-bottomed flask (250 mL) Bunsen burner
thermometer (−10°C to   ring stand, iron ring, wire gauze
  110°C)   2 adjustable clamps
distilled water   50 mL of antifreeze (ethylene
boiling chips   glycol)

CAUTION: Wear safety goggles during this activity.

Procedure

a. Set up the apparatus as shown in Figure 9-4, page 85. Begin with 100 mL of water in the flask.
b. Heat the water until it is boiling. Note the temperature.
c. Add 10 mL of antifreeze to the water. Continue heating the mixture. Note the new boiling point.
d. Repeat step (c) until all the antifreeze has been used.
e. Record your results in a table.

Discussion

1. Describe the effect of antifreeze on the boiling point of water.
2. Use the particle theory to explain this effect.
3. Why do car makers recommend leaving the antifreeze in a car radiator during the summer?
4. Compare the effects of antifreeze on the boiling point and freezing point of water.
5. Does antifreeze help or hinder boiling? Does antifreeze help or hinder melting?

Explanation

The normal boiling point of pure water is 100°C. But, when ethylene glycol is mixed with water, the boiling point goes up. Why is this so? There are strong attractive forces between ethylene glycol molecules and water molecules. Therefore the water molecules need extra energy to evaporate. As a result, more heat must be added before boiling will occur. Thus the boiling point rises.

Car makers tell us to leave the antifreeze in our car radiators in the summer. You probably know why now. On a hot summer day pure water might start to boil in the radiator. But water with antifreeze in it won't boil so easily.

11.7   ACTIVITY   Selecting Motor Oil for a Car

The term **viscosity** [vis-KOS-i-tee] describes how easily a liquid "runs" when it is poured. Some liquids such as corn syrup are thick. They run slowly when poured. They are called **viscous** [VIS-kus]. They are said to have a high viscosity. Other liquids such as gasoline and water are thin and splashy. They have a low viscosity. As a result, they run easily through a rubber hose. Imagine trying to run corn syrup through a garden hose!

What makes corn syrup more viscous than gasoline? According to the particle theory, both liquids are made of particles. Also, the particles have attractive forces that pull them together. Apparently the forces between corn syrup particles are stronger. Viscous liquids thin out when heated. How does the particle theory explain that?

In this activity you will compare the viscosities of different oils.

You will also study the effect of temperature on viscosity. The results should help you decide on the best oil to use in different seasons. Using the right oil increases the fuel efficiency of a car. It also increases engine life.

Problem

What grade of motor oil should be used in different seasons?

Materials

tall glass containers (at least 20 cm)
single grade motor oil of several viscosities (SAE 10; SAE 20; SAE 30; SAE 40; SAE 50)
multigrade motor oil (SAE 10W-40; SAE 20W-50)
hot-plate (on teacher's bench)
plastic sphere
thermometer ($-10°C$ to $110°C$)
large tin can
watch or clock with second hand

CAUTION: Avoid open flames near the oil. Do not heat oil above 45°C. Hot oil can cause serious burns.

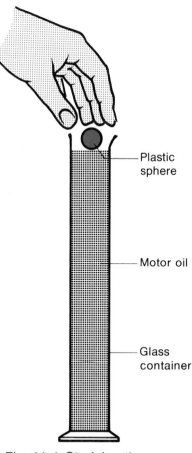

Fig. 11-4 Studying the viscosity of motor oil

Procedure A Comparing Different Grades of Motor Oil

Your teacher has placed several tall glass containers at various stations in the lab. Each container contains motor oil of a certain grade. The grade is marked on the container. At each station you are to do the following:
a. Take the temperature of the oil.
b. Hold a plastic sphere so it just touches the surface of the oil (Fig. 11-4). Release it. Time how long it takes for the sphere to reach the bottom. Do not try to remove the spheres. Your teacher will do this.
c. Record your results in a table with the column headings "Grade" and "Time".

Procedure B Viscosity and Temperature

a. Prepare a data table with the headings "Temperature" and "Time".
b. Your teacher has samples of SAE 50 motor oil at several temperatures on the demonstration desk. Form a group of 6 to 8 students and go to the demonstration desk. Repeat steps (a) and (b) of Procedure A using each of the samples. Your teacher will remove the spheres. Record all your results.

Procedure C What is Multigrade Motor Oil?

a. A multigrade motor oil is described by two numbers, for example, SAE 10W-40. Write in your notebook what you think this means. If you have no idea at all, discuss the matter with a classmate who

does. What you write in your notebook is a hypothesis concerning the nature of multigrade motor oil.
b. Design an experiment to test your hypothesis. Write the procedure in your notebook.
c. Discuss your procedure with your teacher. Then get the materials to do the experiment.
d. Record your results in a table.

Discussion

1. Make a general conclusion that summarizes the results of Procedure A. Keep in mind that the oil has to flow around the sphere as it drops.
2. Compare the attractive forces between the particles in each of the oil samples used in Procedure A.
3. How does temperature affect the viscosity of a single grade motor oil? Explain your answer using the particle theory.
4. Compare the results from Procedures B and C. What are the advantages of a multigrade motor oil?
5. If you are using a single grade motor oil in your car, why must you change the oil when the season changes?
6. What do 10W and 40 mean for 10W-40 oil?

Main Ideas

1. Evaporation requires heat. Therefore it will cool things on which it occurs.
2. A refrigerator uses an evaporation-condensation cycle.
3. Salt lowers the melting point of ice.
4. Antifreeze lowers the freezing point of water.
5. Antifreeze raises the boiling point of water.
6. The viscosity of a single grade oil changes with the temperature. The viscosity of a multigrade oil changes only slightly with temperature.
7. The particle theory explains evaporation and viscosity.

Glossary

ethylene glycol	ETH-i-leen GLY-kol	antifreeze used in car radiators
freon	FREE-on	coolant in a refrigerator
viscosity	vis-KOS-i-tee	stickiness of a liquid
viscous	VIS-kus	sticky liquid that will not pour easily

Study Questions

A. True or False

Decide whether each of the following sentences is true or false. If the sentence is false, rewrite it to make it true. (Do not write in this book.)
1. Rubbing alcohol cools your skin because it is a cold liquid.
2. All liquids evaporate at about the same rate.
3. Salt lowers the freezing point of water.
4. The freezing point and melting point of a substance are the same.
5. Antifreeze should be drained from a car in the summer.

B. Completion

Complete each of the following sentences with a word or phrase which will make the sentence correct. (Do not write in this book.)
1. A refrigerator uses two changes of state, ▓▓▓▓ and ▓▓▓▓.
2. Salt melts ice because it lowers the ▓▓▓▓ of the ice.
3. Antifreeze raises the ▓▓▓▓ of water.
4. A liquid that pours slowly is said to be ▓▓▓▓.
5. The viscosity of ▓▓▓▓ motor oil changes little with temperature.

C. Multiple Choice

Each of the following statements or questions is followed by four responses. Choose the correct response in each case. (Do not write in this book.)
1. What change of state does freon undergo in the cooling coils of a refrigerator?
   a) condensation **b)** freezing **c)** evaporation **d)** liquefaction
2. The addition of antifreeze to water causes
   a) an increase in both the boiling point and the freezing point of water
   b) a decrease in both the boiling point and the freezing point of the water
   c) an increase in the boiling point and a decrease in the freezing point of water
   d) an increase in the freezing point and a decrease in the boiling point of the water
3. Water will evaporate most rapidly on
   a) a cold calm day   **c)** a warm calm day
   b) a cold windy day   **d)** a warm windy day
4. An SAE-50 motor oil is quite viscous at 0°C. At 30°C it will be
   a) slightly more viscous   **c)** much more viscous
   b) slightly less viscous   **d)** much less viscous
5. A multigrade motor oil has SAE 20W-50 printed on the top.
   a) This oil will not get thicker in the winter than an SAE-20 oil; nor will it get thinner in the summer than an SAE-50 oil.

b) This oil will not get thicker in the winter than an SAE-50 oil; nor will it get thinner in the summer than an SAE-20 oil.
c) This oil will not get thicker in the summer than an SAE-20 oil; nor will it get thinner in the winter than an SAE-50 oil.
d) This oil will not get thicker in the summer than an SAE-50 oil; nor will it get thinner in the winter than an SAE-20 oil.

D. Using Your Knowledge

1. Why is rubbing alcohol used on the skin of bed-ridden persons?
2. Imagine you are swimming in a swimming pool. The water and air are warm. But it is a windy day. You will feel cold when you come out of the pool. Why?
3. In some warm countries, drinking water is stored in porous jugs. (Porous means "having many tiny holes".) Why is this done?
4. A desert nomad often carries his drinking water in a leather bag. The bag leaks slowly over its entire surface. The nomad knows this. Why does he use a leather bag?
5. How does sweating help us keep cool on a hot day?
6. Pioneers used to store vegetables and fruits in their basements. On a cold night these foods could get frozen. To prevent this, the pioneers put pans of water in the basement. The water would freeze, but the foods would not freeze. Why is this so?
7. An air conditioner is basically a refrigerator that is partly outdoors. What part is outdoors? Why?
8. Why is salt not used in the radiators of cars to prevent freezing?

E. Investigations

1. Dogs do not sweat. Find out what adaptation they have for keeping cool on a hot day. Explain how it works.
2. Does a fan make things cooler? Use a fan to blow air on a thermometer. See if the temperature changes. Now blow the air on yourself. Do you feel cooler? Explain your results.
3. Check the owner's manual of the family car. What grade(s) of oil does it tell you to use? Is the right grade being used?
4. Refrigerators are made that burn natural gas or propane. They have a flame in them. How can they cool the food? Find out how gas refrigerators work.
5. Find out what a heat pump is. Explain how it can both heat and cool a building.
6. "Air conditioners are needed in many homes and stores. Many people who drive a lot also need them in their cars." Do you agree or disagree? If you agree, prepare a list of the benefits of air conditioning. If you disagree, prepare a list of the "costs" to society of air conditioning. Your teacher will set up a debate on this matter.

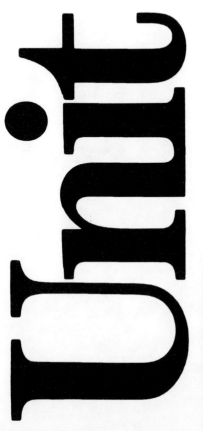

Solutions and Mixtures

CHAPTER 12
Solutions and Other Mixtures

CHAPTER 13
Separation of Substances: Filtration and Distillation

CHAPTER 14
Separation of Substances: Sedimentation and Floc Formation

In Unit 3 you studied physical properties. Changes of state, freezing point, and boiling point are some of them. Then you used the particle theory to explain your results. You used water in some activities. You made solutions in others. You even made some mixtures. But you didn't find out much about the properties of solutions and mixtures.

In this unit you will investigate solutions and mixtures in detail. You probably know very little about solutions and mixtures. But you will learn much useful information in this unit.

Fig. 12-0 These students are heating a solution. A solution is a kind of mixture. The heat will separate the mixture into its parts.

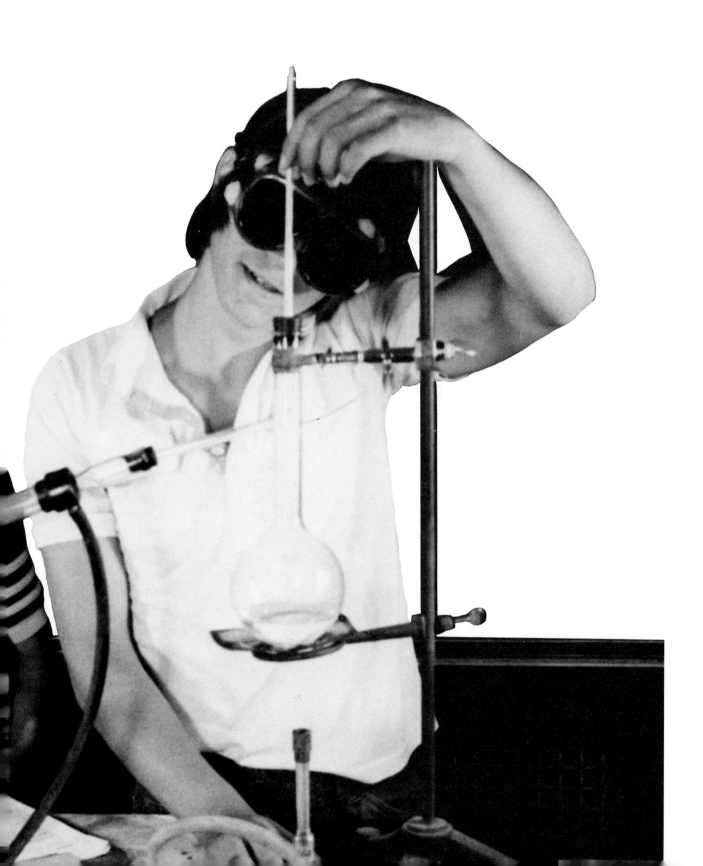

12 Solutions and Other Mixtures

12.1 Solutions
12.2 Other Mixtures
12.3 Activity: Classifying Some Mixtures
12.4 Activity: Temperature Changes During Dissolving
12.5 Activity: Volume Changes During Dissolving
12.6 Activity: What Factors Affect the Rate of Dissolving?
12.7 Classes of Solutions
12.8 Activity: Properties of a Supersaturated Solution
12.9 Solubility
12.10 Activity: Factors Affecting Solubility

You are familiar with many solutions. Soft drinks, tea, and vinegar are solutions. So is tap water. Solutions are mixtures. Ketchup and milk are also mixtures. But they are not solutions. This chapter will show you how these mixtures differ.

12.1 Solutions

This section reviews things you probably know about solutions. It also defines terms you should know. Then it describes other types of mixtures. Learn this section well. You need to know it to do the activities in this chapter.

Fig. 12-1 Solutions. What do all of these have in common?

What is a Solution?

Soft drinks are solutions. Vinegar and shaving lotion are solutions. So are perfumes, tea, and mouthwash. If you stir sugar into tea you make a solution. If you shake some salt with water, you make a solution. What are solutions? What do all of these have in common (Fig. 12-1)?

Some solutions are coloured. Others are colourless. But all solutions, except one type, are clear. (Clear is the opposite of cloudy.)

Solutions are mixtures. They are made of two or more substances mixed together. Yet they look like only one substance. We say a solution has only one **phase** [FAYZ]. That is, you can see only one form of matter in it. Matter with only one phase is **homogeneous** [ho-mo-GEE-nee-us]. It is of uniform composition throughout. Solutions, then, are homogeneous. For example, a bottle of pop is the same composition throughout. It is not thicker or darker near the bottom.

A solution can be defined this way: *A **solution** is a homogeneous mixture of two or more substances.*

A solution is made by **dissolving** one substance in another. The substance which dissolves is the **solute** [SOL-yoot]. The substance which does the dissolving is the **solvent** [SOL-vent]. The solute is said to be **soluble** [SOL-you-bil] in the solvent. Thus:

$$\text{SOLUTE + SOLVENT} \xrightarrow{\text{dissolving}} \text{SOLUTION}$$

For example, suppose a sugar cube is stirred in a glass of water. The sugar is the solute. The water is the solvent. The sugar is said to be soluble in the water. The sweet liquid formed is a solution.

Not all solutions are made by dissolving a solid in a liquid. Nor is the liquid always water. For example, soft drinks have carbon dioxide gas dissolved in water. Vinegar is acetic acid (a liquid) dissolved in water. Shaving lotion and perfume are both made of several solids and liquids dissolved in alcohol.

Some solutions consist of solids dissolved in solids! Brass is an example. It looks like one substance. Yet it is really a homogeneous mixture, or solution, of copper and zinc. Of course, solid solutions are not clear.

Particle Explanation of Dissolving

Suppose you put a sugar cube in a cup of water but did not stir it. A few days later the sugar will be gone. It has dissolved in the water. Or, it has formed a solution. How does this happen?

The particle theory tells us that sugar is made of particles. They are held together by attractive forces. They need extra energy to overcome those forces. Water particles are always bumping into sugar particles. Sometimes they give the sugar particles extra energy. Then the sugar particles break away and move into the water.

Section Review

1. Define solution.
2. What does a solution look like?
3. Define these terms: phase, homogeneous, solute, solvent.
4. Use the particle theory to explain dissolving.

12.2 Other Mixtures

All mixtures are not solutions. Dirty water, ketchup, and oil in water are examples. These mixtures have more than one phase. Also, their composition is not uniform. Such mixtures are said to be **heterogeneous** [he-ter-o-GEE-nee-us] (Fig. 12-2). At first glance, some of these mixtures may look like solutions. But this section explains how they are different.

Fig. 12-2 Heterogeneous mixtures. What do they have in common?

The Tyndall Effect

Have you ever seen the beam of light when a movie projector shines through dusty or smoky air? The particles in the air are large enough to reflect light. This makes the beam of light visible. We call this the **Tyndall effect**. There are three different kinds of heterogeneous mixtures that can be confused with solutions. These are suspensions, emulsions, and colloidal dispersions. All three show the Tyndall effect. Solutions do not. The particles in a solution are too small to reflect the light.

Suspensions

A **suspension** [sus-PEN-shun] is a heterogeneous mixture of two or more substances. Some of its particles are large enough to be seen with the naked eye. It is cloudy. Some of the particles will settle out on standing. It will show the Tyndall effect. Clay in water is a suspension. Can you think of any other examples of suspensions?

Emulsions

An **emulsion** [ee-MUL-shun] is a suspension of one liquid in another. Again, some particles are large enough to see. These particles are tiny liquid droplets. They are suspended in another liquid. An emulsion is cloudy. It will show the Tyndall effect. The two liquids may separate on standing. Oil and water often form an emulsion if they are mixed. Can you think of any other examples of emulsions?

Colloidal Dispersion

A **colloidal dispersion** [kol-LOID-al di-SPER-shun] is part way between a solution and a suspension. It is slightly cloudy. It may appear homogeneous, but it is actually heterogeneous. Its particles are large enough to show the Tyndall effect. But they are not large enough to settle out on standing. You cannot see the particles, even with your microscope. Yet they are much larger than the particles of solute in a solution. You probably cannot think of an example. But you will meet one in the activity that follows.

Section Review

1. What is a heterogeneous mixture?
2. What is the Tyndall effect?
3. Name three types of heterogeneous mixtures.
4. Describe a suspension.
5. Describe an emulsion.
6. Describe a colloidal dispersion.
7. "An emulsion is a special type of suspension." Explain this statement.
8. What is the main difference between a colloidal dispersion and a suspension?

12.3 ACTIVITY Classifying Some Mixtures

In this activity you will make some mixtures. Some are homogeneous (solutions). Others are heterogeneous (suspensions, emulsions, colloidal dispersions). After you have made them, you will decide what each mixture is. Use the information in Sections 12.1 and 12.2.

Problem

Can you classify some mixtures as solutions, suspensions, emulsions, or colloidal dispersions?

Materials

6 test tubes
test tube rack
marking pen
ammonium chloride
copper sulfate

cobalt chloride
liquid laundry starch
fine soil
kerosene
light source (e.g. flashlight)

Procedure

a. Number the test tubes from 1 to 6.
b. Fill all 6 test tubes about two-thirds full of water. Stand them in the test tube rack.
c. Add the materials to the test tubes as described in Table 12-1. (Note: The amounts need not be exact.)

Table 12-1

Test tube number	Material to be added
1	1 g of ammonium chloride
2	1 g of copper sulfate
3	1 g of cobalt chloride
4	1 mL of laundry starch
5	1 g of fine soil
6	1 mL of kerosene

Fig. 12-3 Can you see the Tyndall effect?

d. Make a mixture in each case. Shake each test tube vigorously for 2-3 min.
e. Write a description of each mixture.
f. Shine a beam of light through each mixture. Note whether or not the Tyndall effect shows (Fig. 12-3).
g. After 10 min write a new description of each mixture. Also, repeat step (f) for each one.

Section 12.3   123

Discussion

1. Which mixtures are solutions? Why?
2. Which mixtures are suspensions? Why?
3. Which mixture is an emulsion? Why?
4. Which mixtures are colloidal dispersions? Why?
5. How do your studies of mixtures support the particle theory?

12.4 ACTIVITY Temperature Changes During Dissolving

In this activity you will dissolve a solid in a liquid. The purpose is to find out if heat is required or given off. Let's begin by making a prediction.

The particles in a solid are held together by attractive forces. When the solid dissolves, these particles break away from one another. Will this require heat or release heat? Why? Will the liquid get warmer or cooler?

Problem

Is heat given off or required when a substance dissolves?

Materials

test tube   thermometer ($-10°C$ to $110°C$)
ammonium chloride, potassium nitrate, sodium thiosulfate, sodium sulfate, and other soluble solids

Procedure

a. Add 5 mL of water to the test tube. Take the temperature of the water.
b. Add 1 g of ammonium chloride to the water. Shake the mixture vigorously for about 30 s. Take the temperature of the resulting solution.
c. Repeat steps (a) and (b) with the other solids provided by your teacher. Try to begin each test with water at the same temperature.
d. Record all of your data in a table.

Discussion

1. Make a conclusion regarding heat changes that occur during the dissolving of a solid in a liquid.
2. Explain your conclusion using the particle theory.
3. Are the heat changes the same for all solids? Explain your answer.

12.5 ACTIVITY Volume Changes During Dissolving

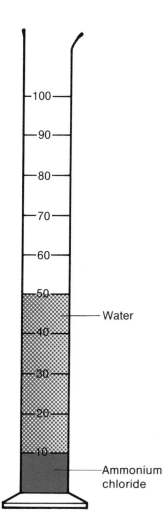

Fig. 12-4 Does the total volume change when a solution forms?

The particle theory tells us that there are spaces between the particles in any type of matter. Ammonium chloride has spaces between its particles. Liquid water also has spaces between its particles. Imagine that 10 mL of ammonium chloride is in a graduated cylinder with 90 mL of water as shown in Figure 12-4. The total volume is 100 mL. Now, suppose you dissolve the ammonium chloride by shaking. What will happen to the total volume? (Use the particle theory to make your prediction.)

Problem

Does the volume change when a solution forms?

Materials

graduated cylinder (100 mL)
ammonium chloride
alcohol (coloured)
100 mL gas measuring tube (if available)

Procedure A Ammonium Chloride and Water

a. Place 50 mL of water in the graduated cylinder.
b. Add 5 g of ammonium chloride to the water without stirring.
c. Note the total volume of water plus ammonium chloride.
d. Shake the mixture until the ammonium chloride dissolves. Be careful not to lose any of the solution.
e. Note the volume of the resulting mixture.

Procedure B Alcohol and Water

a. Place 50 mL of water in the graduated cylinder. (Use a gas measuring tube if one is available. It will give a more accurate result.)
b. *Slowly* pour about 50 mL of coloured alcohol down the side of the graduated cylinder. It should form a layer on top of the water.
c. Measure the total volume of alcohol and water.
d. Mix the two liquids. Do not lose any of the solution.
e. Note the volume of the resulting solution.

Discussion

1. What happens to the volume when a solution forms?
2. Use the particle theory to explain the results of your experiment in Procedure A.
3. Use the particle theory to explain the results of your experiment in Procedure B.

12.6 ACTIVITY What Factors Affect the Rate of Dissolving?

Suppose you want to dissolve a sugar cube in water as fast as possible. What would you do? Can you use the particle theory to explain why?

Problem

What factors will speed up the rate of dissolving?

Materials

2 test tubes
copper sulfate crystals
powdered copper sulfate
Bunsen burner

CAUTION: Wear safety goggles during this activity.

Procedure A Effect of Shaking or Stirring

a. How will shaking or stirring affect the rate of dissolving of a solid in a liquid? Write your prediction in your notebook.
b. Half fill the two test tubes with water.
c. Add 1 g of copper sulfate crystals to each test tube.
d. Let one tube stand undisturbed. Shake the second one vigorously for 2-3 min.

Procedure B Effect of Size of Pieces

a. How will grinding the solid into small pieces affect the rate of dissolving of a solid in a liquid? Write your prediction in your notebook.
b. Half fill the two test tubes with water.
c. Add 2 g of copper sulfate crystals to one test tube. Then add 2 g of powdered copper sulfate to the other.
d. Shake each tube vigorously for 2 min.

Procedure C Effect of Temperature

a. How will an increase in temperature affect the rate of dissolving? Write your prediction in your notebook.
b. Half fill the two test tubes with water.
c. Add 2 g of copper sulfate crystals to each test tube.
d. Let one test tube stand at room temperature. Heat the other one carefully with a Bunsen flame for 2-3 min (Fig. 12-5).

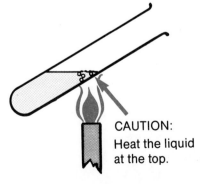

CAUTION: Heat the liquid at the top.

Fig. 12-5 Always heat a liquid at the top. If you don't, the liquid may shoot out of the test tube.

Discussion

1. What was the effect of shaking on the rate of dissolving? Explain

the effect using the particle theory.
2. Will the copper sulfate in the undisturbed test tube eventually dissolve? Explain.
3. State the effect of size of pieces on the rate of dissolving. Explain the effect using the particle theory.
4. State the effect of temperature on the rate of dissolving. Explain the effect using the particle theory.

12.7 Classes of Solutions

You may have heard of dilute and concentrated solutions. Or you may have heard of a saturated solution. These and other important terms are explained in this section.

Dilute and Concentrated

Suppose you dissolve just 2 or 3 crystals of table salt in 100 mL of water. The solution will taste slightly salty. You have made a **dilute** [di-LOOT] solution. But, suppose you dissolve a spoonful of table salt in the same amount of water. The solution will taste very salty. Compared to the first solution, this one is **concentrated** [CON-sen-tray-ted]. Dilute and concentrated are relative terms only. They don't tell you exactly how much salt is in the water. Imagine that you and your classmates were asked to make up dilute solutions of salt. You would not all use the same amount of salt, but all of you would probably put only a few crystals of salt in 100 mL of water. Now imagine that each person in your class was asked to make up concentrated solutions of salt. Again, you would not all add the same amount, but each of you would use more salt than you did to make a dilute solution.

Concentration

Dilute and concentrated are relative terms. They only tell us in a general way whether there are small or large amounts of solute in the solvent. They do not tell us exactly *how much* solute is in a certain amount of solvent. Sometimes we need to know that. That is, we need to know the **concentration** [con-sen-TRAY-shun] of the solution.

The **concentration** *of a solution is the mass of solute in grams dissolved in 100 g of solvent.* Here is an example: In Activity 12.5 you dissolved 5 g of ammonium chloride in 50 mL (50 g) of water. This is the same as dissolving 10 g of ammonium chloride in 100 g of water. Therefore the concentration of the solution is said to be 10 g per 100 g of water.

Unsaturated, Saturated, and Supersaturated Solutions

Three other terms are used to classify solutions. They are unsaturated, saturated, and supersaturated. Let's see what they mean.

You proved in Activity 12.5 that at least 10 g of ammonium chloride will dissolve in 100 g of water. Suppose you tried 20 g. Would all of it dissolve? Or 30 g? Or 40 g? What is the most ammonium chloride that will dissolve in 100 g of water? Experiments show that, at room temperature (20°C), 37 g is all that can dissolve. If you put in 38 g, 1 g will remain undissolved. It doesn't matter how long you shake it. If time permits, your teacher will let you prove this. Now, let us use this information to explain the three terms.

A **saturated** [SATCH-er-ay-ted] *solution is one in which no more solute can dissolve in the same amount of solvent at the same temperature.* Thus 37 g of ammonium chloride in 100 g of water at 20°C makes a saturated solution. No more will dissolve.

An **unsaturated** [un-SATCH-er-ay-ted] *solution is one in which more solute can dissolve in the same amount of solvent at the same temperature.* Thus 2 g of ammonium chloride in 100 g of water at 20°C is an unsaturated solution. In fact, any amount up to 37 g of ammonium chloride is unsaturated.

A **supersaturated** [soo-per-SATCH-er-ay-ted] *solution contains more solute than can normally be dissolved in the solvent at that temperature.* In other words, it contains more solute than a saturated solution. How can this be? You will find out in the next activity.

Section Review

1. What is the difference between a dilute and concentrated solution?
2. Define the term concentration.
3. A solution has a concentration of 45 g per 100 g of water at 20°C. What does this mean?
4. How would you make a salt solution with a concentration of 2 g per 100 g of water at 20°C?
5. What is the difference between a saturated and an unsaturated solution?
6. A saturated solution of potassium nitrate at 60°C has a concentration of 110 g per 100 g of water. What does this mean?
7. What is a supersaturated solution?

12.8   ACTIVITY   Properties of a Supersaturated Solution

Most solutions can hold more solute at high temperatures. Suppose that you cool a warm saturated solution. Some solute will settle out. The cooler solution cannot hold all of it. This happens with most solutes.

However, not all solutes behave in this way. Some solutes will stay in the solution when cooled. The cool solution now holds more solute than a saturated solution. It is said to be **supersaturated** Do this activity and see if you can make a supersaturated solution.

Problem

Can you make a supersaturated solution?

Materials

test tube (20 x 150 mm)   sodium acetate crystals
adjustable clamp   Bunsen burner
CAUTION: Wear safety goggles during this activity.

Procedure

a. Make sure the test tube is clean and dry.
b. Fill the test tube half full of the crystals.
c. Add 10 drops of water to the crystals. Shake the mixture. Note any change in temperature with your hand.
d. Gently heat the mixture until all of the crystals have dissolved.
e. Hold the test tube in a stream of running water from a tap. Keep it there until the solution has been cooled to about room temperature. If no crystals form, you have made a supersaturated solution. If crystals do form, repeat steps (d) and (e). But be more gentle with your handling of the test tube.
f. Add one crystal of sodium acetate to the supersaturated solution. Describe what happens. Note any changes in temperature with your hand.

Discussion

1. Is the solution after step (c) unsaturated or saturated?
2. Which word best describes the solution after step (d) — concentrated or saturated?
3. Is the solution saturated or unsaturated after step (f)? Explain.
4. Is heat required or given off when the crystals dissolve? Explain your answer using the particle theory.
5. Is heat required or given off when the crystals crystallize out of solution? Explain your answer using the particle theory.

12.9 Solubility

What is Solubility?

Solubility *is the mass of solute in grams that dissolves in 100 g of solvent to form a saturated solution at a given temperature.* That's a long definition. Let's see what exactly it means. For example, the solubility of ammonium chloride is 37 g per 100 g of water at 20°C. This means that 37 g of ammonium chloride in 100 g of water makes a saturated solution at 20°C.

At 20°C only 11 g of potassium sulfate will dissolve in 100 g of water. Therefore its solubility is 11 g per 100 g of water at 20°C.

Potassium sulfate is not as soluble as ammonium chloride. It takes less of it to make a saturated solution.

Section Review

1. Define solubility.
2. The solubility of ammonium chloride at 50°C is 50 g per 100 g of water. What does this mean?

12.10   ACTIVITY   Factors Affecting Solubility

Three factors can affect solubility. One is the **nature of the solute**. Suppose the particles in the solute have strong forces between them. They will not dissolve easily. And the solubility will be low.

A second factor is the **nature of the solvent**. One solvent, like water, may help to pull solute particles apart. Another, like alcohol, may not be able to do so.

The third factor is **temperature**. High temperatures give extra energy to particles. This should help them dissolve.

Proceed with the activity to learn more about the three factors that can affect solubility.

Problem

How do the nature of the solute, the nature of the solvent, and the temperature affect solubility?

Materials

5 test tubes   Bunsen burner   adjustable clamp
sugar, common salt, calcium carbonate, kerosene, alcohol, iodine
   crystals, ammonium chloride
CAUTION: Wear safety goggles during this activity.

Procedure A   Nature of the Solute

a. Half fill the 5 test tubes with water. To each test tube add one of the following solutes: 1 g common salt, 1 g sugar, 1 g calcium carbonate, 2 mL kerosene, 2 mL alcohol.
b. Shake each tube vigorously for 2-3 min. Let it stand for 2-3 min. Record the results.

Procedure B   Nature of the Solvent

a. Half fill 1 test tube with water and another with alcohol.
b. Add an iodine crystal of the same size to each test tube. CAUTION: Do not touch or smell the crystal. It will burn flesh and clothing.
c. Shake each tube vigorously for 2-3 min. Record the results.

Procedure C Effect of Temperature

a. Place 5 mL of water in a test tube.
b. Obtain 5 g of ammonium chloride from your teacher. Divide it into 5 portions that are about 1 g each.
c. Add one portion of the ammonium chloride to the water. Shake the test tube vigorously until the ammonium chloride dissolves.
d. Repeat step (c) with a second portion of ammonium chloride. Then use a third portion. Continue until some of the ammonium chloride will not dissolve.
e. Heat the mixture to 50°-60°C. Record what happened to the undissolved ammonium chloride.
f. Add another portion of ammonium chloride. Shake the test tube vigorously.
g. Heat the mixture to the boiling point. Did all the ammonium chloride dissolve?
h. Add any remaining ammonium chloride. Shake the tube vigorously after each addition.
i. Cool the solution by running cold water over the outside of the test tube. Record your observations.

Discussion

1. Which of the solutes in Procedure A were soluble? Which were insoluble?
2. Describe and explain the results of Procedure B.
3. Describe and explain the effect of increased temperature on the solubility of ammonium chloride. What happened when the hot concentrated solution was cooled? Why?

Main Ideas

1. A solution is a homogeneous mixture of two or more substances.
2. A solute dissolves in a solvent to form a solution.
3. The particle theory explains dissolving.
4. Suspensions, emulsions, and colloidal dispersions are heterogeneous mixtures.
5. The temperature often decreases when a solid dissolves in a liquid.
6. The total volume decreases during dissolving.
7. Shaking, size of pieces, and temperature affect the rate of dissolving.
8. A saturated solution can dissolve no more solute at the same temperature.
9. Solubility is the mass of solute in grams that dissolves in 100 g of solvent to form a saturated solution at a given temperature.
10. Solubility is affected by the nature of the solute, the nature of the solvent, and the temperature.

Glossary

colloidal dispersion	kol-LOID-al di-SPER-shun	a heterogeneous mixture with no particles visible, even with a microscope
concentration	con-sen-TRAY-shun	the mass of solute in 100 g of solvent
emulsion	ee-MUL-shun	a suspension of liquid droplets in another liquid
heterogeneous	he-ter-o-GEE-nee-us	having two or more phases and non-uniform composition
homogeneous	ho-mo-GEE-nee-us	having one phase and uniform composition
phase	FAYZ	a sample of matter that appears to be all the same
saturated	SATCH-er-ay-ted	no more solute can dissolve in the same amount of solvent at the same temperature
soluble	SOL-you-bil	able to dissolve
solute	SOL-yoot	the substance which dissolves
solution		a homogeneous mixture of two or more substances (solute in a solvent)
solvent	SOL-vent	the substance that does the dissolving
supersaturated	soo-per-SATCH-er-ay-ted	contains more solute than a saturated solution at the same temperature
suspension	sus-PEN-shun	a heterogeneous mixture of two or more substances; some particles visible

Tyndall effect		a beam of light made visible by particles
unsaturated	un-SATCH-er-ay-ted	more solute can dissolve in the same amount of solvent at the same temperature

Study Questions

A. True or False

Decide whether each of the following sentences is true or false. If the sentence is false, rewrite it to make it true. (Do not write in this book.)
1. All solutions are colourless.
2. A solute is the substance that dissolves.
3. Solutions show the Tyndall effect.
4. An emulsion is a suspension of liquid droplets in another liquid.
5. A supersaturated solution contains less solute than a saturated solution under the same conditions.

B. Completion

Complete each of the following sentences with a word or phase which will make the sentence correct. (Do not write in this book.)
1. A homogeneous mixture of two or more substances is called a _____.
2. When a solid dissolves in a liquid the temperature often _____.
3. The total volume _____ when a solution is formed.
4. The mass of solute dissolved in 100 g of solvent is called the _____.

C. Multiple Choice

Each of the following statements is followed by four responses. Choose the correct response in each case. (Do not write in this book.)
1. 10 g of salt were dissolved in 100 g of water. What is the concentration of this solution?
   a) 10 g/110 g **b)** 10 g/100 g **c)** 100 g/110 g **d)** 110 g/100 g
2. The solubility of sodium nitrate at 25°C is 90 g/100 g. How much sodium nitrate will dissolve in 500 g of water at 25°C?
   a) 7.2 g **b)** 7200 g **c)** 180 g **d)** 450 g
3. The solubility of ammonium chloride at 20°C is 37 g/100 g. How much water is needed to dissolve 111 g of ammonium chloride?
   a) 300 g **b)** 333 g **c)** 11.1 g **d)** 33.3 g
4. A substance formed a slightly cloudy mixture with water. The

mixture showed the Tyndall effect. It did not settle on standing. This mixture is
a) a solution
b) a suspension
c) an emulsion
d) a colloidal suspension

5. The solubility of a substance at 20°C is 40 g/100 g of water. A solution of this substance was found to contain 50 g of solute in 100 g of water. This solution is
a) supersaturated   b) saturated   c) unsaturated   d) dilute

D. Using Your Knowledge

1. What type of mixture is each of the following? (Some may be more than one type.) Vinegar, vanilla extract, ketchup, house paint, shaving lotion, skin lotion, salad dressing, freshly squeezed orange juice, homogenized milk.
2. Table 12-2 shows the solubility of carbon dioxide gas in water.
   a) How does the solubility differ from that for solids in water?
   b) Soft drinks contain carbon dioxide. It gives them the stinging taste. Explain why a soft drink goes "flat" if you let it get warm in a glass.

Table 12-2 Solubility of Carbon Dioxide

Temperature (°C)	Solubility (g/100 g)
0	0.34
20	0.18
40	0.12
60	0.086

E. Investigations

1. Find 10 different mixtures in your home. Classify each one as a solution, suspension, emulsion, or colloidal dispersion.
2. The ocean is a solution. Find the names and amounts of the main substances in sea water.
3. Do an experiment to show that the solubility of carbon dioxide in water decreases as the temperature increases.

13 Separation of Substances: Filtration and Distillation

13.1 Activity: Filtration
13.2 Uses of Filtration
13.3 What is Distillation?
13.4 Demonstration: Simple Distillation
13.5 Demonstration: Fractional Distillation
13.6 Fractional Distillation of Petroleum

The last chapter was about solutions, suspensions, and other mixtures. You learned how to make these mixtures and you studied their properties. Now, you will learn how to separate mixtures.

We often find it necessary to separate mixtures. For example, air is a mixture of many gases. But it also contains a suspension of dust. This dust will damage a car engine if it gets inside it. Therefore the dust must be separated from the air. Our drinking water supplies are also mixtures. Some substances must be separated from the water before we drink it (Fig. 13-1). You will learn how to separate mixtures in this chapter and the next.

Fig. 13-1 A large city gets its drinking water from this lake. But it puts its sewage and other wastes into the same lake. Many substances must be separated from the water before people drink it.

Chapter 13 135

13.1 ACTIVITY **Filtration**

Solids can form three types of mixtures in a liquid. These are solutions, colloidal dispersions, and suspensions. In this activity you will try to remove the solid from each of these types of mixtures. You will do this by **filtration** [fil-TRAY-shun].

In these filtrations you will use filter paper. Filter paper has tiny holes in it. Particles larger than these holes will stay on the filter paper. The substance on the filter paper is called the **residue** [REZ-i-doo]. Particles smaller than the holes will pass through the paper. This material is called the **filtrate** [FIL-trayt].

Problem

What kinds of mixtures can be separated by filtration?

Materials

funnel	salt solution
filter paper	laundry starch
8 test tubes	clay suspension
4 microscope slides	dropper
copper sulfate solution	hot plate

Procedure

a. Prepare the filter paper and funnel for filtration. Your teacher will tell you how.
b. Obtain a test tube full of each of the 4 mixtures.
c. Filter each mixture as shown in Figure 13-2.
d. Look at the filter paper in each case. Is there any residue on it?
e. In each case compare the filtrate to the original mixture.
f. Place a drop or two of each filtrate on a microscope slide. Carefully evaporate the water from the slides by placing them on a hot plate.

Discussion

1. Explain the terms filtrate and residue.
2. Describe the effect of filtration on the solutions.
3. Describe the effect of filtration on the colloidal dispersion.
4. Describe the effect of filtration on the suspension.
5. What kind(s) of mixtures can be separated using filter paper? Why can they be separated while the others cannot?
6. Use the particle theory to explain the results of this activity.

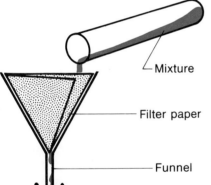

Fig. 13-2 What types of mixtures can be separated by filtration?

13.2 Uses of Filtration

Filtration is widely used in our daily lives. Several filters are used on a car engine. Hot air heating systems use filters. Filtration is also used in air conditioners, clothes dryers, vacuum cleaners, and swimming pools. Even your nose has a filtration system. Let's see how some of these work.

How Filters Work

Filters work on a very simple principle. Every filter has holes in it. Particles larger than the holes cannot pass through the filter. Particles smaller than the holes can pass through (Fig. 13-3). We say that the large particles are filtered out or, they have been separated from the mixture by filtration.

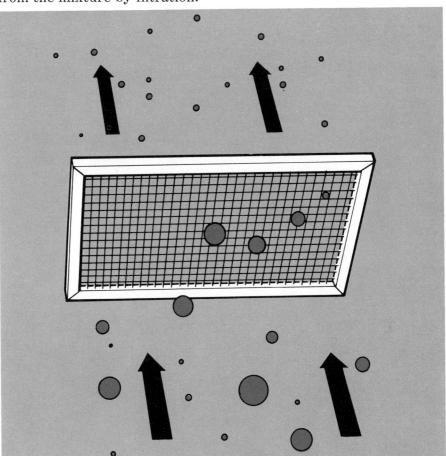

Fig. 13-3 A filter stops particles that are larger than the holes in it.

A car engine has three main filters. These are the air filter, oil filter, and fuel filter (Fig. 13-4).

a) *Air Filter*
A car engine needs air. The oxygen in the air helps the fuel to burn. But the air must be clean. If it isn't, the dust will ruin the engine.

Fig. 13-4 Three types of filters used in car engines. What does each one do?

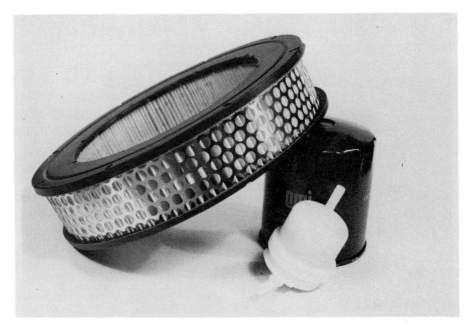

Therefore, the air is passed through an **air filter** before it enters the engine. Look at an air filter when you get a chance. You will see that the holes in it are very small. They will stop even the smallest dust particles but air particles can still get through.

b) *Oil Filter*

A car engine also has an **oil filter**. It filters particles from the engine oil. This oil lubricates the moving parts in the engine. Therefore it must not have dirt in it. Dirt will damage the moving parts.

c) *Fuel Filter*

Finally, the fuel goes through a **fuel filter** before entering the engine. Dirt may get in the fuel at the service station. It must not be allowed to get to the engine. If it does, it can plug the carburetor or fuel injection system.

d) *Changing Filters*

An air filter must be changed regularly. Otherwise too little air enters the engine. This wastes fuel and pollutes the air.

An oil filter must also be changed regularly. If not, the oil will not be cleaned. And the engine could be damaged.

A fuel filter must also be changed before it gets clogged. Otherwise the engine will not run properly. It will not get enough fuel.

Home Filters

a) *Hot Air Furnace Filter*

Many homes are heated with hot air. Hot air is blown from the furnace to the rooms. Cold air returns from the rooms to the furnace. This circulation of air blows dust around. Therefore the dust must be removed from the air. A **hot air furnace filter** does this (Fig. 13-5). This filter must be cleaned and changed regularly. If it gets clogged, the air cannot circulate properly. More important, fuel is wasted. A clogged filter can increase fuel consumption by over 25%.

Fig. 13-5 A hot air furnace filter

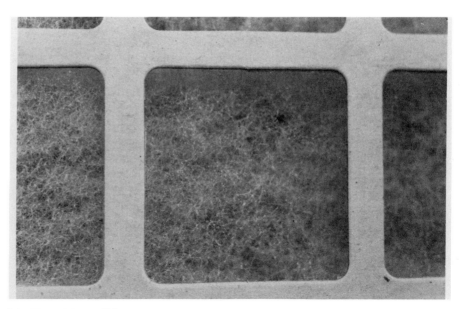

b) *Two More Filters*

A clothes dryer has a **lint filter** in it. This removes tiny pieces of cloth from the air. The bag of a vacuum cleaner is also an air filter. Do you know how it works?

Swimming Pool Filters

A swimming pool has several filters (Fig. 13-6). The water first goes through a filter called the **skimmer basket**. It has rather large holes in it. It removes leaves and other large floating things.

The water then passes through a **lint filter**. It has smaller holes in it. It removes objects like small pieces of leaves, grass clippings, and hair.

Finally the water enters the **main filter**. It has very small holes in it. This filter is filled with fine sand. The holes are the spaces between the grains of sand. This filter removes small particles like dust, dirt, and sand.

All three filters must be kept clean. If not, they will not do their job. The skimmer basket and lint filter are cleaned by simply emptying them. The main filter is cleaned by **backwashing**. Water is run through the filter backwards. This washes the dirt out.

Cigarette Filters

Tobacco companies spend millions of dollars every year telling us how good their filters are. Yet, if these filters were perfect, no smoke would pass through them at all! But they are far from perfect. They do remove traces of a few substances. But they do not remove dangerous gases. The particles of gases are very small. No filter can stop them. One of the gases is **nitrogen dioxide**. It causes lung cancer. Another gas is **carbon monoxide**. It causes headaches and drowziness. It also lowers a person's reaction time. It can even kill.

About 50% of all fatal poisonings in Canada are caused by carbon monoxide. Many of these are caused by car exhaust.

Cigarette filters will also not remove tiny solid particles. These smoke particles enter deeply into the lungs. They take with them hundreds of chemicals. Many of these cause lung diseases. Three of these diseases are found mainly in smokers. They are **lung cancer**, **chronic bronchitis**, and **emphesema**. If you smoke, you should know about these diseases. They are all killers. And filters do little to protect you from them. If you are a non-smoker, remember that filters do not protect you at all. Your best protection is to avoid smoky places. Second-hand smoke can harm you.

Section Review

1. Explain how all filters work.
2. a) Name the three main filters in a car.
   b) Explain what each type does.
   c) Why must these filters be changed regularly?
3. Name three filters used in the home.
4. Explain why a swimming pool has three filters.
5. a) What kinds of substances are not removed by a cigarette filter?
   b) Why should these substances be kept out of human lungs?

Fig. 13-6 The filters of a swimming pool. Why are three types needed?

13.3 What is Distillation?

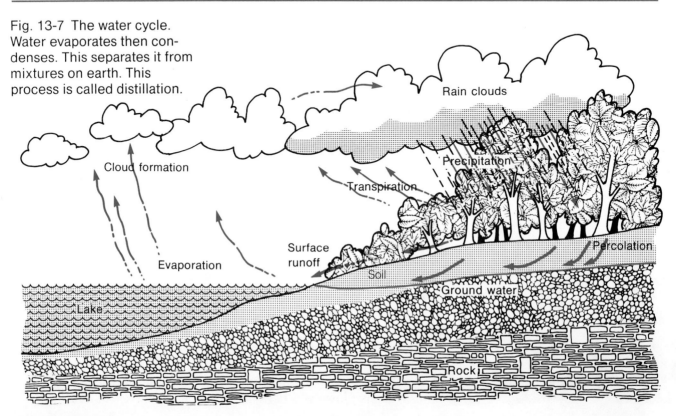

Fig. 13-7 The water cycle. Water evaporates then condenses. This separates it from mixtures on earth. This process is called distillation.

You can separate a suspension by filtration. You discovered that in Activity 13.1. But you cannot separate a solution by filtration. How do you separate a solution? Why would we want to separate a solution?

The Process of Distillation

Distillation is a process that can separate many mixtures. It can separate most solutions. It can also separate suspensions.

The water cycle in nature separates mixtures by distillation (Fig. 13-7). Water **evaporates** from the soil and bodies of water. Some of the bodies of water are dirty but the dirt does not evaporate. Only the water goes into the air. It goes into the air as a vapour. The vapour **condenses** in the upper atmosphere. The water returns to the earth as precipitation (rain or snow). The water has been separated from the mixture in the soil and bodies of water. It has been separated by a process called distillation. **Distillation** [dis-til-LAY-shun] *is the separation of a mixture by evaporation and condensation.*

How Distillation Works

Imagine a mixture of two substances, A and B, in a flask. Suppose A has a lower boiling point than B. If the mixture is warmed up, A will evaporate first. The vapour of A can now be condensed. A is now separate from B.

The condensed substance (the product that is collected) is called the **distillate** [DIS-til-late]. The substance that remains in the flask is called the **residue** [REZ-i-doo]. If the distillate is water, we call it distilled water.

The separation of the liquid from a solid-liquid mixture is called **simple distillation**. A solution of salt in water is a solid-liquid mixture. The solid is salt. The liquid is water. The water can be separated from the salt by simple distillation.

Many countries lack fresh water for drinking but they have lots of sea water nearby. Drinking water can be made by distilling the sea water. However, heat is needed to distil the sea water. Some countries now use solar energy to supply the heat.

Section Review

1. What is distillation?
2. What kinds of mixtures can be separated by distillation?
3. Define distillate and residue.
4. Which has the lower boiling point, the distillate or the residue?
5. What is simple distillation?

13.4 DEMONSTRATION Simple Distillation

In this demonstration we will separate a mixture by simple distillation. That is, we will separate a solid from a liquid. We will try to do this for both a solution and a suspension.

Problem

Will simple distillation separate solutions and suspensions into their parts?

Materials

ring stand (2), iron ring, wire gauze
Bunsen burner
boiling chips
adjustable clamp (2)
two-hole rubber stopper
Erlenmeyer or round-bottom flask
thermometer ($-10°C$ to $110°C$)
beaker
solution of copper sulfate in water (Mixture 1)
suspension of clay in water (Mixture 2)
Liebig condenser

CAUTION: Wear safety goggles during this activity.

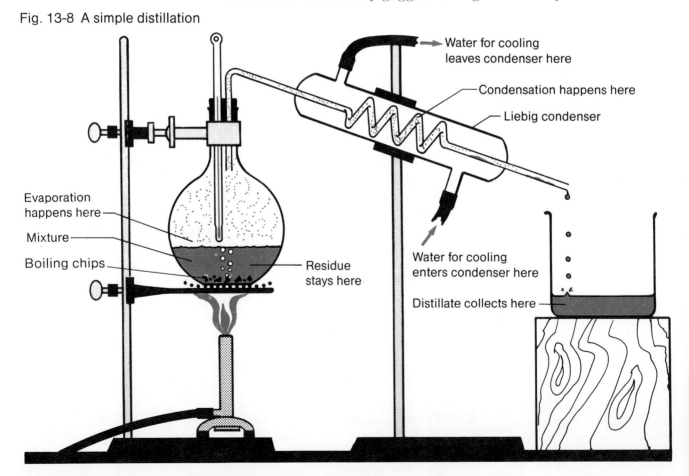

Fig. 13-8 A simple distillation

142   Chapter 13

Procedure

a. Set up the apparatus as shown in Figure 13-8. Put 30-40 mL of mixture 1 in the flask. Then start the water running through the condenser.
b. Heat the mixture slowly and carefully until it is boiling gently. CAUTION: Do not heat too strongly. The vapour may blow the stopper from the flask.
c. Turn off the Bunsen burner when a few millilitres of distillate have been collected.
d. Repeat the entire procedure with mixture 2.

Discussion

1. Describe the original mixture, distillate, and residue for mixture 1.
2. Describe the mixture, distillate, and residue for mixture 2.
3. Does distillation separate the substances in both solutions and suspensions?
4. Explain how your distillation worked.

13.5 DEMONSTRATION Fractional Distillation

In the last demonstration we used distillation to separate a liquid and a solid. It works well because the liquid and solid differ greatly in boiling points. The boiling point of the liquid is much lower than that of the solid. Therefore the liquid vapourizes easily, leaving the solid behind.

Distillation can also be used to separate a mixture of two or more liquids. However, the liquids must have different boiling points. When the mixture is heated, the liquid with the lowest boiling point distils off first. When it is nearly all gone, the liquid with the next lowest boiling point distils off. Then the next distils, and so on. Each separate liquid that distils off is called a **fraction**. The process is called **fractional distillation** [FRAK-shun-al dis-til-LAY-shun].

The mixture that we will use in this demonstration has just two liquids in it. One is water (b.p. = 100°C). The other is alcohol (b.p. = 78.5°C). Let's see if we can separate these liquids by fractional distillation.

Problem

Can alcohol and water be separated by fractional distillation?

Materials

Same as Demonstration 13.4, except the mixture is alcohol and water. CAUTION: Wear safety goggles during this activity.

Procedure

a. Set up the apparatus as shown in Figure 13-8. Start the cooling

water running. Make sure the thermometer bulb is well above the liquid.
b. Place 50 mL of the alcohol-water mixture in the flask.
c. Heat the mixture slowly and carefully until it is boiling gently.
d. Record the temperature every minute.
e. The temperature will start to change quickly after a few minutes. When it does, start collecting the distillate in another beaker.
f. Continue heating and recording temperatures for about 5 min.

Discussion

1. Explain the terms fraction and fractional distillation.
2. How must the liquids in a mixture differ before they can be separated by fractional distillation?
3. What liquid is most of the first fraction made of? Why is this so?
4. What liquid is most of the second fraction made of? What proof do you have of this?
5. The alcohol fraction always contains some water. How could you make the alcohol fraction more pure?

13.6 Fractional Distillation of Petroleum

The most important use of fractional distillation is the separation of **petroleum** [pe-TRO-lee-um]. Petroleum (crude oil) occurs in the ground. It is our main source of energy for heating homes and for transportation. But, petroleum must be separated into its fractions before it can be used. This is done by fractional distillation. Fractional distillation of petroleum produces many substances including gasoline, kerosene, furnace oils, diesel fuel, lubricating oils, greases, and hundreds of other useful substances. Let's see how petroleum is distilled.

The petroleum that comes out of the ground is a black, sticky liquid. It is a mixture of thousands of substances — gases, liquids, and solids. Most of these substances have different boiling points. Therefore they can be separated from one another by fractional distillation. The last demonstration showed us how that is done. However, the boiling points of some of these substances are very similar. Therefore, a simple distillation apparatus like the one we used is not good enough. A **bubble tower** is used for the fractional distillation of petroleum (Fig. 13-9). It separates the petroleum into fractions according to their boiling points.

A bubble tower contains 20 to 30 trays. They are arranged one above the other and about 0.7 m apart. Thus bubble towers are often over 20 m tall. The temperature of the tower is kept high near the bottom. It gets gradually lower up the tower. The petroleum is heated to 370°C. Then it is pumped into the bubble tower 6 or 7 trays from the bottom. At this temperature much of the petroleum

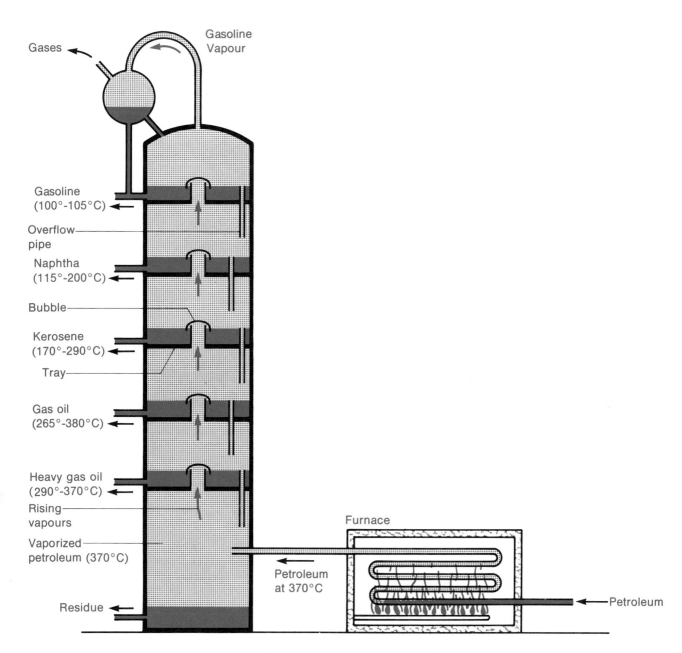

Fig. 13-9 A bubble tower for the fractional distillation of petroleum. Only 5 trays are shown. A tower usually has 20-30 trays.

vapourizes as soon as it enters the tower. The vapours rise through openings in the trays called **bubbles**. Each vapour finally reaches a tray where the temperature is low enough for it to condense into a liquid. Substances with high boiling points such as lubricating oils condense near the bottom. Further up the column, furnace oils condense. (They have lower boiling points.) Still further up, kerosene condenses. Gasoline has a very low boiling point; it will not condense at all in the tower. It leaves the top as a vapour and is condensed by a special condensor. Many other substances pass straight through the column in the gaseous state. Others collect in the bottom as viscous liquids and solids. They are called the residue.

The bubbles and overflow pipes allow part of each condensed liquid to flow down to a lower tray. This gives substances a "second

Section 13.6 145

chance" to stop at the right tray. Often the upward rush of vapour carries some substances too far up the tower.

Section Review

1. Name six substances in petroleum.
2. Explain how a bubble tower is used to separate the substances in petroleum.

Main Ideas

1. Filtration produces a filtrate and a residue.
2. Filtration will separate suspensions. It will not separate solutions and colloidal dispersions.
3. Distillation involves evaporation and condensation.
4. Distillation produces a distillate and a residue.
5. The residue has a higher boiling point than the distillate.
6. Simple distillation will separate a liquid from a solid-liquid mixture.
7. Fractional distillation will separate the liquids in liquid-liquid mixtures.
8. Fractional distillation is used to separate the substances in petroleum.

Glossary

distillate	DIS-til-late	a substance that evaporates and condenses in a distillation
distillation	dis-til-LAY-shun	the separation of a mixture by evaporation and condensation
filtrate	FIL-trayt	a substance passing through a filter
filtration	fil-TRAY-shun	the separation of a suspension by passing the liquid through a filter
fraction		a liquid obtained in a fractional distillation
fractional distillation	FRAK-shun-al dis-til-LAY-shun	the separation of a mixture of liquids by distillation
petroleum	pe-TRO-lee-um	crude oil obtained from the ground

residue	REZ-i-doo	a substance remaining after a filtration or distillation
simple distillation	dis-til-LAY-shun	the separation of a solid and a liquid by distillation

Study Questions

A. True or False

Decide whether each of the following sentences is true or false. If the sentence is false, rewrite it to make it true. (Do not write in this book.)
1. The oil filter of a car removes oil from the fuel.
2. During a distillation, the distillate stays in the flask.
3. Simple distillation separates a liquid from a solid.
4. Fractional distillation separates a mixture of liquids.
5. Petroleum contains gasoline, fuel oil, greases, and diesel fuel.

B. Completion

Complete each of the following sentences with a word or phrase which will make the sentence correct. (Do not write in this book.)
1. A filtration produces a filtrate and a ▓▓▓▓ .
2. A distillation produces a distillate and a ▓▓▓▓ .
3. Three lung diseases caused mainly by smoking are ▓▓▓▓ .
4. Petroleum is found in the ▓▓▓▓ .
5. Petroleum is separated into fractions in a ▓▓▓▓ .

C. Multiple Choice

Each of the following statements is followed by four responses. Choose the correct response in each case. (Do not write in this book.)
1. Filtration with ordinary filter paper will separate the parts of
   - a) solutions and suspensions
   - b) suspensions and colloidal dispersions
   - c) solutions only
   - d) suspensions only
2. 5 g of soil was mixed with 1 L of water. The mixture was then filtered. Which one of the following is most likely to be true?
   - a) The residue is mostly large soil particles. The filtrate is mostly water.
   - b) The residue is mostly large soil particles. The filtrate is mostly fine soil particles.
   - c) The residue is mostly fine soil particles. The filtrate is mostly large soil particles.
   - d) The residue is mostly water. The filtrate is mostly large soil particles.

3. Simple distillation can be used to separate
   a) a mixture of liquids
   b) the substances in solutions ONLY
   c) the substances in suspensions ONLY
   d) the substances in both solutions and suspensions

4. A mixture contains water, antifreeze, and glycerine. These liquids all have different boiling points. They can best be separated by
   a) simple distillation
   b) filtration
   c) fractional distillation
   d) evaporation

5. Heptane and octane are found in gasoline. Heptane has a boiling point of 98°C. Octane has a boiling point of 125°C. If a mixture of heptane and octane is distilled, what will happen?
   a) Heptane will be the distillate. Octane will be the residue.
   b) Heptane will be the residue. Octane will be the distillate.
   c) Nothing will happen. Heptane and octane cannot be separated by distillation.
   d) Both the heptane and the octane will become distillate.

D. Using Your Knowledge

1. Name three places in a car where filters are used. In each case name the filtrate. Also, name the residue that will likely be found on the filter.
2. Would an oil filter weigh more or less after it has been used? Why?
3. Why is petroleum important to all of us?
4. a) Why is petroleum called a non-renewable resource?
   b) Name a renewable resource that can be used instead of petroleum for heating homes.
5. Explain how you would separate a mixture of salt and sand.

E. Investigations

1. Filtration is used to help purify your drinking water. Find out how this is done.
2. Check a car manual to see how often the three main filters should be changed.
3. Examine a new and a used air filter for a furnace or car. Write a report on your observations.
4. Find out how a vacuum cleaner filter works.
5. Research one of the three lung diseases caused by smoking. How is the disease caused? What are its symptoms? Can it be cured?

6. Millions of cigarette filters are thrown on city streets every day. Find out if this is littering (which is against the law). Do an experiment to see if the filters will decay quickly if buried in the earth.
7. Find out what liquid air is. Explain how the substances in liquid air are separated by fractional distillation.
8. Find out how the salt is removed from sea water to make drinking water.
9. Do an experiment to prove that tomato juice is both a suspension and a solution.

14 Separation of Substances: Sedimentation and Floc Formation

14.1 What are Sedimentation and Floc Formation?
14.2 Activity: Effect of Particle Size on Rate of Sedimentation
14.3 Activity: Sedimentation of Soil from Water
14.4 Activity: Floc Formation
14.5 Sedimentation with a Centrifuge
14.6 Uses of Sedimentation and Floc Formation

In the last chapter you studied two methods for separating mixtures. In this chapter you will meet two more methods. They are called sedimentation and floc formation. Both have many important uses. You will study some of these uses. But first you need to know the meaning of sedimentation and floc formation.

14.1 What Are Sedimentation and Floc Formation?

People often wish to separate suspended solids from liquids. You studied two ways to do this in Chapter 13. You can separate the suspension by filtration or by distillation. Sometimes these methods cannot be used because they may be too expensive. Or, the heat of distillation may destroy part of the suspension. As a result, sedimentation or floc formation may be used. Let's see what they are.

Sedimentation

Have you ever noticed the water in a lake or river just after a storm?

It is muddy. Clay and other solid particles were washed into the water by the runoff. But a few days later, the water may be clear again. The suspended particles have settled out. This settling is called **sedimentation** [sed-i-men-TAY-shun].

If you studied Chapter 12, you have seen sedimentation in the lab. In Section 12.2 you learned that the large particles of a suspension settle out on standing. This is sedimentation.

Floc Formation

You also saw in Section 12.2 that very small particles do not settle out on standing. They seem to stay suspended forever. Clay is an example. The smallest particles of clay stay suspended in water for years. They do not easily settle out. Nor can they be easily removed by filtration. They are too small. And distillation costs too much. How can they be removed? The answer is by **floc formation** [FLOK for-MAY-shun]. Here is how it works.

A floc is formed in the water. A **floc** [FLOK] is made of globs or "flocs" of a sticky solid called **aluminum hydroxide** [al-OO-min-um hy-DROX-ide]. These globs are more dense than water. Therefore they settle to the bottom. Suspended clay particles stick to the globs and also settle to the bottom. Floc formation simply speeds up sedimentation. You will try this in Activity 14.4.

Sometimes the floc settles slowly. If you are in a hurry, you can filter the floc from the water.

Section Review

1. What is sedimentation?
2. What is a floc?
3. How does floc formation speed up sedimentation?

14.2 ACTIVITY Effect of Particle Size on Rate of Sedimentation

Which will settle more quickly, large or small particles? (Assume that the particles are the same substance.) Write your prediction in your notebook. Then test it by doing this activity.

Problem

Will large or small particles settle more quickly?

Materials

jar of motor oil
plastic beads (diameters of 12 mm, 7 mm, and 4 mm)

Procedure

a. Hold the largest bead so it just touches the surface of the oil.
b. Release the bead. Time how long it takes to reach the bottom. This is the rate of sedimentation.
c. Repeat steps (a) and (b) with each of the other beads.

Discussion

1. How does particle size affect the rate of sedimentation?
2. A suspension settles. A colloidal dispersion settles slowly, if at all. And a solution never settles. Use your results from this activity and the particle theory to explain these differences in settling.

14.3   ACTIVITY   Sedimentation of Soil from Water

Soil contains particles of many sizes. Suppose a suspension is made of soil in water. Which soil particles will settle out first? Write your prediction in your notebook. Then test it by doing this activity. (The results of the last activity should help you with your prediction.)

Problem

Which soil particles will settle out first?

Materials

100 mL graduated cylinder   loam soil
water

Procedure

a. Put 100 mL of water in the graduated cylinder.
b. Add 2-3 cm^3 of loam soil to the water.
c. Shake the mixture for 15-20 s.
d. Observe the mixture several times during the rest of the period. Record the appearance of the suspension. Also, record the appearance of the **sediment** [SED-i-ment].
e. Let the mixture stand for 24 h. Do not disturb it.
f. Observe the mixture again. Record the appearance of the suspension and sediment.
g. Save the mixture for the next activity. Do not shake it!

Discussion

1. What did the mixture look like when you first shook it?
2. How did the suspension change on standing?
3. Did all the particles settle out? Why?
4. Describe the appearance of the sediment.

5. Which particles settle out first, large ones or small ones?

14.4 ACTIVITY Floc Formation

Very small particles take days to settle out by sedimentation. In fact, colloidal particles may take years. But we can speed up sedimentation by using a floc. In this activity you will make a floc. Then you will see how it settles small particles.

Problem
How does a floc speed up sedimentation?

Materials
the soil-water mixture from Activity 14.3
alum (aluminum sulfate) solution
lime (calcium hydroxide) solution
100 mL graduated cylinder

Fig. 14-1 Pour carefully. Do not disturb the sediment.

Procedure
a. Carefully pour about half of the suspension (from Activity 14.3) into a 100 mL graduated cylinder. Do not disturb the sediment (Fig. 14-1). You now have in the graduated cylinder those particles that did not settle in 24 h. Write a description of this mixture.
b. Add 5-10 mL of alum solution.
c. Now add 5-10 mL of lime solution.
d. Shake the mixture.
e. Let the mixture stand for the rest of the period. Record any changes that occur.

Discussion
1. The lime reacts with the alum to form aluminum hydroxide. The aluminum hydroxide is the floc. Describe it.
2. What effect did the floc have on the appearance of the mixture?
3. Why does the floc speed up the rate of sedimentation?

14.5 Sedimentation with a Centrifuge

The force of gravity causes sedimentation. It pulls the particles down. If the force was stronger, the particles would settle faster. A **centrifuge** [SEN-truh-fuge] provides this stronger force (Fig. 14-2).

Fig. 14-2 A centrifuge whirls the mixture around. This increases the force that settles the particles.

The mixture to be separated is placed in a test tube in the centrifuge. When the centrifuge is turned on, it swings the test tube around. This increases the force pulling the particles to the bottom of the test tube. Thus a centrifuge increases the rate of settling. Your school may have a centrifuge. If so, your teacher will show you how it works.

Section Review

1. What is a centrifuge?
2. Why does a centrifuge separate a suspension faster than sedimentation does?

14.6 Uses of Sedimentation and Floc Formation

This section describes four common uses of sedimentation and floc formation. In all cases, solid particles are being removed from a liquid.

Sewage Treatment

Sedimentation is used to remove solids from sewage. Sewage runs from homes through pipes to the treatment plant (Fig. 14-3). At the plant, the sewage first passes through a screen which removes floating solids. Then the sewage is pumped to a **settling tank**. This is where sedimentation occurs. The sewage stays in the tank until most of the solids settle out.

The solids are removed from time to time. After treatment, they can be used as soil. The liquid goes on for further treatment. Then it is released to a river or lake.

Fig. 14-3 A sewage treatment plant uses sedimentation to remove suspended solids.

Water Purification

Many towns and cities get their drinking water from rivers and lakes. Often the water contains suspended solids. It will contain more of these after a storm. The water treatment plant must remove these solids.

Most solids can be removed by filtration. The water is simply passed through beds of sand which filters out most particles. However, some small particles get through the sand. These are removed by floc formation. Alum and lime are added to the water. A floc of aluminum hydroxide forms. It settles, taking the small particles with it.

154 Chapter 14

Septic Tanks

Most farms and cottages must have their own sewage systems. They are too far from sewage treatment plants.

Most of the sewage systems have a **septic tank**. This is a large underground tank. The solids settle to the bottom by sedimentation. There they are broken down by bacteria. The liquids run off the top. They go through **weeping tiles** into the ground. The weeping tiles must not be near a lake. If they are, the liquids will pollute the lake. Many lakes have been destroyed by poor septic systems in cottages.

Separating Blood

Blood has two main parts. One part is liquid. It is called **plasma** [PLAZ-ma]. It consists of water and dissolved substances. It is clear and straw-coloured. You may have seen it oozing through a reopened cut.

The other part is solid. It is made up of **red blood cells, white blood cells**, and **platelets** [PLATE-eh-lets].

The Red Cross often separates the solids from the plasma (liquid). The plasma is used to treat patients for shock. Red cells may be used to treat anemia. Platelets may be used to help clots to form in a bleeding person.

The solids are removed in this way. First, the blood is placed in a bottle. It is allowed to clot. The platelets form the clot. The clot is then removed. The remaining mixture contains blood cells and plasma. These are separated with a centrifuge. The blood cells end up in the bottom of the tubes. The plasma can be poured off the top.

Section Review

1. How is sedimentation used in sewage treatment?
2. Why is floc formation needed in water purification?
3. How does a septic tank remove solids from sewage?
4. Why is blood centrifuged?

Main Ideas

1. The settling of solids from a liquid is called sedimentation.
2. Floc formation is used to speed up sedimentation.
3. Large particles settle faster than small particles of the same substance.
4. A centrifuge speeds up sedimentation.
5. Sedimentation and floc formation have many important uses.

Glossary

aluminum hydroxide	al-OO-min-um hy-DROX-ide	a floc used in water purification
centrifuge	SEN-truh-fuge	a machine that speeds up sedimentation by whirling a suspension around
floc	FLOK	a sticky solid that occurs in globs
floc formation	FLOK for-MAY-shun	making a floc by mixing alum and lime in water
plasma	PLAZ-ma	the fluid part of blood
platelets	PLATE-eh-lets	a part of the solid in blood; cause clotting
sediment	SED-i-ment	the solid that settles out of a suspension
sedimentation	sed-i-men-TAY-shun	the process of settling a solid from a liquid

Study Questions

A. True or False

Decide whether each of the following sentences is true or false. If the sentence is false, rewrite it to make it true. (Do not write in this book.)
1. The settling of suspended particles is called sedimentation.
2. Large particles settle more slowly than small particles.
3. A centrifuge speeds up floc formation.
4. Sedimentation is used to remove floating solids in sewage treatment plants.
5. Solids leave a septic tank through weeping tiles.

B. Completion

Complete each of the following sentences with a word or phrase which will make the sentence correct. (Do not write in this book.)
1. A floc of aluminum hydroxide is made by mixing ▨▨▨ and ▨▨▨ in water.

2. The solid and liquid parts of blood cells can be separated with a _____.
3. Water purification plants remove very small particles by using _____.
4. Weeping tiles too close to a lake can cause _____.
5. The liquid part of blood is called _____.

C. Multiple Choice

Each of the following statements is followed by four responses. Choose the correct response in each case. (Do not write in this book.)
1. After a storm, the water in a lake was dirty. A few days later it was clear. What happened during those days?
   a) floc formation **b)** filtration **c)** sedimentation **d)** solidification
2. Some pebbles, sand, and clay are mixed with water. The mixture is shaken then allowed to stand. In what order would the particles settle?
   a) clay, sand, then pebbles   **c)** sand, clay, then pebbles
   b) pebbles, sand, then clay   **d)** no particular order
3. The best method for removing very small (colloidal) particles from a city's drinking water is
   a) floc formation **b)** distillation **c)** filtration **d)** sedimentation
4. A floc helps speed up sedimentation because the floc is
   a) more dense than the dirt in the water
   b) less dense than the dirt in the water
   c) sticky and less dense than water
   d) sticky and more dense than water
5. The Red Cross separates blood cells from plasma by means of
   a) a centrifuge **b)** floc formation **c)** filtration **d)** clotting

D. Using Your Knowledge

1. An ultra-centrifuge runs much faster than a regular centrifuge. It is used for separating colloidal dispersions. Why?
2. Use the particle theory to explain why clay particles settle very slowly in water.
3. Use the particle theory to explain why large soil particles settle faster than small soil particles.

E. Investigations

1. Limestone and sandstone are called sedimentary rocks. Find out why they are given that name.
2. Contact the Red Cross and find out more about blood separation.
3. Visit a sewage treatment plant. Find out what happens after the sewage leaves the settling tank.
4. Too many septic systems near a lake can pollute the lake. The humus toilet will not pollute the lake. Find out how it works.
5. Find out what else is done in water purification besides filtration and settling with a floc.

The Chemical Properties of Matter

Your world consists of thousands of different substances. Just look around you. What are all these different forms of matter made of? Why do they behave differently? Why do some substances keep changing?

This unit takes a close look at matter. You will learn how to classify all the different types of substances. Then you will examine chemical changes. You will also study the changing model of the atom. You will learn how chemistry explains some of the mysteries of matter.

CHAPTER 15
The Classification of Matter

CHAPTER 16
Chemical Change

CHAPTER 17
Structure of Matter

CHAPTER 18
Atoms and the Periodic Table

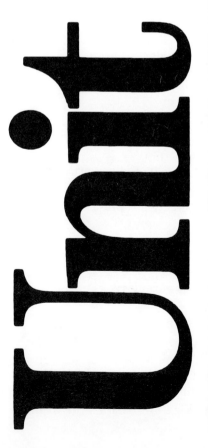

Fig. 15-0 This is a chemical change. How does it differ from a physical change?

15 The Classification of Matter

15.1 The Main Types of Matter
15.2 Pure Substances
15.3 Chemical Symbols and Formulas
15.4 Mixtures
15.5 Activity: Classifying Matter

This chapter explains how matter is classified. It then describes the four main types of matter. You will learn to use chemical symbols. And you will learn to write chemical formulas.

15.1 The Main Types of Matter

Matter *is anything that has mass and takes up space.* Matter can be a solid, liquid, or gas. It comes in all shapes and colours and reacts in many different ways. Chemists study matter and how it behaves. But where does a chemist begin? Matter must first be organized. So chemists use the classification system shown in Figure 15-1. This system does not sort out matter into solids, liquids, and gases. It uses composition instead. The composition of a substance depends upon the kind of particles it is made of.

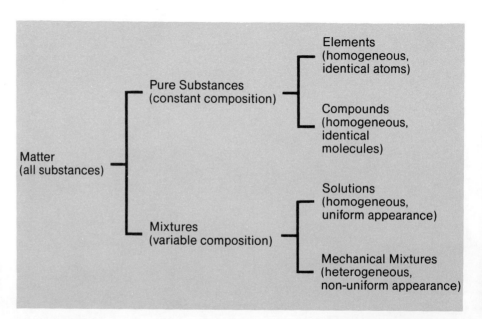

Fig. 15-1 A classification system for matter. This system uses composition to organize the different types of matter.

160 Chapter 15

Homogeneous Matter

Homogeneous [hom-o-GEE-nee-us] means uniform or "the same". Homogeneous matter has only one visible part; it looks exactly the same all over. **Pure substances** are homogeneous. The particles of a pure substance are identical. There are two kinds of pure substances, elements and compounds. **Elements** are made of identical atoms. **Compounds** are made of identical molecules. Oxygen and gold are pure substances. They are elements. Water and salt are also pure substances. But they are compounds.

Pure substances have a constant composition. Any two samples of a pure substance always contain the same kind of particles. Because they are identical, they always behave the same way.

Solutions are also homogeneous. They have only one visible part. They have a uniform appearance. But the particles in a solution are *not* identical. Solutions are made of two or more different kinds of particles. They are blended together. You cannot tell them apart. For example, you cannot see salt particles when they dissolve in water. The solution looks just like pure water. But you can sure taste the difference!

Solutions are homogeneous mixtures. They do not have a constant composition. The composition of a mixture depends on the number and kind of different particles mixed. The amount of each kind of particle is important too. Unless two mixtures are exactly the same, they will not behave the same way.

Compare the two solutions of sugar and water shown in Figure 15-2. Although they both contain the same two kinds of particles, one solution has more sugar particles than the other. They would look exactly the same. But one would taste much sweeter.

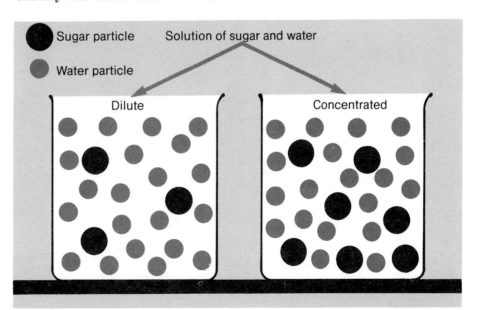

Fig. 15-2 A solution is homogeneous in appearance. Its composition is variable.

Heterogeneous Matter

Heterogeneous [he-ter-o-GEE-nee-us] means non-uniform or different.

Section 15.1

Mechanical mixtures are heterogeneous. They have two or more visible parts. They do not look the same all over. You can see the different kinds of particles. You can see sand mixed with water. You can see gas bubbles mixed with soda pop.

Mechanical mixtures do not have a constant composition. Two mixtures of sand and water can have any amount of water or any amount of sand. One mixture might look like muddy water. The other might look like wet sand (Fig. 15-3).

Section Review

1. Why do chemists use a classification system?
2. What determines the composition of a substance?
3. Describe a homogeneous substance and a heterogeneous substance.
4. Why do samples of the same pure substance always behave the same way?
5. Name three factors that determine the composition of a mixture.
6. Describe the two different types of mixtures.

Fig. 15-3 Mechanical mixtures have a heterogeneous appearance. They have a variable composition.

15.2 Pure Substances

Solutions and pure substances often look alike. But chemists can test their behaviour to tell them apart. These tests can also help us identify any pure substances. There are two types of pure substances — elements and compounds.

Elements

*An **element** is a pure substance made of identical atoms.* Atoms are particles which cannot be broken down during normal physical or chemical changes. Therefore an element cannot be broken down into simpler substances. There are 106 known elements. They are the building blocks of all matter. Ninety-two of these elements occur naturally on our planet; the other elements have been made by scientists.

Some elements such as nitrogen and oxygen are found in the air while others are found in the earth's crust. Figure 15-4 shows the percentage composition by mass of elements found in the earth's crust. You are surrounded by elements. You can recognize many of them. Gold, silver, copper, carbon, and sulfur are examples of elements. Most elements do not exist alone. The atoms of one element join with the atoms of other elements. This forms larger particles called **molecules** [MOL-uh-kyools].

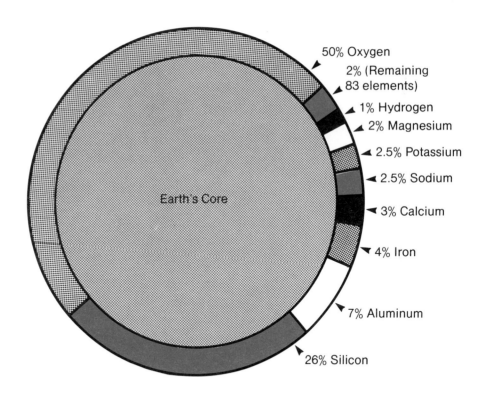

Fig. 15-4 Composition of the Earth's Crust

Compounds

*A **compound** is a pure substance made of identical molecules containing atoms from two or more different elements.* Molecules can be broken down into the atoms which formed them. Similarly, compounds can be broken down into the elements which formed them (Fig. 15-5).

You can think of elements as letters in an alphabet. Our real alphabet has only 26 letters. Yet think of all the words those letters form — and in many languages! Elements combine to form an

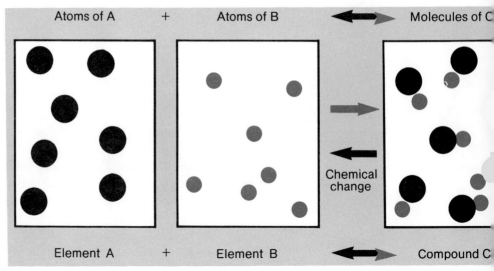

Fig. 15-5 The molecules of compounds can be broken down into the atoms of elements.

endless variety of compounds in the same way letters combine to form words.

Each molecule of a compound may contain many atoms of several different elements. For example, your red blood cells contain the compound hemoglobin. It is formed from five different elements. Each hemoglobin molecule contains 8640 atoms!

Section Review

1. Define an element. How many elements have been studied?
2. Explain why an element cannot be broken down into simpler substances.
3. What are the two most abundant elements found in the earth's crust?
4. How are molecules formed?
5. Define a compound.
6. Why are there so many compounds?

15.3 Chemical Symbols and Formulas

Nearly 2 000 000 (two million) compounds have been studied and new ones are found every day. How do chemists keep track of them all? They use a special naming system. They also use chemical symbols.

Chemical Symbols

Each element is identified by a **chemical symbol**. Modern chemists use letters from our alphabet. The symbol for any element uses a capital letter, usually the first letter in the name of an element. For instance, the symbol for hydrogen is H. If two elements have the same first letter, a second small letter is added. Thus, the symbol for helium is He. The set of elements in Table 15-1 is a good example. Some elements have symbols from their Latin names (Table 15-2).

Table 15-1   Symbols of Some Elements

Element	Symbol
Carbon	C
Calcium	Ca
Cadmium	Cd
Chlorine	Cl
Cobalt	Co
Chromium	Cr

Table 15-2   Symbols Using Latin Names

Element	Latin name	Symbol
Copper	Cuprum	Cu
Gold	Aurum	Au
Iron	Ferrum	Fe
Lead	Plumbum	Pb
Mercury	Hydroargentum (means liquid silver)	Hg
Potassium	Kalium	K
Silver	Argentum	Ag
Sodium	Natrium	Na
Tin	Stannum	Sn

Chemical Formulas

Table 15-3   Compound Names and Formulas

Common name	Elements used to form compound	Chemical name	Chemical formula
Table salt	Sodium (Na), chlorine (Cl)	Sodium chloride	NaCl
Lye	Sodium (Na), hydrogen (H), oxygen (O)	Sodium hydroxide	NaOH
Bluestone	Copper (Cu), sulfur (S), oxygen (O)	Copper sulfate	$CuSO_4$
Rust	Iron (Fe), oxygen (O)	Iron oxide	Fe_2O_3
Ethyl alcohol	Carbon (C), hydrogen (H), oxygen (O)	Ethanol	C_2H_5OH
Table sugar	Carbon (C), hydrogen (H), oxygen (O)	Sucrose	$C_{11}H_{22}O_{11}$

The chemical names of compounds often use the names of the elements which form the compound. So the name of a compound can describe its composition. The symbol for a compound is called a **chemical formula**. It includes the symbol of each element involved. Numbers are added to show how many atoms of each element are present (Table 15-3).

Water is a compound formed from the elements hydrogen and oxygen. One molecule of water consists of two atoms of hydrogen and one atom of oxygen. The chemical formula for water is H_2O.

Carbon dioxide is a compound. One molecule of carbon dioxide consists of one atom of carbon and two atoms of oxygen. The chemical formula for carbon dioxide is CO_2.

Molecules of Elements

In some elements, the identical atoms combine to form molecules. When you inhale oxygen gas, you are breathing in molecules. Each molecule of oxygen is made of two identical oxygen atoms. The symbol for oxygen is O. Oxygen gas molecules are shown as O_2. Nitrogen gas contains the molecules N_2. Hydrogen gas contains the molecules H_2. But these molecules differ from those of a compound. The molecules of a compound contain at least two different kinds of atoms (Fig. 15-6).

Fig. 15-6 The molecules of a compound contain two or more different kinds of atoms.

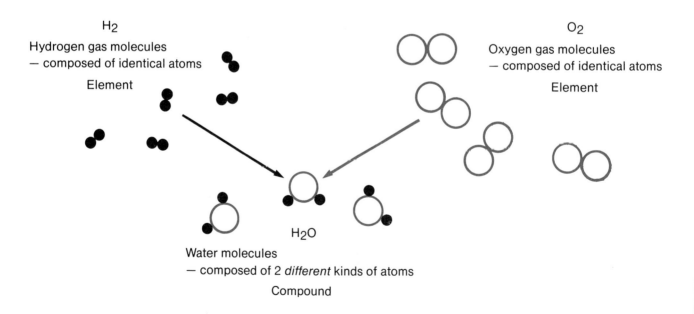

Using Chemical Symbols

Every pure substance has a chemical symbol or formula. This can tell you:
a) whether the pure substance is an element or compound;
b) which elements are present in a compound;
c) how many atoms of each element are present in a compound.

Section Review

1. What is a chemical symbol?
2. Why does gold have the symbol Au instead of G?
3. What is a chemical formula?
4. A molecule of sulfur dioxide contains one sulfur atom and two oxygen atoms. Write the chemical formula.
5. How do the molecules of an element differ from those of a compound?
6. What does a chemical formula tell you?

15.4 Mixtures

You have to use chemical symbols to distinguish an element from a compound. But it is easy to tell the difference between a mechanical mixture and a solution.

Mechanical Mixtures

Mechanical mixtures are heterogeneous in appearance. The different kinds of particles are visible. They can be separated in many ways. Sand stirred up in water will slowly settle to the bottom. It can also be filtered out. A magnet can pick out iron filings mixed with chalk dust. The method of separation depends on the kinds of particles in the mixture (Fig. 15-7).

Fig. 15-7 There are many ways of separating the different kinds of particles in a mechanical mixture.

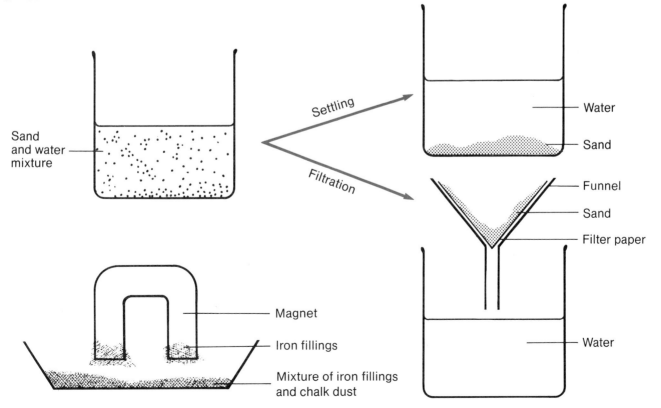

Solutions

Solutions are homogeneous in appearance. They look just like pure substances. But a solution contains at least two different kinds of substances. The substance which dissolves is called the **solute** [SOL-yoot]. The substance it dissolves in is the **solvent** [SOL-vent]. Most solutions you work with have a liquid solvent — usually water.

Consider a salt water solution. Salt is the solute. Water is the solvent. The formula for salt is NaCl. A salt water solution is labelled NaCl(aq). The symbol (aq) means "dissolved in water". Bluestone or copper sulfate is $CuSO_4$. A solution of bluestone is labelled $CuSO_4$(aq).

The particles of a solution are hard to separate. They must be heated until the particles of one substance evaporate from the solution. This is called **distillation** [dis-til-LAY-shun] (Fig. 15-8).

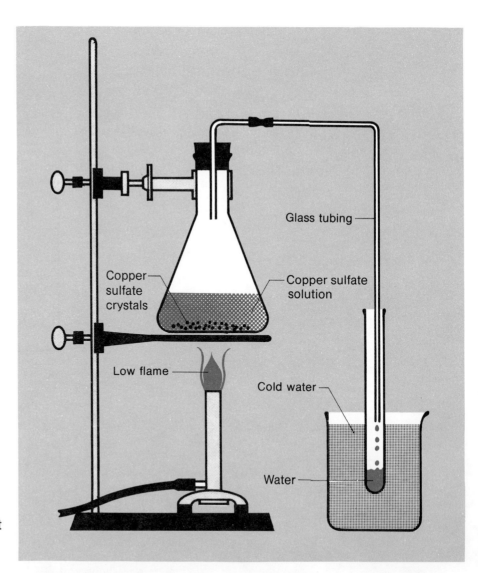

Fig. 15-8 Distillation of a solution. As the water solvent evaporates, copper sulfate crystals are left.

Section Review

1. How can you recognize a mechanical mixture?
2. What does the separation method for a mechanical mixture depend upon?
3. Define a solute and a solvent.
4. How would you label a solution of glucose sugar ($C_6H_{12}O_6$) in water?
5. How could you separate the different particles of a solution?

15.5 ACTIVITY Classifying Matter

In this activity, your teacher will display several different substances. You have studied the four main types of matter. Now test yourself. See how well you can use this classification system.

Problem

Can you classify several different substances?

Materials

Samples of several different substances. Each sample has an identification number. It is also labelled with the name and chemical symbol or formula of the contents.

Procedure

a. Make a copy of Table 15-4 in your notes. Find out the number of samples. Number a space in the table for each sample.
b. Observe each sample carefully. Find the sample number in your observation table. Place a check mark in the correct column for appearance.
c. Read the label on the sample. Then classify it as an element, compound, solution, or mechanical mixture.

Table 15-4 Classifying Matter

Sample number	Appearance		Classification
	Homogeneous	Heterogeneous	
1			
2			
3			
4			
5			

Main Ideas

Discussion

1. How did you distinguish between the elements and compounds?
2. How did you recognize the mechanical mixtures?
3. How did you identify the solutions?
4. Solutions can be confused with pure substances. Why?
5. Why are proper labels important?

Main Ideas

1. Substances are classified according to their composition. Composition depends upon the kinds of particles in a substance.
2. Homogeneous matter has only one visible part. Heterogeneous matter has two or more visible parts.
3. Elements and compounds are pure substances. They are homogeneous. They have a constant composition.
4. Every element has a chemical symbol. When elements combine, the compounds formed have a chemical formula.
5. Mechanical mixtures are heterogeneous. They have a variable composition. They can be separated in many different ways.
6. Solutions are homogeneous mixtures. They have a variable composition. They can only be separated by distillation.

Glossary

compound		a pure substance made of identical molecules containing atoms from two or more different elements
distillation	dis-til-LAY-shun	a process of separating the parts of a solution
element		a pure substance made of identical atoms
heterogeneous	het-er-o-GEE-nee-us	non-uniform; 2 or more visible parts
homogeneous	hom-o-GEE-nee-us	uniform; seems to be one part
molecule	MOL-uh-kyool	a particle made by the combination of 2 or more atoms
solution		a homogeneous mixture

Study Questions

A. True or False

Decide whether each of the following sentences is true or false. If the sentence is false, rewrite it to make it true. (Do not write in this book.)
1. Homogeneous matter has at least two visible parts.
2. Molecules can be broken down into atoms.
3. Compounds are pure substances which cannot be broken down.
4. The symbol N_2 represents a molecule of a compound.
5. Solutions are easier to separate than mechanical mixtures.

B. Completion

Complete each of the following sentences with a word or phrase which will make the sentence correct. (Do not write in this book.)
1. Matter is classified according to its _____ .
2. The two types of pure substances are _____ and _____ .
3. Solutions and pure substances are _____ in appearance.
4. A compound is represented by a _____ .
5. Solutions must be separated by a process called _____ .

C. Multiple Choice

Each of the following statements or questions is followed by four responses. Choose the correct response in each case. (Do not write in this book.)
1. Two samples of the same pure substance always behave the same because
   a) they are both homogeneous
   b) they are both heterogeneous
   c) they contain identical particles
   d) they can be broken down into simpler substances
2. Sugar dissolved in water is classified as
   a) a homogeneous mixture   c) a homogeneous pure substance
   b) a heterogeneous mixture   d) a heterogeneous pure substance
3. A solution has a variable composition because
   a) it is homogeneous   c) it can be separated
   b) it is heterogeneous   d) the amount of solute can differ
4. Carbon dioxide (CO_2) is classed as a compound because
   a) it is a pure substance   c) it is homogeneous
   b) it is formed from two different elements d) it contains 3 atoms
5. Which one of the following is a mechanical mixture?
   a) salt water   c) water and oil
   b) gasoline and oil   d) water and alcohol

D. Using Your Knowledge

1. Only 92 elements occur naturally on our planet. Explain how they can form thousands of different compounds.
2. Suppose you are testing a pure substance. As you heat the orange powder, a colourless gas is given off. A shiny silver liquid remains. Was this powder an element or a compound? Explain.
3. How would you separate each of the following mixtures?
    - a) gravel and water
    - b) salt and sand
    - c) iron filings and sawdust
    - d) sugar dissolved in water
4. For each of the following formulas
    - a) state the number of different elements in the compound
    - b) state the total number of atoms in each molecule
    - c) try to identify the different elements
        - (i) $NaHCO_3$ (baking soda)
        - (ii) $HC_2H_3O_2$ (vinegar)
        - (iii) $C_{55}H_{72}O_5N_4Mg$ (chlorophyll a)
        - (iv) $C_{254}H_{377}N_{65}O_{75}S_6$ (insulin)

E. Investigations

1. Air is a mixture of gases. Research the composition of air. How many different elements and compounds do you breathe?
2. A gas mask can separate harmful substances from the air you inhale. Find out how a gas mask works.
3. Research information on water treatment plants. Find out how harmful pollutants are separated from our drinking water.

16 Chemical Change

16.1 Physical and Chemical Properties
16.2 Word Equations
16.3 Qualitative Tests
16.4 Activity: The Decomposition of a Compound
16.5 Activity: Collecting and Testing a Gas Product
16.6 Activity: The Action of a Catalyst

This chapter explains how you can identify substances. You will learn how to use characteristic properties. Then you will test the products of several chemical reactions.

16.1 Physical and Chemical Properties

You are identified by your physical characteristics. Some characteristics are **qualitative** [KWAL-i-tat-iv]; they can be sensed and described. The colour of your hair and eyes are qualitative characteristics. Some characteristics are **quantitative** [KWAN-ti-tat-iv]; they can be measured. Your height and mass are examples of quantitative characteristics.

Physical Properties

Pure substances can also be identified by their physical characteristics. These characteristics are called **properties**. Qualitative physical properties can be sensed and described. Quantitative physical properties can be measured. Table 16-1 lists the physical properties of water.

Table 16-1 The Physical Properties of Water

Qualitative properties	Quantitative properties
Liquid at room temperature	Boiling point = 100°C
Clear	Freezing point = 0°C
Colourless	Density = 1000 kg/m^3
Odourless	
Tasteless	

Chemical Properties

A picture of you shows many of your physical characteristics. But it does not reveal how you behave with other people. Are you a loner? Or do you prefer being part of a group?

Similarly, the appearance of a substance does not reveal how it reacts with other matter. Some substances are highly reactive. For example, oxygen combines with many materials. Other substances do not react at all. For example, helium is inert. It does not combine with any other substance. **Chemical properties** *describe how a substance behaves with other matter.*

Physical Changes

You can change your physical appearance. A simple haircut can make quite a difference. But you are still the same person. Your personality has not changed.

A substance can also have a physical change. This involves a change of state or appearance. The particles of the substance do not change. They simply spread out or move closer together. *A **physical change** is a rearrangement of identical particles* (Fig. 16-1).

When sugar dissolves in tea, it seems to disappear. The molecules of the sugar crystals have simply separated and spread out. They are too small to see, but they are still sugar molecules. They have the same properties. They make the tea taste sweet. If the tea is evaporated, the same sugar molecules are left behind.

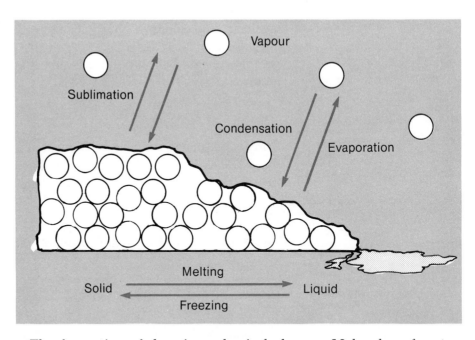

Fig. 16-1 A physical change is a rearrangement of identical particles.

The formation of dew is a physical change. Molecules of water vapour group together. They form droplets of liquid water. A glass shatters. This is a physical change because the shape and appearance changes. But the broken pieces are made of the same particles as the original glass.

Chemical Changes

When one substance reacts with another, a chemical change occurs.

A new and different substance forms. It is made of different particles. It has different properties.

Carbon is a black solid. Oxygen is a clear colourless gas. Carbon burns in oxygen to form carbon dioxide gas (Fig. 16-2). The properties of carbon dioxide are very different from those of carbon or oxygen. This is a chemical change. *A **chemical change** involves the formation of a new substance.* Chemical changes occur within you and all around you. Glucose sugar reacts with oxygen in your cells. Your food digests. Fuels burn to provide your heat and transportation. Green plants produce your food and oxygen. Dead matter decays. The list is endless.

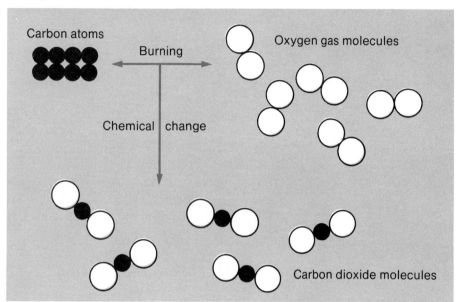

Fig. 16-2 A chemical change involves the formation of a new substance.

Section Review

1. What is a physical property?
2. Define and give two examples of a qualitative physical property.
3. Define and give two examples of a quantitative physical property.
4. What are chemical properties?
5. Define and give two examples of a physical change.
6. Define and give two examples of a chemical change.

16.2 Word Equations

A chemical change can be represented by a **word equation**. This equation shows the substances reacting and the substances produced.

Reactants and Products

Glucose sugar reacts with oxygen in living cells. This chemical change is called respiration. The glucose and oxygen are called the reactants [ree-AK-tents]. Glucose molecules are broken down into carbon and hydrogen atoms. These combine with oxygen atoms to form new molecules called **products**. The products of respiration are carbon dioxide and water molecules. The word equation for respiration is:

glucose + oxygen ⟶ carbon dioxide + water + energy
(reactants) (products)

The reactants are shown on the left of the arrow. The products are shown on the right. Respiration releases energy. It is shown as a product of this chemical change.

Sulfur tarnishes silver. The sulfur atoms react with silver atoms. The products of this chemical change are molecules of silver sulfide. The word equation is written:

silver + sulfur ⟶ silver sulfide
(reactants) (product)

Section Review

1. What does a word equation do?
2. What are reactants?
3. What are products?
4. Why is energy sometimes shown as part of a word equation?
5. What separates the reactants and products in a word equation?

16.3 Qualitative Tests

You can use physical and chemical properties to identify the products of a chemical change. Suppose you have just reacted substances containing carbon, hydrogen, and oxygen atoms. One of the products is collected in a test tube. It is a clear, colourless, odourless gas. It might be oxygen, hydrogen, or carbon dioxide. These three gases have similar qualitative physical properties. How can you identify the gas? You must test its chemical properties.

Standard Test for Oxygen Gas

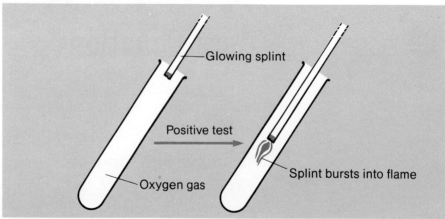

Fig. 16-3 The standard test for oxygen gas

Oxygen gas is very reactive. Fuels burn more intensely in pure oxygen than in air. Light a wooden splint, then blow on the splint until it just glows. Lower the splint into the gas. If the gas is oxygen, the splint will burst into flame again (Fig. 16-3).

Standard Test for Hydrogen Gas

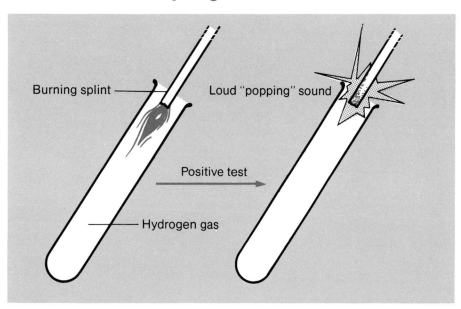

Fig. 16-4 The standard test for hydrogen gas

Hydrogen gas burns explosively when heated with the oxygen in air. Insert a burning splint into the mouth of the test tube. If the gas is hydrogen, you will hear a loud "pop" (Fig. 16-4).

Standard Test for Carbon Dioxide Gas

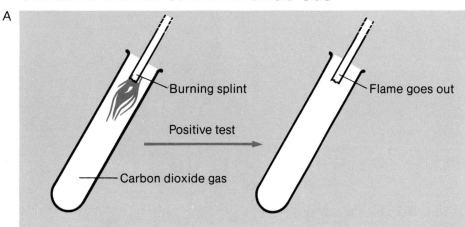

Fig. 16-5 The standard tests for carbon dioxide gas

Carbon dioxide gas does not burn. In fact, it will smother a flame. Lower a burning splint into the gas. If the gas is carbon dioxide, the flame will go out (Fig. 16-5).

There is a better standard test for carbon dioxide. It will react with limewater solution. One of the products is chalk which does not dissolve. Bubble the gas you are testing through the clear colourless

Section 16.3 177

limewater. If the gas is carbon dioxide, the limewater turns cloudy white (Fig. 16-5).

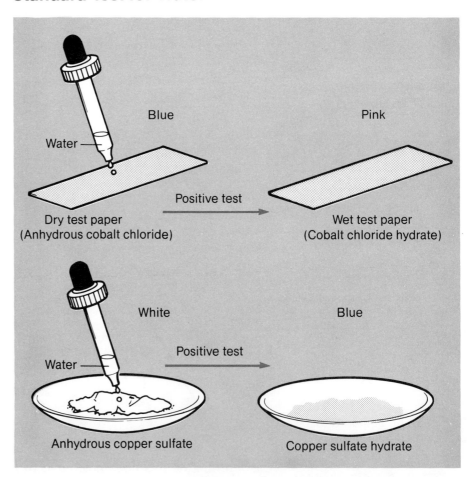

Fig. 16-6 Standard tests for water

Water is another product which is easy to identify. Water combines with certain compounds to form hydrates. This changes the colour of the original compound. The original compound is called anhydrous which means "without water". For example, anhydrous copper sulfate is white. It combines with water. The hydrate formed is blue (Fig. 16-6). You can also test for water with a piece of blue cobalt chloride paper. It turns pink when exposed to water, forming a hydrate.

Section Review

1. How can you identify the products of a chemical change?
2. Describe the standard test for oxygen gas. What property are you testing?
3. Describe the standard test for hydrogen gas. What property are you testing?
4. Describe the standard tests for carbon dioxide gas.
5. Describe the standard tests for water.

16.4 ACTIVITY The Decomposition of a Compound

A compound can be broken down into simpler substances. Some compounds, like baking soda, break down easily. In this activity you will heat baking soda. Then you will use qualitative tests to identify the products.

Problem

What is formed when baking soda decomposes [dee-kom-POZ-es]?

Materials

baking soda
limewater
Bunsen burner
test tubes (2)
glass elbow with rubber stopper

ring stand
test tube clamp
cobalt chloride paper
or
anhydrous copper sulfate

CAUTION: Wear safety goggles during this activity.

Procedure

a. Place a spoonful of baking soda in a clean test tube. Insert the glass elbow.
b. Clamp the test tube into position as shown in Figure 16-7.
c. Fill a clean test tube ⅓ full with limewater. Hold this test tube in position at the end of the glass elbow.
d. Move the Bunsen burner back and forth as you *gently* heat the baking soda. Record any changes you observe.

e. When no further changes occur, remove the test tube with limewater. Then stop heating the baking soda.

CAUTION: You must remove the limewater before you stop heating. If you don't, the limewater may back up into the hot test tube and break it.

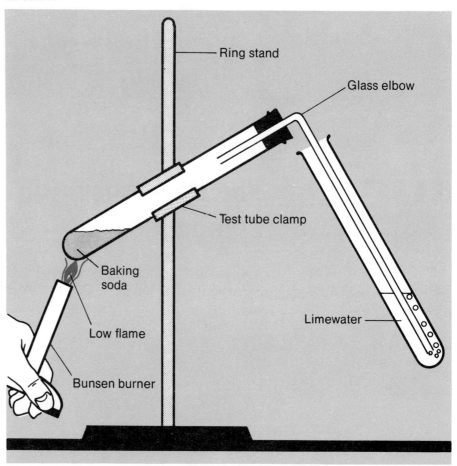

Fig. 16-7 Heating baking soda

f. Test the product which formed near the mouth of the test tube. Use a piece of dry cobalt chloride paper or a small quantity of anhydrous copper sulfate.

Discussion

1. Describe the baking soda before heating and during heating.
2. Describe the gas which bubbled through the limewater.
3. Describe any changes in the limewater. Identify the gas.
4. How would this gas affect a burning splint?
5. Identify the product formed near the mouth of the test tube. What is your evidence?
6. The chemical name for baking soda is sodium bicarbonate. Heating baking soda releases three products. One is a white solid called sodium carbonate. You identified the other two products. Now write a word equation for this reaction.
7. Explain how baking soda could be used as a fire extinguisher.

8. Fresh bread should be light and spongy. Why would a baker mix baking soda with dough?

16.5 ACTIVITY Collecting and Testing a Gas Product

In many chemical changes two different substances react. In this activity, calcium metal reacts with water. One of the products is a gas. You will collect this gas using a method called displacement of water. Then you will test the gas to find out what it is.

Problem

What gas is formed when calcium metal reacts with water?

Materials

chips of calcium metal
water
beaker
test tubes (2)

wooden splint
Bunsen burner
paper towel

CAUTION: Wear safety goggles during this activity.

Procedure

a. Fill a clean beaker ½ full of water.
b. Completely fill a clean test tube with water. Cover the rim of the test tube with a small piece of paper towel. Pat it firmly against the edge of the rim. Let it soak up water (Fig. 16-8). Now you can turn the full test tube upside down.

Fig. 16-8 Inverting the test tubes used to collect the gas

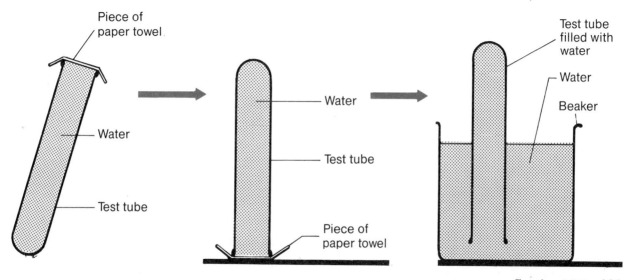

c. Place the test tube upside down in the beaker of water. Remove the paper towel (Fig. 16-8).
d. Repeat this method with the other test tube.
e. Place a small piece of calcium metal on a dry paper towel. CAUTION: Do not touch calcium with wet fingers. It may burn your skin.
f. Record a description of calcium metal.
g. Add the calcium to the water in the beaker. Observe the reaction.
h. Cover the reacting calcium with one of the test tubes. The gas produced will displace the water inside the test tube (Fig. 16-9).

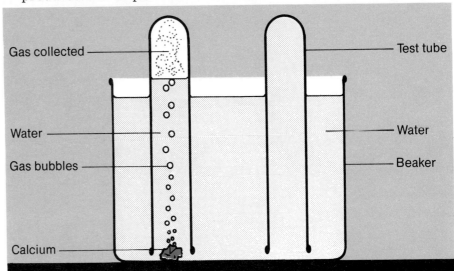

Fig. 16-9 The gas produced displaces the water in the test tube.

i. Use this method to fill a second test tube with gas.
j. Light a wooden splint to test the gas.
k. Keep each test tube upside down as you remove it from the beaker. Then hold it horizontally to test it (Fig. 16-10). Quickly place the burning splint in the mouth of the test tube. Record your observations.

Fig. 16-10 Testing the collected gas

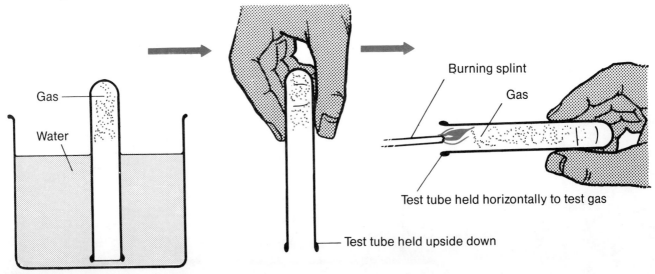

l. If your test results were uncertain, you may wish to collect more gas.

Discussion

1. Describe the calcium metal.
2. Describe the reaction of calcium and water.
3. Describe the gas produced. What is the gas?
4. Why is the water displaced in each test tube as the gas bubbles in?
5. Why must you hold each test tube upside down until you test the gas?
6. Why is this gas too dangerous to use in large balloons? Which gas is used instead? Why?
7. Two products form in this reaction. One is a base called calcium hydroxide. The other is the gas you tested. Write the word equation for this reaction.
8. Explain how the piece of paper towel can hold the water in the test tube.

16.6 ACTIVITY The Action of a Catalyst

A **catalyst** *[KAT-uh-list] is a substance which speeds up a chemical reaction without being used up.* In this activity, you will react two substances. One is a catalyst. The other is a substance that decomposes. You will test to identify one of the products. Observe this reaction carefully. See if you can tell which substance is the catalyst.

Problem

Which substance is the catalyst: manganese dioxide or hydrogen peroxide?

Materials

dilute hydrogen peroxide (10%) wooden splint
manganese dioxide Bunsen burner
test tube spatula
CAUTION: Wear safety goggles during this activity.

Procedure

a. Fill a clean test tube ¼ full of hydrogen peroxide. Observe.
b. Add a small amount of manganese dioxide using a spatula. Observe.
c. Place your thumb over the end of the test tube to trap any gas produced.
d. Lower a glowing splint into the test tube (Fig. 16-11). Record your observations.

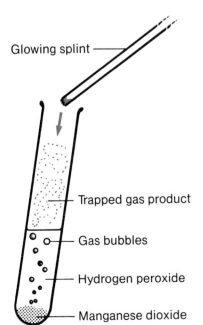

Fig. 16-11 Testing the gas produced

e. Wait until all reaction stops in the test tube. Then add a small amount of hydrogen peroxide. Observe. Trap and test any gas produced.
f. Now add a small amount of manganese dioxide to the test tube. Observe. Trap and test any gas produced.

Discussion

1. Describe hydrogen peroxide.
2. Describe manganese dioxide.
3. What happens when you add manganese dioxide to hydrogen peroxide?
4. Describe and identify the gas produced.
5. Which substance is the catalyst? How can you tell?
6. Which substance is decomposed? How can you tell?
7. A catalyst is not changed in a chemical reaction. It appears among the products in its original form. One of the products of this reaction is water. The other is the gas you tested. Write the word equation for this reaction.

Main Ideas

1. Pure substances can be identified by their physical and chemical properties.
2. Qualitative physical properties describe a substance. Quantitative physical properties can be measured.
3. Chemical properties describe how a substance reacts with other matter.
4. A physical change is a rearrangement of identical particles.
5. A chemical change involves the formation of a new substance.
6. A chemical change can be represented by a word equation.
7. Standard qualitative tests can often identify the products of a reaction.
8. A catalyst speeds up a chemical reaction without being used up.

Glossary

catalyst	KAT-uh-list	a substance which speeds up a reaction without being used up
decompose	dee-kom-POZE	to break down a compound into simpler substances
reactants	ree-AK-tents	the substances which enter a chemical change

Study Questions

A. True or False

Decide whether each of the following sentences is true or false. If the sentence is false, rewrite it to make it true. (Do not write in this book.)
1. Colour is a quantitative physical property.
2. New kinds of particles form during a physical change.
3. Reactants appear on the left hand side of a word equation.
4. A catalyst is decomposed as it speeds up a chemical reaction.
5. Carbon dioxide gas turns cloudy limewater clear.

B. Completion

Complete each of the following sentences with a word or phrase which will make the sentence correct. (Do not write in this book.)
1. A physical change is a ▭ of identical particles.
2. The density of water is a ▭ physical property.
3. ▭ properties describe how a substance reacts with other matter.
4. A ▭ splint is used to test for oxygen gas.
5. Anhydrous copper sulfate has a ▭ colour.

C. Multiple Choice

Each of the following statements or questions is followed by four responses. Choose the correct response in each case. (Do not write in this book.)
1. A clear colourless gas "pops" when tested with a burning splint. The gas tested is
   a) oxygen **b)** hydrogen **c)** carbon dioxide **d)** helium
2. Which of the following examples is a chemical property of hydrogen gas? Hydrogen gas
   a) is less dense than air **c)** burns in air
   b) will dissolve in water **d)** is clear and colourless
3. Which of the following is a chemical change?
   a) decomposition **b)** freezing **c)** melting **d)** boiling
4. Water changes the colour of anhydrous copper sulfate to
   a) pink **b)** white **c)** blue **d)** colourless
5. Which of the following can be broken down during a chemical change?
   a) all pure substances **c)** elements only
   b) compounds only **d)** catalysts only

D. Using Your Knowledge

1. State whether each of the following is a physical or a chemical property of the element sulfur. Sulfur
   - a) is a yellow solid
   - b) tarnishes silver
   - c) burns with a blue flame
   - d) does not dissolve in water
   - e) melts at 118.9°C
   - f) combines with hydrogen
   - g) reacts to form sulfuric acid
   - h) boils at 444.6°C

2. Classify each of the following as a physical or a chemical change:
   - a) Wood is ground into sawdust
   - b) Bread bakes in the oven
   - c) Milk turns sour
   - d) Frost forms on the window
   - e) A car rusts
   - f) Mothballs grow smaller in a closet
   - g) Grapes ferment into wine
   - h) Tea dissolves in water
   - i) Ice cream melts
   - j) Copper tarnishes

3. Which properties could you use to distinguish between unidentified samples of each of the following pairs:
   - a) Sugar and salt
   - b) Water and gasoline
   - c) Diamond and glass
   - d) Air and pure oxygen
   - e) Pure gold and fool's gold
   - f) Glass and clear plastic

E. Investigations

1. Rusting is a major problem for most car owners. What causes rust? What are the reactants? What are the products? Is there a catalyst? How does rustproofing work?
2. Respiration in living cells needs a constant supply of oxygen. Burning fuels use oxygen. Why don't we ever run out of oxygen? How is oxygen returned to the air?
3. Why do some substances burn more easily than others? What is spontaneous combustion? What causes it? What causes explosions?

17 The Structure of Matter

17.1 The Beginning of Atomic Theory
17.2 Activity: Does Chemical Change Affect Mass?
17.3 Dalton's Atomic Theory
17.4 Inside the Atom
17.5 The Particles of the Atom

This chapter deals with the changing story of the atom. For centuries, curious people have wondered about the nature of matter. What is matter made of? Why do various substances behave in different ways? What causes chemical changes?

Many clever minds have suggested answers. Most have tried to prove their ideas. Countless experiments have been performed. The results indicate that matter is made up of particles. But they are incredibly small! We cannot see them, even using the most powerful microscope. So no one has been able to describe them in detail. We only have theories about these particles. As we learn more about the particles, these theories change.

17.1 The Beginning of Atomic Theory

In Chapter 9, you studied the particle theory of matter. Let us review the major points before we go on.

The Particle Theory

1. All matter is made of small particles.
2. All particles of the same substance are identical.
3. The particles in matter attract one another. The "attractive forces" get stronger as the particles get closer together.
4. The spaces between the particles are large compared to the sizes of the particles themselves.
5. The particles in matter are always in motion.

This theory is used to explain *physical* properties and changes. But it cannot explain *chemical* properties and changes. What are these particles? What are they made of? Why do they behave in different ways?

The Democritus Theory

Long ago, in ancient Greece, students discussed questions about matter with their teachers. One famous teacher was Democritus. He taught that all matter is made of tiny, hard, indestructible particles. He believed that you can divide any substance into smaller and smaller pieces. Finally you will come to a piece so small that you cannot divide it any further. It is indivisible. The Greek word for indivisible is "atom". So Democritus called these particles **atoms**.

But Democritus was a philosopher, not a scientist. He never bothered to test his ideas with experiments. So scientists did not accept his ideas for over two thousand years. Then certain experiments made these ideas important again. But before this story proceeds, you should perform Activity 17.2.

Section Review

1. Why is the particle theory so useful?
2. Why is this theory not sufficient for scientists?
3. Describe Democritus' theory of matter.
4. What does "atom" mean?
5. Why was this idea not accepted by scientists for many centuries?

17.2 ACTIVITY Does Chemical Change Affect Mass?

In this activity, substances react with one another in a sealed flask. Particles cannot enter or escape. You will measure the mass of the reactants. Then, after the reaction, you will measure the mass of the products. How should they compare? Write down your prediction. Explain your reasoning. Now test your theory.

Problem

Does the mass change during a chemical reaction?

Materials

balance	solutions of each of the following:
Erlenmeyer flask and stopper	Pair 1: lead nitrate, potassium iodide
	Pair 2: copper sulfate, sodium hydroxide
small test tube	Pair 3: barium chloride, sodium sulfate

Procedure

a. Copy Table 17-1 into your notebook. Record your results in it. Leave enough space to describe the reactants and the products.

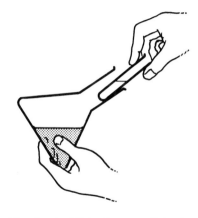

Fig. 17-1 Slide the small test tube carefully into the flask. Do not let the solutions mix.

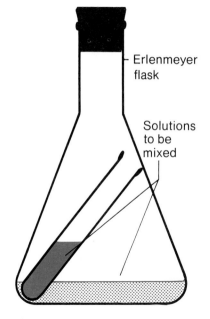

Fig. 17-2 Find the mass of the stoppered flask and contents before mixing the solutions.

Each reaction uses two solutions. You perform the reactions one at a time.

b. Pour about 10 mL of one solution into the flask.
c. Half-fill the small test tube with the other solution of the pair.
d. Carefully slide the small test tube containing solution into the flask. Do not let the solutions mix (Fig. 17-1).
   CAUTION: Avoid spilling the solutions on you or your clothing. Wash thoroughly with water if you accidentally spill some.
e. Place the stopper in the flask. Describe the reactants in your observation table (Fig. 17-2).
f. Place the flask on the balance. Record the mass of the stoppered flask and its contents. Use the "Mass of Reactants" column.
g. Remove the flask from the balance pan. Do not adjust the balance. Tip the flask to mix the solutions. Swirl the flask carefully.
h. Return the stoppered flask to the balance pan. Record the mass. Use the "Mass of Products" column.
i. Describe the reaction and products in your table.
j. Wash the flask and test tube thoroughly. Then continue with the other two reactions.

Table 17-1 Table of Observations

Reaction number	Names of reactants	Descriptions of reactants	Mass of reactants (g)	Description of reaction and products	Mass of products (g)
1	Lead nitrate (aq) Potassium iodide (aq)				
2	Copper sulfate (aq) Sodium hydroxide (aq)				
3	Barium chloride (aq) Sodium sulfate (aq)				

Discussion

1. For each reaction, explain how you know that a chemical change took place.
2. Your teacher will identify any products which formed. Write the word equation for each chemical reaction.
3. For each reaction, compare the mass of the reactants with the mass of the products. Was your prediction correct?
4. Your teacher can help you compare your results with the rest of the class. Do most of the results agree?
5. The **Law of Conservation of Mass** states that *the total mass of the products formed in a chemical reaction is equal to the total mass of the reactants.* Do the results of your class experiment support this law? Do the results of this experiment *prove* this law? Explain.

6. A chemical reaction occurs when a flash bulb flashes. How could you prove that mass is conserved?
7. When magnesium metal burns, a white ash forms. The mass of solid increases. Study the word equation:

   Magnesium + oxygen ⟶ magnesium oxide
   (metal)   (gas)   (white ash)

   Explain why the mass of the product is greater than the mass of the magnesium metal.
8. When a candle burns, it becomes shorter. Its mass decreases. Study the word equation:

   wax + oxygen ⟶ carbon dioxide + water
   (solid) (gas)   (gas)   (vapour)

   Why does the mass of the solid decrease? How could you prove that the mass of the products is equal to the mass of the reactants?
9. 20 g of sulfur react with 30 g of oxygen. Predict the mass of the product, sulfur trioxide. Write the word equation for the reaction.

17.3 Dalton's Atomic Theory

By 1800, many experiments had proven the Law of Conservation of Mass. Other experiments had proven a second important law.

The Law of Constant Composition

The **Law of Constant Composition** states that *every chemical compound always contains the same elements in the same proportions by mass*. For example, every sample of carbon dioxide gas contains the elements carbon and oxygen. Carbon is always 27% of the total mass of the carbon dioxide. Oxygen is always 73% of the total mass. Similarly, table salt always contains the elements sodium and chlorine. Sodium is always 39% of the total mass. Chlorine is always 61% of the total mass. Can you explain why?

The Dalton Theory

John Dalton deserves the credit for the first scientific theory of the atom. Dalton was an English school teacher who was very interested in science. He carefully studied how matter behaved. Dalton did many experiments, especially with gases. He tried to explain the two important laws we have just studied. In 1803, Dalton wrote his theory about atoms:

1. All matter is made of tiny invisible particles called atoms.
2. Atoms cannot be broken down into smaller particles.
3. The atoms of an element are all identical. They all have the same mass and properties.
4. Atoms of different elements have different masses and properties.

5. The atoms of different elements combine to form compounds.
6. Atoms cannot be created or destroyed during any physical or chemical change.

Dalton did not *prove* that atoms do exist. But the points of his theory seemed to explain the two basic laws we discussed. Choose the ideas from Dalton's theory which could explain the Law of Conservation of Mass. Then choose the ideas which could explain the Law of Constant Composition. Discuss these ideas with your class.

Some of Dalton's ideas about atoms are still accepted. Some of his ideas have been changed as we learn more about the atom. His theory still left many questions unanswered. So the story continues.

Section Review

1. Why was Dalton's theory accepted by scientists?
2. Summarize the points of Dalton's atomic theory in your notes.
3. Which of Dalton's ideas explain the Law of Conservation of Mass? Explain your choice.
4. Which of Dalton's ideas explain the Law of Constant Composition? Explain your choice.
5. Why have some of these ideas been changed since Dalton wrote his theory?

17.4 Inside the Atom

Dalton's theory was very important to science. But it left many questions unanswered. What are atoms made of? Why do atoms react? Why do some kinds of atoms conduct electricity? More scientists searched for answers. More experiments were done. More facts were discovered. And slowly, the theory of the atom changed.

The Thomson Model

J. J. Thomson was an English science professor. He studied electric particles. His experiments showed that all atoms gave off negatively charged particles. These particles were all identical. They had the same mass. They had the same negative electric charge. These particles were called **electrons** [e-LEK-trons]. Thomson concluded that all atoms contain electrons.

Atoms do not have a negative charge. If they did, they would repel one another. Thomson knew that atoms are electrically neutral. All atoms contain particles with negative charges. They must also contain an equal number of positive charges. But Thomson was never able to cause atoms to give off positively charged particles. So he suggested the following model of the atom: Atoms were made of positively charged matter. The negative electrons were scattered all through this matter (Fig. 17-3). The electrons were just like the raisins in a raisin bun.

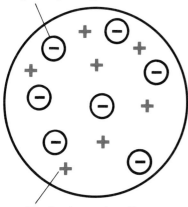

Fig. 17-3 Thomson's model of the atom. Negatively charged electrons were scattered throughout positively charged matter.

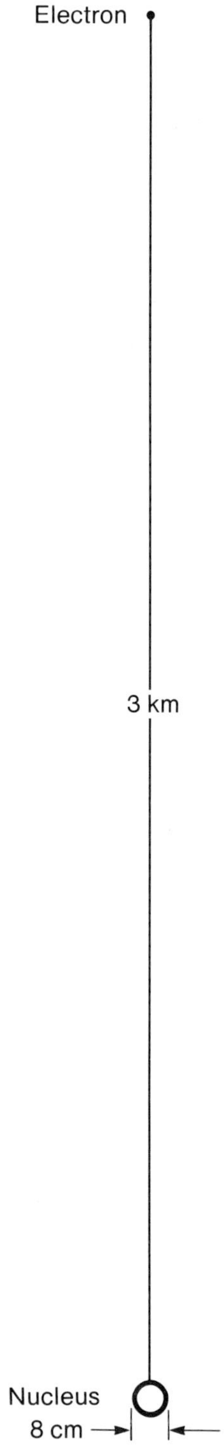

The Rutherford Model

J. J. Thomson taught a very gifted student named Ernest Rutherford. This student later became a very famous scientist. Rutherford tried new experiments to learn more about the atom. His results surprised him greatly! He found that:
1. Atoms were not solid spheres. They were mostly empty space.
2. The positive matter was concentrated in the centre of the atom. Rutherford named this part the **nucleus** [NEW-klee-us] (Fig. 17-4).

So, in 1911, Rutherford suggested a new model for the atom. Negatively charged electrons moved in the empty space around a tiny, positively charged nucleus. The nucleus contains nearly the total mass of the atom. However, other scientists with other experiments would soon change this model too.

The Bohr Model

Rutherford's model of the atom did not explain the movement of electrons. Did they follow a certain path? Did they all have the same energy? And why were they not attracted into the positive nucleus?

Rutherford taught a brilliant young Danish student named Neils Bohr. Bohr became one of the most famous scientists of our century. His studies showed that each electron had a certain amount of energy. In 1913, Bohr suggested the "solar system" model of the atom. Each planet in our solar system moves around the sun in a definite path or orbit. Bohr suggested that the electrons moved around the nucleus in a definite orbit. The model of the atom shown in Figure 17-5 shares two ideas. These are Rutherford's theory of the central nucleus and Bohr's theory of electron orbits.

Bohr's entire theory is too difficult to explain here. His model of electron orbits could explain many facts about the smallest atom, hydrogen. But it did not work well for atoms with many electrons. A more complex model was needed. But Bohr's work was very important; his exciting ideas had a great effect on modern atomic theory.

Fig. 17-4 The atom is mostly empty space. Suppose the nucleus of a hydrogen atom were as big as a tennis ball. The electron would most likely be found about 3 km away!

Section Review

1. Why did J. J. Thomson believe that all atoms contain electrons?
2. Why did Thomson also believe that atoms contain positive charges?
3. Describe Thomson's model of the atom.
4. How did Ernest Rutherford change Thomson's model of the atom?
5. How did Neils Bohr change the model of the atom?
6. Describe the Bohr-Rutherford model of the atom.
7. Why does the model of the atom keep changing?

17.5 The Particles of the Atom

Dalton believed that atoms were the smallest particles of matter. But experiments proved that each atom is made of even smaller particles.

The Electron

J. J. Thomson showed that all atoms contain electrons. An electron is a very tiny particle. It has a negative electric charge. An electron moves rapidly through the space around the central nucleus. An atom can lose or gain electrons during a chemical change.

The Proton

Other experiments showed that atoms also contain positively charged particles. Rutherford called these particles **protons** [PRO-tons]. A proton has a charge equal but opposite to that of an electron. But a proton has a mass nearly 2000 times greater than that of an electron (Fig. 17-6). Protons are found in the nucleus of all atoms. Atoms do not lose or gain protons during normal chemical reactions.

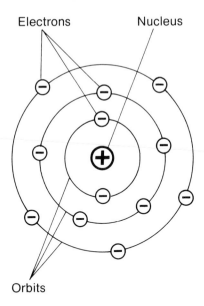

Fig. 17-5 The Bohr-Rutherford model of the atom. Negatively charged electrons travel in orbits around the central positively charged nucleus. This is often called the "solar system" model.

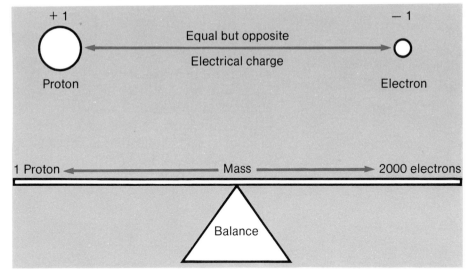

Fig. 17-6 Comparing the proton and the electron

The Neutron

Rutherford showed that nearly all of the mass of the atom is found in the **nucleus**. But the total mass of the nucleus is much greater than the total mass of the protons. Rutherford predicted that the nucleus also contained particles with a neutral charge. These neutral particles made up the remaining mass of the nucleus. Twenty years later, these particles were discovered by James Chadwick, a British scientist. These particles were called **neutrons** [NEW-trons].

Neutrons are found in the nucleus of every atom except the smallest hydrogen atom. A neutron has almost the same mass as a proton. It is electrically neutral. Atoms do not lose or gain neutrons during normal chemical reactions.

The Changing Model

The model of the atom is still changing. Scientists now use computers to plot the motion of electrons. Table 17-2 compares three important particles found in the atom. But modern scientists have found many other kinds of tiny particles in the nucleus. Many of these tiny particles are made of even smaller particles. And scientists are still searching for more!

Table 17-2 The Three Main Particles of the Atom

Particle	Electrical charge	Location in atom	Mass
Electron	−1	outside nucleus	almost zero
Proton	+1	inside nucleus	almost 2000 times greater than mass of electron
Neutron	0	inside nucleus	same as proton

Radioactive Atoms

Pierre and Marie Curie were French scientists who discovered the **radioactive** [RAY-dee-o-AK-tiv] element radium. The nucleus of a radium atom breaks down naturally to release smaller particles. Some of these particles have positive charges; others have negative charges. High energy gamma rays are also released (Fig. 17-7). This is why radioactive atoms are dangerous. There are other radioactive elements, such as uranium. These elements can be detected using a Geiger counter (Fig. 17-8).

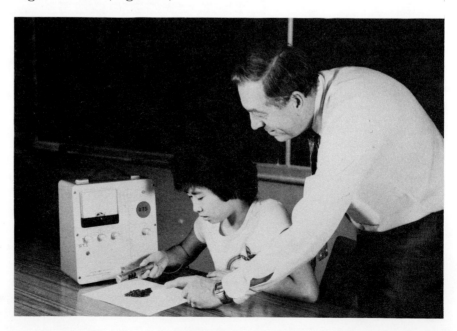

Fig. 17-8 The radioactivity of a substance can be measured with a Geiger counter.

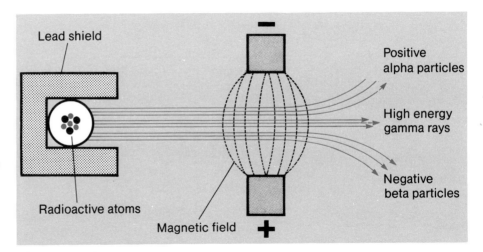

Fig. 17-7 The nucleus of a radioactive atom breaks down automatically. It releases positive and negative particles. It also releases dangerous high energy gamma rays.

Section Review

1. Describe an electron.
2. Describe a proton.
3. Describe a neutron.
4. Show the location of each of these particles on a simple drawing of an atom.
5. Which of these particles can be lost or gained during a chemical change?
6. How have modern scientists changed the model of the atom?
7. How do radioactive atoms behave? Why are they dangerous?

Main Ideas

1. All matter is made of tiny invisible particles called atoms.
2. As scientists learn more facts, the model of the atom keeps changing.
3. Atoms are mostly empty space. They have a tiny dense central nucleus.
4. Negatively charged electrons move through the space around the nucleus.
5. The nucleus contains neutrons and positively charged protons.
6. The nucleus also contains many other tiny particles.
7. The nucleus of a radioactive atom breaks up. It releases electrically charged particles and high energy rays.

Glossary

electron	e-LEK-tron	a negatively charged particle found in all atoms
neutron	NEW-tron	a neutral particle found in the nucleus
nucleus	NEW-klee-us	the positively charged centre of an atom

proton	PRO-ton	a positively charged particle found in the nucleus
radioactive	RAY-dee-o-AK-tiv	a nucleus which breaks down naturally

Study Questions

A. True or False

Decide whether each of the following sentences is true or false. If the sentence is false, rewrite it to make it true. (Do not write in this book.)
1. Democritus wrote the first scientific theory of the atom.
2. Dalton stated that atoms could be broken down into smaller particles.
3. Electrons and protons have the same mass.
4. Neutrons have a mass almost 2000 times greater than electrons.
5. Protons and neutrons are found in the nucleus of atoms.

B. Completion

Complete each of the following sentences with a word or phrase which will make the sentence correct. (Do not write in this book.)
1. The word "atom" means ▭ .
2. Every chemical compound always contains the same ▭ in the same proportions by mass.
3. The nucleus contains nearly the total ▭ of the atom.
4. The nucleus of a ▭ atom can break down naturally.
5. James Chadwick discovered the ▭ .

C. Multiple Choice

Each of the following statements or questions is followed by four responses. Choose the correct response in each case. (Do not write in this book.)
1. J. J. Thomson discovered
    a) protons   b) neutrons   c) electrons   d) the nucleus
2. Ernest Rutherford discovered
    a) protons   b) neutrons   c) electrons   d) the nucleus
3. The scientists who suggested that electrons travel in orbits was
    a) John Dalton   c) Ernest Rutherford
    b) J. J. Thomson   d) Neils Bohr
4. In a chemical reaction
    a) total mass stays the same   c) total mass is lost
    b) total mass is gained   d) mass changes to energy
5. In a chemical change, atoms may lose or gain
    a) electrons   b) protons   c) neutrons   d) mass

D. Using Your Knowledge

1. Explain why most of the mass of an atom is contained in the nucleus.
2. An oxygen atom contains 8 electrons. How many protons are found in the nucleus of an oxygen atom?
3. Why is the mass of an oxygen atom twice as great as the mass of the protons?

E. Investigations

1. Neils Bohr had a very exciting career. Research the story of his life. Why was his work so important during World War II?
2. Modern scientists have discovered many smaller particles inside the nucleus. What kind of particles have they found? How do they trace these tiny particles?
3. Radioactive elements are often used in scientific research. How are these elements used? What precautions must be taken when working with radioactive substances?
4. Find out how a Geiger counter works. How are Geiger counters used?

18 Atoms and the Periodic Table

18.1 The Atoms of Elements
18.2 Atoms and Ions
18.3 Metals and Non-metals
18.4 Activity: Flame Tests
18.5 The Periodic Table
18.6 From Atoms to Molecules

This chapter deals with the modern theory of the atom. The scientists of this century have made many important discoveries. They now have better equipment for studying atoms. They have even learned how to use nuclear energy. We will take a simple look at the atom their work has revealed.

18.1   The Atoms of Elements

The nucleus of an atom does not change during a chemical reaction. The number of protons and neutrons remains the same. This fact is used to identify atoms. Let us see how.

Atomic Number

The **atomic number** *is the number of protons in the nucleus of an atom.* Every atom of an element has the same atomic number. Therefore atoms of an element are identified by their atomic number.

In a neutral atom, the number of protons is equal to the number of electrons. Therefore atoms of the same element have the same number of protons and electrons.

Isotopes

Nearly all of the mass of an atom is found in the nucleus. The mass of the electrons is not important because they are so much smaller. (It is similar to comparing the mass of a dog with or without a few fleas!) But atoms of the same element are not entirely identical. **Isotopes** [EYE-so-topes] *are atoms of the same element which have different numbers of neutrons.* This means they do not have the same mass.

Mass Number

*The **mass number** of an atom is the number of protons and neutrons in the nucleus.* The isotopes of an element have different mass numbers. Then how can scientists determine the **atomic mass** of any element? They must use an average value. (This is similar to finding the class average on a test.)

The Isotopes of Chlorine

The atomic number of chlorine is 17. This means that every chlorine atom has 17 protons and 17 electrons. Almost 75% of chlorine atoms have 18 neutrons. The mass number of these atoms is (17 + 18) = 35. Nearly 25% of chlorine atoms have 20 neutrons. The mass number of these atoms is (17 + 20) = 37. The average atomic mass of chlorine is 35.5.

The Isotopes of Hydrogen

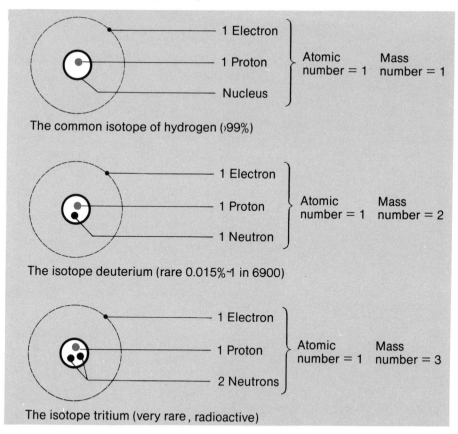

Fig. 18-1 The isotopes of hydrogen

The smallest atoms belong to the element hydrogen. The atomic number of hydrogen is 1. Every hydrogen atom has 1 proton and 1 electron. Hydrogen has 3 isotopes. More than 99% of hydrogen atoms do not have a neutron. The mass number of this isotope is (1 + 0) = 1. The second isotope is called **deuterium** [doo-TEER-e-um]. Deuterium atoms have 1 neutron. The mass number is (1 + 1) = 2. A

Section 18.1

very rare isotope of hydrogen is called **tritium** [TRIT-e-um]. Tritium atoms have 2 neutrons. Their mass number is (1 + 2) = 3 (Fig. 18-1). Because the first hydrogen isotope is so abundant, the average atomic mass is almost 1. The exact value is 1.008.

Section Review

1. Define atomic number.
2. Does the atomic number of an atom change during a chemical reaction? Explain.
3. How is the atomic number related to the number of electrons in a neutral atom?
4. What are isotopes? How does the number of neutrons affect an atom?
5. Define mass number.
6. How do scientists determine the atomic mass of an element?

18.2 Atoms and Ions

The nucleus of an atom does not change during a chemical reaction. Although atoms can lose or gain electrons, this has no measureable effect on the mass of the atom. But the atom is no longer electrically neutral. It has become an ion.

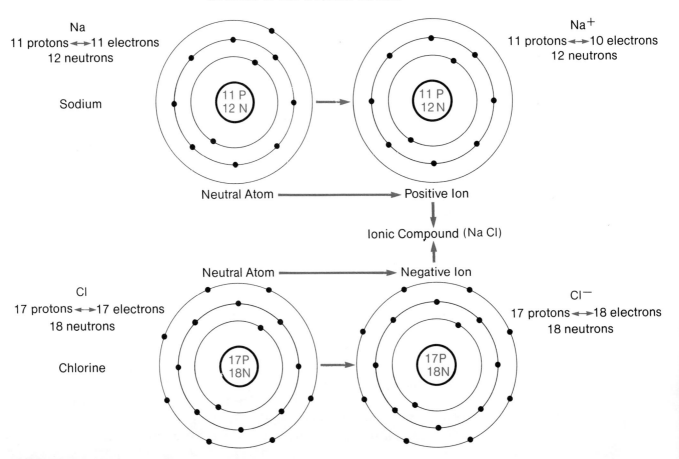

Fig. 18-2 Atoms form ions. Ions form ionic compounds.

Ions

*An **ion** is an atom which has lost or gained electrons.* If an atom loses electrons, it forms a **positive ion**. If an atom gains electrons, it forms a **negative ion**. Ions with opposite charges attract each other. This is how some compounds form.

Ionic Compounds

Ionic compounds [I-on-ik] *consist of pairs or groups of ions.* They are held together by electrostatic attraction. Table salt (NaCl) is an ionic compound.

Sodium atoms (Na) lose electrons to form positive sodium ions (Na^+). Chlorine atoms (Cl) gain electrons to form negative chloride ions (Cl^-). These ions attract each other to form crystals of table salt, NaCl (Fig. 18-2).

When ionic compounds dissolve, ions are released. Solutions of these charged particles are good conductors of electricity (Fig. 18-3). A car battery contains ions in solution.

Section Review

1. Define an ion.
2. How does a positive ion form?
3. How does a negative ion form?
4. How does an ionic compound form? Give an example.
5. Explain why solutions of ionic compounds conduct electricity.

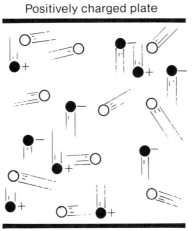

Fig. 18-3 Ions in solution conduct electricity.

Positive ions move toward the negatively charged plate. Negative ions move toward the positively charged plate. Neutral atoms move randomly. This is how electric current passes through a solution of ions.

18.3 Metals and Non-Metals

The properties of different elements have been studied by many scientists. They have divided the elements into two main groups: metals and non-metals.

Metals

The majority of elements are classed as metals. Metals share many properties.
1. *State.* Mercury is a liquid. All other metals are solids.
2. *Colour.* Except for copper and gold, all metals have a silver or grey colour. Most metals are shiny.
3. *Conductivity.* All metals conduct heat and electricity (Fig. 18-4).
4. *Chemical Properties.* The atoms of metals tend to lose electrons. Metals form positive ions. Most metals are fairly reactive with other substances. Platinum and gold are exceptions.

Fig. 18-4 Metals are good conductors of heat. The metal spoon in this pot is too hot to touch. The handle of the wooden spoon is cool.

Non-metals

The small group of non-metals have very different properties. Non-metals are not alike.
1. *State.* Half of the non-metals are gases. Bromine is a liquid. The rest are solids.
2. *Colour.* Most of the gases are colourless. Bromine is brown. The solids have a variety of colours.
3. *Conductivity.* Non-metals are poor conductors of heat and electricity.
4. *Chemical Properties.* The atoms of many non-metals tend to gain electrons. They form negative ions. Some non-metals will form both types of ions. Some will not react at all. They are called **inert gases** [in-URT].

Section Review

1. Name the two main groups of elements.
2. Which is the largest group?
3. Describe the common properties of this group.
4. How are the elements of the smaller group different?
5. Which group would each of the following elements belong to: oxygen, silver, sulfur, nitrogen, tin, and lead? Explain your choices.

18.4 ACTIVITY Flame Tests

When elements are heated strongly, they give off light. This light energy can be analyzed using a **spectroscope** [SPEK-tro-skope] (Fig. 18-5). This instrument separates the light energy into lines of colour.

These coloured lines are called a **spectrum** [SPEK-trum]. Each element can be identified by its spectrum. This is a very useful property. Spectroscopes are used to analyze the light from distant stars and planets. The element helium was discovered on the sun before it was found on earth!

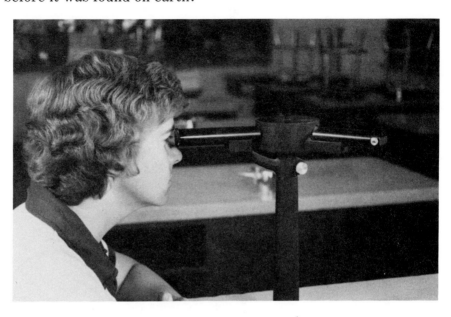

Fig. 18-5 The spectroscope can analyze light energy. Separate lines of colour can be seen and photographed. These coloured lines identify elements.

The ions of metals give off characteristic colours when heated. These colours can be used to identify the element. In this activity you will heat elements and compounds with a Bunsen burner flame. The light they give off colours the Bunsen flame. This is called a **flame test**. Before you begin, make the following prediction: Does the colour of light change when the element is part of a compound? Explain your answer.

Problem

Can flame tests be used to identify elements?

Materials

the elements and compounds listed in Table 18-1
a mystery powder
dilute hydrochloric acid
nichrome wire and handle
Bunsen burner, tongs

CAUTION: Wear safety goggles during this activity.

Procedure

a. Copy Table 18-1 into your notebook.
b. Light the Bunsen burner. Turn the gas on full. Open the air inlet until you can see two cones in the flame. The tip of the inner blue cone is the hottest point in the flame.

Section 18.4

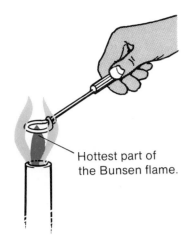

Fig. 18-6 Hold the wire loop in the hottest part of the Bunsen flame.

c. *Flame tests for elements.* Hold the calcium with the tongs. Place it in the hottest part of the Bunsen flame. Record the flame colour in your table. Repeat for copper.
d. The wire must be cleaned before each test. Heat the wire in the hottest part of the Bunsen flame. Dip it in hydrochloric acid. Then heat it again. Keep doing this until the wire is clean. When the wire is clean it glows white hot. It no longer colours the flame.
e. *Flame tests for compounds.* Dip the hot clean wire into dilute hydrochloric acid. Then dip the wire into the solid you wish to test. Particles of the solid will stick to the wet loop in the wire. Hold the wire loop in the hottest part of the Bunsen flame (Fig. 18-6). Record the colour of the flame in your table. Repeat for each of the other compounds.
f. Your teacher will give you a mystery compound. Record the flame colour. Try to identify the metal in the compound.

Table 18-1 Flame Colours

Trial	Pure substance	Flame colour	Trial	Pure substance	Flame colour
1	Barium nitrate		8	Lithium chloride	
2	Calcium		9	Potassium carbonate	
3	Calcium acetate		10	Potassium nitrate	
4	Calcium oxide		11	Strontium chloride	
5	Copper		12	Strontium nitrate	
6	Copper II chloride		13	Sodium chloride	
7	Copper II nitrate		14	Sodium hydrogen carbonate	

Discussion

1. Compare the flame colours in Trials 2, 3, and 4. What do these pure substances have in common?
2. Compare the flame colours in Trials 5, 6, and 7. What do these pure substances have in common?
3. Does the flame colour of an element change when it forms a compound? Explain your answer. Was your prediction correct?
4. Predict the flame colours of barium, lithium, potassium, sodium, and strontium. Explain your answers.
5. Which metal is present in the mystery compound? Explain.
6. Review Dalton's atomic theory. Which of Dalton's ideas explain your observations?
7. Pine cones and rolled newspaper were soaked in a solution. Then they were allowed to dry. When they were burned in a fireplace,

the flamed was coloured with crimson and emerald green. Which metal ions were present in the solution?

18.5 The Periodic Table

The periodic table is a chart. It organizes much of the information about elements. It is very useful to scientists.

The Work of Mendeleef

Dmitri Mendeleef [Men-del-AY-ef] was a Russian scientist. His dream was to organize all the knowledge about elements. Sixty-three different elements had been discovered by 1869. He arranged these elements by atomic mass. Then he looked for a pattern in their properties. He found that similar elements could be placed into groups. These groups formed a regular pattern. His work led to the first "Periodic Table of the Elements".

There were three empty spaces in this table. Mendeleef was sure that his theory was right. He predicted that three new elements would be discovered. He even predicted their properties. His predictions were very accurate. The missing elements were later found. And their properties fit the pattern!

The Modern Periodic Table

The modern periodic table is shown in Figure 18-7. One hundred and six different elements have now been identified. The elements are now arranged by atomic number instead of atomic mass. Hydrogen is the first element in the table.

The groups of similar elements are arranged in vertical columns. Each **group** is identified by a Roman numeral and a letter. The elements of each group share similar properties. You can study one element in a group. Then you can predict the properties of the others.

There are seven horizontal rows in the table. These rows are called **periods**. Each space in the table tells you: 1) the name of the element; 2) the element symbol; 3) the atomic number; 4) the atomic mass. On some tables, the spaces contain much more information about each element.

Notice the line dividing metals and non-metals. Some elements found along this line are called **metalloids** [MET-ul-oyds]. They share properties of both metals and non-metals.

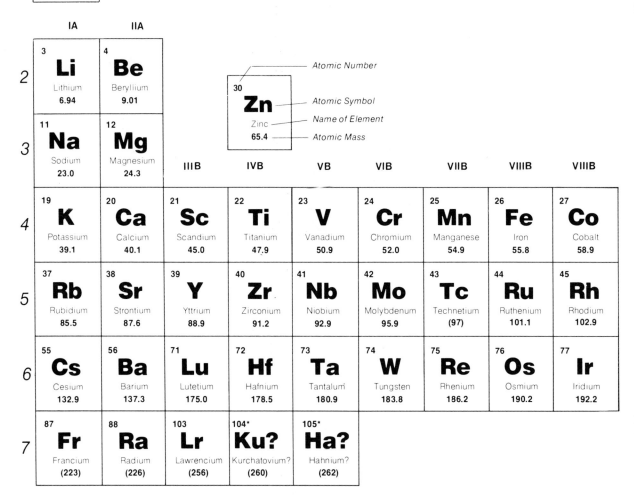

Fig. 18-7 The Modern Periodic Table

								O
			IIIA	IVA	VA	VIA	VIIA	2 **He** Helium 4.00
			5 **B** Boron 10.8	6 **C** Carbon 12.01	7 **N** Nitrogen 14.01	8 **O** Oxygen 16.00	9 **F** Fluorine 19.0	10 **Ne** Neon 20.2
VIIIB	IB	IIB	13 **Al** Aluminum 27.0	14 **Si** Silicon 28.1	15 **P** Phosphorus 31.0	16 **S** Sulfur 32.1	17 **Cl** Chlorine 35.5	18 **Ar** Argon 39.9
28 **Ni** Nickel 58.7	29 **Cu** Copper 63.5	30 **Zn** Zinc 65.4	31 **Ga** Gallium 69.7	32 **Ge** Germanium 72.6	33 **As** Arsenic 74.9	34 **Se** Selenium 79.0	35 **Br** Bromine 79.9	36 **Kr** Krypton 83.8
46 **Pd** Palladium 106.4	47 **Ag** Silver 107.9	48 **Cd** Cadmium 112.4	49 **In** Indium 114.8	50 **Sn** Tin 118.7	51 **Sb** Antimony 121.8	52 **Te** Tellurium 127.6	53 **I** Iodine 126.9	54 **Xe** Xenon 131.3
78 **Pt** Platinum 195.1	79 **Au** Gold 197.0	80 **Hg** Mercury 200.6	81 **Tl** Thallium 204.4	82 **Pb** Lead 207.2	83 **Bi** Bismuth 209.0	84 **Po** Polonium (209)	85 **At** Astatine (210)	86 **Rn** Radon (222)

64 **Gd** Gadolinium 157.2	65 **Tb** Terbium 158.9	66 **Dy** Dysprosium 162.5	67 **Ho** Holmium 164.9	68 **Er** Erbium 167.3	69 **Tm** Thulium 168.9	70 **Yb** Ytterbium 173.0
96 **Cm** Curium (247)	97 **Bk** Berkelium (247)	98 **Cf** Californium (251)	99 **Es** Einsteinium (254)	100 **Fm** Fermium (253)	101 **Md** Mendelevium (257)	102 **No** Nobelium (255)

Section Review

1. What did Mendeleef discover about elements?
2. Why did he predict that three new elements would be discovered?
3. How are elements arranged in the modern periodic table?
4. What is a group?
5. What is a period?
6. What does each space in the table tell you?
7. What is a metalloid?

18.6 From Atoms to Molecules

By sharing electrons atoms combine to form molecules. A molecule behaves as a single particle.

Molecules and Compounds

A molecule is the smallest particle of a compound. Molecules are not broken down during a physical change. But they can be broken down into atoms during chemical changes. The molecules of a compound are identical. They all share the same physical and chemical properties.

Molecular Size

Molecules, like atoms, are very small. A billion atoms in a row would only be 2.5 cm long. In one thimble of water there are 60 000 000 000 000 (sixty thousand thousand million) water molecules. Suppose a single drop of water could be magnified. If that drop were as large as the earth, each water molecule would be the size of a golf ball. It would take you more than a million years to count them all!

Molecular Mass

The mass of a molecule depends upon:
1. The number of different atoms it contains;
2. The mass of each atom it contains.

Section Review

1. What are molecules?
2. How do molecules form?
3. How are molecules affected by i) physical changes, and ii) chemical changes?
4. Describe the molecules of a compound.
5. What determines the mass of a molecule?

Main Ideas

1. The atoms of an element have the same number of protons and electrons.
2. Isotopes are atoms of the same element which have different numbers of neutrons. They do not have the same atomic mass.
3. Ions are atoms which have lost or gained electrons.
4. Most of the elements are classed as metals. Metals share many properties.
5. Elements can be identified by the light energy they give off.
6. The periodic table is a chart which organizes the elements.
7. A molecule is the smallest particle of a compound.

Glossary

isotopes	EYE-so-topes	atoms of the same element which have different masses
Mendeleef	Men-del-AY-ef	the scientist who designed the periodic table
metalloid	MET-ul-oyd	an element which shares the properties of a metal and a non-metal

Study Questions

A. True or False

Decide whether each of the following sentences is true or false. If the sentence is false, rewrite it to make it true. (Do not write in this book.)

1. Most elements are classed as non-metals.
2. The atoms of metals always form positive ions.
3. Ions in solution conduct electricity.
4. Most metals are gases.
5. Non-metals are good conductors of heat and electricity.

B. Completion

Complete each of the following sentences with a word or phrase which will make the sentence correct. (Do not write in this book.)

1. The elements in the periodic table are arranged by _____ .
2. A horizontal row of element is called a _____ .
3. A vertical column of elements is called a _____ .
4. _____ share properties of both metals and non-metals.
5. Ions are atoms which have lost or gained _____ .

C. Multiple Choice

Each of the following statements is followed by four responses. Choose the correct response in each case. (Do not write in this book.)

1. The atomic number of an element is
   a) the number of isotopes
   b) the number of protons in the nucleus
   c) the number of neutrons in the nucleus
   d) the number of protons and neutrons in the nucleus
2. Isotopes of an element have the same
   a) atomic number
   b) mass number
   c) atomic mass
   d) number of neutrons
3. The mass number of an element is
   a) the number of protons in the nucleus
   b) the number of neutrons in the nucleus
   c) the number of protons and electrons
   d) the number of protons and neutrons
4. When an atom gains an electron it becomes
   a) an isotope
   b) a positive ion
   c) a negative ion
   d) electrically neutral
5. The elements of a group on the periodic table
   a) have the same atomic number
   b) have the same atomic mass
   c) give the same flame test
   d) have similar properties

D. Using Your Knowledge

1. Use the periodic table to complete the following chart:

Element	Element symbol	Atomic number	Mass number	Number of protons	Number of electrons	Number of neutrons
Helium			4			
	K					20
		6				6
	N		14			

2. A calcium atom (Ca) loses 2 electrons.
   a) What kind of ion is formed?
   b) What is the atomic number of this ion?
   c) How many positive charges does this ion contain?
   d) How many negative charges does this ion contain?
   e) What is the resulting electrical charge on this ion?
   f) What kind of ion would this calcium ion attract?
   g) What type of compound would form? Why?

3. Match each element in column A to the element in column B which has similar properties:

Column A
Fluorine (F)
Argon (Ar)
Magnesium (Mg)
Carbon (C)
Sodium (Na)

Column B
Silicon (Si)
Potassium (K)
Neon (Ne)
Bromine (Br)
Calcium (Ca)

E. Investigations

1. The elements of group IA are called the alkali metals. Your teacher may demonstrate some of their properties. What properties do these elements share? Why are they classed as metals? How do they differ from metals such as aluminum, iron, or tin? None of the alkali metals is found in its pure form in nature. Why?
2. What properties do the metals of group IB share? Why are they often called "precious" metals? How are they used?
3. The elements of group VIIA are called the halogens. They are very reactive. Research the properties of these elements. Find out how they are used.
4. The elements of group O are called the "noble" or inert gases. Why are they special? Where are they found? How are they used?

Applying Your Knowledge of Chemistry

CHAPTER 19
Air Pollution

CHAPTER 20
Water Pollution

CHAPTER 21
How to be a Thinking Consumer

In Unit 5 you studied the chemical properties of matter. You learned what a chemical change is. And you studied theories that explain why chemical changes occur.

In this unit you will use what you learned in Unit 5. You will study air and water pollution. We hope that what you learn will do two things: help you avoid mistakes that harm the environment; help you correct mistakes others have made.

This unit ends with consumer chemistry. In this chapter you will get a chance to use your knowledge and skills to solve everyday problems.

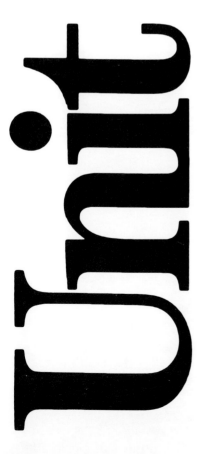

19 Air Pollution

19.1 What is Air Pollution?
19.2 Airborne Particles
19.3 Activity: Examining Particles
19.4 Sulfur Dioxide
19.5 The Oxides of Carbon
19.6 Activity: Combustion of Fuels
19.7 Nitrogen Oxides and Hydrocarbons
19.8 Other Air Pollutants

Air is mostly nitrogen and oxygen. But it also contains small amounts of other gases. Some of these are harmful to living things. You will study those gases in this chapter. Air also contains dust particles. You will learn how they affect life in this chapter.

This chapter is about air pollution. Learn the content well. As you study this chapter ask yourself this question: What can I do to help stop air pollution?

19.1 What is Air Pollution?

Fig. 19-1 Chimneys belch tonnes of waste into the air. Nature cannot recycle it fast enough.

Everyone talks about air pollution. But very few people know what it is. What is polluted air? *It is air containing substances harmful to living things or their habitats.* What are these substances?

Common Pollutants

Some common pollutants are tiny solid particles, sulfur dioxide, carbon monoxide, carbon dioxide, nitrogen oxides, hydrocarbons, ozone, and lead. Many of these are a natural part of air. Solid particles include pollen grains from plants. Wind carries dust into the air. Volcanoes emit sulfur dioxide. And every living thing gives off carbon dioxide.

Concentrations of Pollutants

Nature used to be able to absorb and recycle these substances. They did no harm. In fact, they were not called pollutants. But we changed all that. Our industries, homes, and cars produce too much of these substances (Fig. 19-1). Because they enter the air too quickly, nature can no longer absorb them all. When high enough concentrations are formed to harm living things, they are called pollutants.

Sulfur dioxide is a good example. Volcanoes emit sulfur dioxide. It spreads through the atmosphere. Much of it drifts over the oceans. It gradually dissolves in the water. It usually does not harm living things or their habitats, so it is not called a pollutant. Besides, there are not many active volcanoes; certainly there are none within large cities. Large eruptions can, of course, cause pollution for a time.

However, there are many coal-burning power plants near cities. They emit sulfur dioxide constantly. It drifts over nearby cities and farmland. Human health is affected; soil and crops are damaged. When sulfur dioxide is harmful, it is called a pollutant.

Summary

The degree of air pollution depends upon three factors:
1. The kind of substances that enter the air.
2. How much of these substances enters the air.
3. How fast these substances enter the air.

Section Review

1. What is polluted air?
2. Name the common pollutants.
3. Which of these are a natural part of air?
4. How does nature control substances entering the air?
5. How does a substance become a pollutant?

19.2 Airborne Particles

Take a deep breath. Do you live in a large city? If so, you probably just inhaled about 70 000 solid particles. If you live in the country, then you "only" inhaled about 40 000 particles! Does that sound unbelievable? Just look at a beam of light in a dark room. You can see hundreds of specks floating around in the "clean" air.

Types of Particles

What are these particles? They are tiny pieces of many of the substances around us. Some are dust. Others are fly ash or soot. Some are grease, oil, or even metal. Some may be asbestos.

Sources of Particles

Where do all these particles come from? The wind stirs up dirt from the ground. Burning produces fly ash and soot. Friction is always at work; watch a construction site. Sawing, drilling, and grinding materials release millions of tiny particles. Tiny pieces of asphalt and rubber tires come from every road. The sources are endless.

Effects of Particles

How do all these particles affect you? People are constantly cleaning their homes, cars, and clothing. Particles invade your nose and throat. Tiny ones even reach the deepest parts of your lungs. Studies show that some of these may cause cancer.

Particles in the air cause a haze over most cities. You cannot see very far; pilots cannot see as well when landing airplanes. This haze gradually spreads into the countryside. It blocks out some of the sunlight. Scientists fear such haze may gradually lower the average temperature of the earth. The next ice age may happen sooner than expected!

Section Review

1. What kinds of particles are in the air?
2. Name several common sources of particles.
3. State three effects of airborne particles.
4. Why are certain types of particles dangerous to inhale?
5. Why are some scientists concerned about constant haze?

19.3   ACTIVITY Examining Particles

You can often identify particles when you magnify them. You can study their size, shape, colour, and texture. For example, sand grains and metal shavings are easy to recognize. In this activity, you will collect particle samples from different sites. Then you will look at them with a microscope.

Problem

What do airborne particles look like?

Materials

compound microscope
a stereo microscope
   (if available)

several glass slides
petroleum jelly (or other clear,
   sticky material)

Procedure

a. Coat several glass slides with a *thin* film of petroleum jelly.
b. Select several sites which should have many airborne particles. Choose construction sites, factories, busy streets, and smoky rooms. Place one slide at each site. Place each slide where dirt on the ground will not blow over it. You only want to collect particles from the air. Expose each slide long enough to collect several particles. Often one day will do.

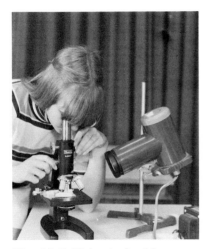

Fig. 19-2 This student is using oblique incident light to examine particles. Note the position of the light.

c. If possible, use a stereo microscope to examine the particles on each slide. It will help you study the largest particles.
d. Examine each slide with a compound microscope. Use oblique light as shown in Figure 19-2. Do not use light underneath the slide. This method will help you view the smaller particles.
e. The very smallest particles are seen best when you use a light underneath the slide as usual. If the film on the slide is not too thick, you should be able to focus on the smallest specks. Do not forget to move the slide around. You can only study a small section at a time.

Discussion

1. Draw and describe the different types of particles you collect. Try to identify different materials.
2. Suggest the most likely source of each type of particle.

19.4 Sulfur Dioxide

You cannot see sulfur dioxide. It is a clear, colourless gas. But you can certainly smell it. It has a sharp, choking odour. You may have noticed it after a wooden match was lit.

Sources of Sulfur Dioxide

Almost 80% of sulfur dioxide in air comes from natural sources. About 16% comes from burning coal and oil. These fuels contain sulfur. 4% comes from smelters and refineries. Human sources may not seem very important. But remember that most human sources release sulfur dioxide into city air. The concentration of this gas in cities often becomes very dangerous.

Effects of Sulfur Dioxide

When sulfur dioxide meets water, it forms sulfurous acid. Then the sulfurous acid changes to sulfuric acid. This is a very powerful acid which damages people and other living things. Your lungs and other tissues contain moisture. When you inhale sulfur dioxide, it forms sulfuric acid. Even very small amounts of this acid are harmful. As the sulfur dioxide level in air rises, some people die. Heavy smokers and people with asthma are most affected.

Rain forms acid with sulfur dioxide in the air. You may have heard of **acid rain**. It washes into the soil. It kills many plants (Fig. 19-3). Acid rain also harms life in rivers and lakes where it kills fish and other aquatic life.

Fig. 19-3 The stacks in the distance emit sulfur dioxide. Acid rain results. This soil is poisoned. Notice the effects on plant life.

Sulfur dioxide corrodes metals such as steel railway tracks and cars. It also reacts with brick, stonework, and even granite. It discolours many materials. Plastics and rubber become brittle. Almost every kind of material is affected.

Controlling Sulfur Dioxide

Many cities measure sulfur dioxide levels in air daily. They try to alert major sources when these levels get too high. Burning more coal and oil will make things worse. Science must find a way to cheaply remove sulfur from these fuels. Better yet, we must stop burning so much fuel. Sulfur dioxide in air is a critical problem. We must solve it!

Section Review

1. Name the major human sources of sulfur dioxide.
2. Why are these human sources so important?
3. Why is sulfur dioxide harmful to people?
4. How does "acid rain" form?
5. How does acid rain affect the life of soil and water?
6. State three other effects of sulfur dioxide.
7. How can we reduce sulfur dioxide levels in the air?

19.5 The Oxides of Carbon

Fossil fuels (coal, oil, and gas) contain carbon. When they burn, the carbon reacts with oxygen. If the combustion is complete, carbon dioxide forms. If the combustion is not complete, carbon monoxide forms. Greater amounts of fossil fuels are burned each year. The products are pollutants.

Carbon Dioxide

Carbon dioxide is a clear, colourless gas. It is a natural part of air. Plant and animal respiration releases carbon dioxide. Without it, plants could not photosynthesize. All life on earth would eventually die. Then why is carbon dioxide called a pollutant? Burning so much fossil fuel adds carbon dioxide to the air faster than nature can recycle it. More than 6 000 000 t of carbon dioxide are accumulating in the air each year. Green plants absorb carbon dioxide. But cities keep growing. Much of our forest and farmland has been lost.

Effects of Carbon Dioxide

Carbon dioxide is not directly harmful to humans. You can breathe high concentrations without harm. The main concern is the temperature effect.

This gas forms a "heat blanket" around the earth. Carbon dioxide in the air acts much like the glass in a greenhouse. It traps heat energy. This is called the "greenhouse effect". It helps to keep the earth warm. Without carbon dioxide, most of the earth's heat would radiate back into space. The earth would get very cold.

But too much carbon dioxide could raise the average global temperature. The polar icecaps could melt. Then ocean levels would rise. Coastal cities would be flooded! The whole climate of the earth could change. This is only a theory, but many scientists agree with it. Although this would happen some time in the future, we must think ahead. Our actions today affect the future of our planet.

Carbon Monoxide

About 50% of the air pollution in North America is carbon monoxide. There are few natural sources; it is nearly all man-made. Most of it comes from the incomplete burning of gasoline and oil. Cars produce about 80% of the earth's carbon monoxide. The rest comes from power plants, homes, incinerators, and industries.

Effects of Carbon Monoxide

You cannot detect carbon monoxide. It is a colourless, odourless gas. This makes it especially dangerous. Carbon monoxide can kill you. Carbon monoxide slows down the rate of oxygen transfer in your body. Red blood cells contain **hemoglobin** [HEEM-uh-glow-bin]. It carries oxygen from the lungs to all other body cells. But carbon monoxide is greatly attracted to hemoglobin. So the hemoglobin carries carbon monoxide instead. A person exposed to carbon monoxide suffers from lack of oxygen. The brain is affected. Coordination decreases. Reactions are slower. You cannot think clearly. You get dizzy and sleepy. If oxygen is not restored, you die.

Overweight people and those with heart or respiratory problems are most affected by carbon monoxide. These people need all the oxygen they can get. Heavy smokers also suffer first when carbon monoxide levels rise. Cigarette smoke has a great deal of carbon monoxide in it. Smokers have already tied up 5-10% of their hemoglobin. So carbon monoxide reaches a poisonous level in their bloodstreams faster.

Smoking just one cigarette in a car with the windows up can be dangerous. It increases the carbon monoxide level. This affects the driver and passengers alike. Imagine the carbon monoxide level when two or three people smoke in a car in heavy downtown traffic!

Controls of Carbon Monoxide

The obvious remedy is to reduce traffic and to ban smoking. Tokyo, Japan has huge traffic jams. Policemen who direct rush hour traffic breathe a lot of carbon monoxide. They must stop every half hour to breathe from oxygen tanks. In 1970, cars were banned from downtown Tokyo every Sunday. Carbon monoxide levels quickly dropped. People could enjoy strolling through the city streets. It was a great success — the ban is now permanent.

Section Review

1. How do carbon dioxide and carbon monoxide form?
2. Why is carbon dioxide important to green plants?
3. Explain why high levels of carbon dioxide could be harmful.
4. What is the major source of carbon monoxide?
5. Why do heavy smokers suffer more from carbon monoxide?
6. How can carbon monoxide levels be reduced?

19.6 ACTIVITY Combustion of Fuels

Acetylene gas is a fuel used in welding torches. Its combines with oxygen to produce a very hot flame. Because it is an organic compound, it contains carbon. When it burns in pure oxygen, it should produce carbon dioxide. This gas is the product of **complete combustion**. If there is not enough oxygen, other products form. We call this **incomplete combustion**.

This activity involves complete and incomplete combustion. You will compare the products. You will then discuss combustion in car engines and furnaces.

Problem

What is the difference between complete and incomplete combustion of a fuel?

Materials

250 mL beaker   forceps
4 pyrex test tubes   wooden splint
a few lumps of calcium carbide   Bunsen burner
CAUTION: Wear safety goggles during this activity.

Procedure

a. Calcium carbide reacts with water to form acetylene. Collect a test tube of acetylene as follows: Half fill a beaker with water. Invert a test tube full of water into the beaker. Use forceps to drop a small piece of calcium carbide into the water. Place the

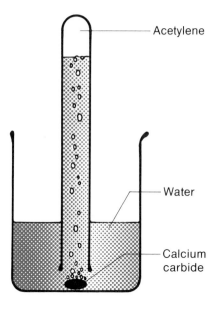

Fig. 19-4 Collecting acetylene gas

inverted test tube over the calcium carbide (Fig. 19-4). Let the acetylene gas completely fill the test tube. Lift the test tube from the water. Place it mouth down on top of the desk.

b. Repeat step a three more times, but each time collect less acetylene. Collect one half of a test tube of acetylene in one tube. Collect one third of a test tube in another. Collect one twelfth of a test tube in the last.

c. Lift each tube from the water. Remember to keep them inverted. Allow air to replace the water that runs out. Put your thumb over the mouth of each tube. Shake it hard for 20-30 s to mix the acetylene and air. Then place each test tube mouth down on top of the desk.

d. Hold each test tube in a horizontal position. Bring a burning splint to its mouth. Wait until each reaction is complete. Then place each test tube mouth down on the desk top. Record all observations. Use a diagram to compare the four test tubes after reaction.

e. Add 3-4 mL of limewater to each test tube. Shake it hard for several seconds. Record observations.

Discussion

1. Which test tube held the most oxygen? the least?
2. In which test tube did complete combustion occur? What product was formed?
3. In which test tubes did incomplete combustion occur? What products formed?
4. Where did the flame burn in the first test tube? Why?
5. If a car engine is not tuned, incomplete combustion occurs. What product will form in the engine? How might it affect the engine?
6. The burner of a gas or oil furnace must be serviced properly. Otherwise incomplete combustion of the fuel occurs. Describe the likely results.
7. Carbon monoxide is another product of incomplete combustion. This deadly gas has killed many people. Why does an idling car engine produce carbon monoxide?

19.7 Nitrogen Oxides and Hydrocarbons

Nitrogen oxides and hydrocarbons can unite to form **photochemical** smog [fow-tow-KEM-i-kul] (Fig. 19-5). Sunlight provides the energy for the reaction. Sunny Los Angeles is famous for this unhealthy brown haze. It blankets the city much of the time. It causes great damage to people, plants, and animals.

Section 19.7 221

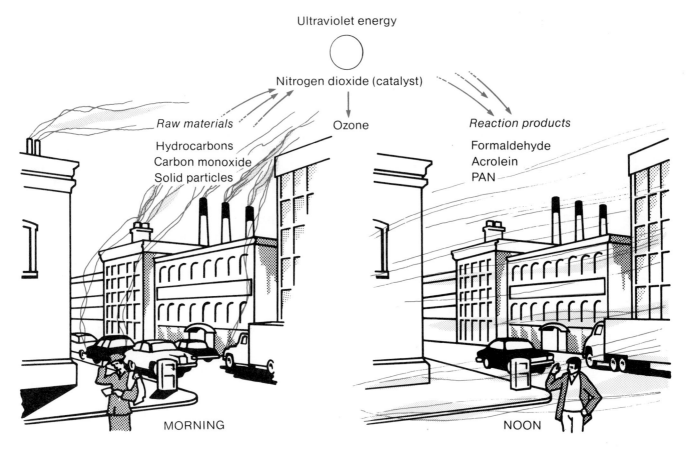

Fig. 19-5 City air often has high levels of pollutants. With energy from the sun, they can react together. This "factory in the sky" forms photochemical smog.

Nitrogen Oxides

Traces of nitrogen oxides occur naturally in the air. But the concentrations of these gases are very high in cities. Cars, furnaces, and power plants are sources. They pour millions of tonnes of these gases into the air each year.

Effects of Nitrogen Oxides Nitrogen oxides react with water to form acids. One of these is nitric acid. It is one of the most damaging acids known. Nitrogen oxides react with the moisture in your nose, throat, lungs, and eyes forming stinging acids. High levels of these gases can kill you. Long exposure to low levels can cause cancer.

Controls of Nitrogen Oxides These dangerous oxides come mainly from cars. The government has forced car makers to improve car engines. This should lower the levels of nitrogen oxides. But smoking is possibly a greater problem than cars. Smoke curling from a cigarette has about 160 times more nitrogen oxides than is safe to breathe!

Hydrocarbons

Hydrocarbon molecules contain only two kinds of atoms — hydrogen and carbon. They are organic compounds.

Sources of Hydrocarbons Most hydrocarbons come from natural sources. Decay of organic matter forms the hydrocarbon methane. It is called natural gas. Forests and other vegetation release many other hydrocarbons. Human sources only form about 15% of the hydrocarbons in air. But nearly all of it forms in city air. Almost half of these emissions [e-MI-shuns] come from car exhausts. They produce about 200 different kinds of hydrocarbons. Industries, planes, and evaporation of gasoline are other sources.

Effects of Hydrocarbons Many hydrocarbons can cause cancer. One example is benzopyrene. It comes from cars and coal-burning furnaces. It causes lung cancer. Benzopyrene also occurs in cigarette smoke; in fact, almost all lung cancer victims were heavy smokers!

Section Review

1. What is photochemical smog?
2. Why are nitrogen oxides dangerous?
3. What is the main source of nitrogen oxides?
4. What are hydrocarbons? Why are they dangerous?
5. What is the main source of hydrocarbons in cities?
6. Discuss tobacco smoke as a form of air pollution.

19.8 Other Air Pollutants

We should discuss three other pollutants. All are dangerous. All are problems because of human activities.

Ozone

Ozone is a necessary and natural part of the air. It occurs at high altitudes. And it protects us from harmful radiation from the sun. But humans produce high levels of ozone at ground level. And nature cannot absorb it fast enough.

An ozone molecule is made of three oxygen atoms. It is a strong oxidizing agent. It attacks or "burns" other substances. Even very low levels cause stinging eyes, coughing, and chest pains. Only 1 µL/L breathed in during 8 h every day for a year is damaging. It produces bronchitis and other respiratory problems. Ozone is very harmful to plants. Even very low levels cause leaf damage. Farmers can no longer grow good bean crops in parts of southwestern Ontario. Ozone from distant Detroit factories harms the plants. Ozone also damages rubber and most fabrics.

Lead

Lead is also a natural part of the air. But human sources now put most of the lead into the air. Sources include coal burning, pesticide spraying, and the burning of waste. But the most important source is leaded gasoline.

Lead is a cumulative poison. Every bit you inhale stays with you. It gradually builds up to dangerous levels. At first, lead poisoning gives you headaches and dizziness. Then you cannot sleep. You become weak and anemic. Finally, you fall into a coma and die.

Government laws are gradually getting rid of leaded gasoline. New cars run on non-leaded gasoline (Fig. 19-6). Since leaded gas is cheaper, people still buy it. People often care more about their money than their health!

Fig. 19-6 Most new cars run on non-leaded gas. But many car owners prefer cheaper leaded gas.

Asbestos

Asbestos is a fireproof material. It is widely used in buildings and industry. Asbestos is found in pipe and electric insulations. It is used in brake linings of cars. Asbestos fibres are valuable. But in the air they can also be deadly. If you inhale asbestos fibres they can lodge in your lung tissue. This irritates your lung and can cause a lung tumor or lung cancer. Many asbestos handlers have been affected. But most victims are people who live close to an asbestos industry. People are now aware of the hazard. Conditions should be made safer for both the asbestos workers and their neighbours.

Section Review

1. Why is ozone dangerous?
2. How does ozone affect plants?
3. What is the major source of lead in the air?
4. Why is lead dangerous?
5. What is a cumulative poison?
6. What is asbestos used for?
7. Why is asbestos dangerous?

Main Ideas

1. Pollutants are substances which nature cannot recycle fast enough.
2. Friction and burning are major sources of airborne particles.
3. The acid rain formed from sulfur dioxide is harmful to living things.
4. Increasing carbon dioxide levels may change the climate of our planet.
5. Cars produce most of the air pollution.
6. Smoking is the worst form of air pollution as far as human health is concerned.

Glossary

emissions	e-MI-shuns	products released
hemoglobin	HEEM-uh-glow-bin	the oxygen-carrying part of blood
photochemical	fow-tow-KEM-i-kul	a reaction using light energy
pollution		the presence of harmful substances

Study Questions

A. True or False

Decide whether each of the following questions is true or false. If the sentence is false, rewrite it to make it true. (Do not write in this book.)
1. Haze from solid particles may increase the average temperature of the earth.
2. Acid rain forms when sulfur dioxide reacts with water in the air.
3. More than 50% of the air pollution is carbon monoxide.
4. Carbon monoxide has a sharp, choking odour.
5. Ozone improves plant growth.

B. Completion

Complete each of the following sentences with a word or phrase which will make the sentence correct. (Do not write in this book.)
1. Sulfur dioxide inhaled into your moist lungs forms _____ .
2. Carbon monoxide attracts the _____ in your blood. Your body cells do not obtain enough _____ .
3. A hydrocarbon molecule contains _____ and _____ atoms.
4. Lead is a _____ poison. It builds up in your system.
5. Asbestos fibres can cause _____ .

C. Multiple Choice

Each of the following statements or questions is followed by four responses. Choose the correct response in each case. (Do not write in this book.)
1. Carbon monoxide gas is produced by
    a) friction
    b) complete combustion
    c) incomplete combustion
    d) the "greenhouse effect"
2. Increasing levels of carbon dioxide could eventually
    a) increase global temperatures
    b) block out sunlight
    c) decrease global temperatures
    d) stop plant photosynthesis

3. Nitrogen oxides and hydrocarbons react in sunlight to form
   a) the "greenhouse effect"   c) incomplete combustion
   b) acid rain   d) photochemical smog
4. The major source of airborne lead is
   a) power plants   c) leaded gasoline
   b) home furnaces   d) metal factories
5. Asbestos is used
   a) as a fireproof material   c) in unleaded gas
   b) as a fuel   d) to produce ozone

D. Using Your Knowledge

1. Decay in a marsh releases gases such as methane. These are not called pollutants. But methane gas released from an oil refinery is called a pollutant. Explain.
2. What processes release carbon dioxide? What processes use up carbon dioxide? How has the natural balance of carbon dioxide been upset?
3. Cars are a major source of air pollutants. Suggest several ways of reducing these pollutants. How would these change your lifestyle?
4. List some of the pollutants in cigarette smoke. Describe their effects on humans.

E. Investigations

1. How much dirt falls near your home or school in a month? Are the nearest factory areas dirtier? Select a number of test sites in your community. Design an experiment to collect and measure dustfall. Figure 19-7 shows a sample collection jar.
2. Car engines release about 0.3 kg of carbon monoxide for every litre of gas burned. Calculate the amount of carbon monoxide gas released by your family car: a) during an average working day; b) during one week; c) during one month; d) during one year. You must first know the average distance driven and the amount of gasoline used.
3. Many communities have banned smoking in some public places. Do you agree or disagree with this policy? Has the ban gone far enough? Has it gone too far? Select one side of this issue. Collect all the information you can. Then prepare for a class debate on the topic.

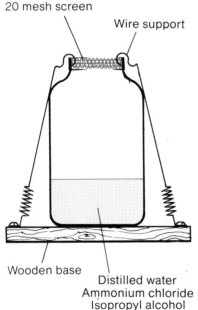

Fig. 19-7 The sample collection jar should be secured to a support base.

20 Water Pollution

20.1 Oxygen
20.2 Carbon Dioxide and pH
20.3 Activity: Measuring the pH of Water Samples
20.4 Nitrogen and Phosphorus
20.5 Activity: Studying Suspended Solids in Water
20.6 Demonstration: Effect of Organic Matter on Water Quality

Many of our lakes and rivers are polluted (Fig. 20-1). They are unsafe for drinking or swimming. Most of us don't like this; we want clean water. But our towns and cities still put sewage in the water. And runoff from farms continues to pollute streams. Why?

Fig. 20-1 This river gets sewage from five towns. All this sewage ends up in a lake.

Perhaps we haven't complained enough. Have you ever told anyone you don't like water pollution? When you do complain, you should know what you are talking about. So let's learn about water pollution.

20.1 Oxygen

Most aquatic organisms need oxygen. They use oxygen that is dissolved in the water. But some kinds of pollution use up this oxygen. For example, suppose sewage is put into the water. Bacteria feed on this sewage. As they feed, they use oxygen. They must respire. Therefore they take oxygen from the water. Then fish and other animals don't have enough oxygen. So they die.

Section 20.1

How Much Oxygen is Needed?

Most trout need at least 10 µg/g of oxygen. (That's ten micrograms of oxygen in one gram of water.) However, chub need only 7 µg/g. And carp can live with only 1-2 µg/g of oxygen. In general, water should contain at least 5 µg/g of oxygen. If it doesn't, many kinds of animals will die (Fig. 20-2).

The amount of oxygen an animal needs depends on many factors. One factor is activity. An animal needs more oxygen when it is active than when it is resting. Another factor is the temperature. As the water temperature goes up, most animals need more oxygen. For example, a resting trout with a mass of 1 kg needs 55 mg of oxygen per hour at 5°C. But, at 25°C it needs 285 mg per hour.

Fig. 20-2 This water has 4 ug/g of oxygen in it. Is this a trout stream?

Solubility of Oxygen in Water

The amount of oxygen in water depends on the temperature. Table 20-1 shows this dependence. Note that water holds less oxygen at high temperatures.

Table 20-1 Solubility of Oxygen in Water (when air is the only source)

Temperature of water (°C)	Solubility µg/g
0	14.6
10	11.3
20	9.1
30	7.5

Sources of Oxygen

In streams and rivers the oxygen comes mainly from the air. As the water splashes over rocks, it picks up oxygen from the air.

Lakes and ponds get some of their oxygen from the air. It enters the water by diffusion. But most of the oxygen comes from plants and algae. They make oxygen by photosynthesis.

Thermal Pollution

Power generating plants and some industries put warm water into lakes and rivers. This is called **thermal pollution** [THER-mal].

Table 20-1 tells us that the warm water will not hold as much oxygen. Yet we know that animals need extra oxygen in warm water. As a result, the warm water kills some species of fish and other animals.

Sewage Pollution

Much of **sewage** [SU-aj] was once part of living organisms. Sewage is a serious form of pollution. Sewage tends to remove oxygen from the water. Bacteria feed on the sewage and respire as they do so. Often they will reduce the oxygen level in the water to zero.

Sewage also adds nutrients to the water. **Nutrients** [NU-tree-ents] are elements that living things need for growth. **Nitrogen** [NI-tro-jen] and **phosphorus** [FOS-for-us] are examples. Extra nutrients make algae grow very quickly. Large amounts of algae build up in the lake. This is called an **algal bloom** [AL-gal BLOOM]. When these algae are alive, they are good for the water. They produce oxygen by photosynthesis and are eaten by other organisms. Often, however, algae crowd themselves out of food and space and then die. Bacteria now feed on the dead algae. This uses up oxygen and kills other organisms (Fig. 20-3).

Fig. 20-3 This lake has an algal bloom. When the algae die, the oxygen level of the lake may drop. Then fish may die.

Section Review

1. What is the least amount of oxygen water should contain?
2. How does temperature affect the solubility of oxygen in water?
3. Where does a stream get its oxygen?
4. Where does a lake get its oxygen?
5. What is thermal pollution? How does it affect animals?
6. What is sewage pollution? How does it affect organisms in the water?
7. What is an algal bloom?

20.2 Carbon Dioxide and pH

Two other factors that tell us about water quality are carbon dioxide and pH. Let us see what they are.

Carbon Dioxide

Most of the carbon dioxide in water comes from organisms. Most water has organisms in it. As they respire, they put carbon dioxide into the water.

Water near the surface usually has less than 10 µg/g of carbon dioxide, but water near the bottom often has much more. Bacteria feed on organic matter in the bottom ooze. As they respire, they add carbon dioxide to the water at the bottom.

Polluted water often has over 25 µg/g of carbon dioxide. This harms most living things. In fact, 50-60 µg/g kills many species of animals.

pH

The pH of water is a measure of its acidity. Values for pH run from 0-14 (Fig. 20-4). A pH of 7 is neutral. A pH less than 7 is acidic. And a pH greater than 7 is basic. The lower the pH of water, the more acidic the water is. That is, water gets less acidic as its pH goes from 0 to 14. And it gets more basic at the same time.

A pH between 6.7 and 8.6 supports a good balance of fish types. Few fish species can live for long at a pH below 5.0 or above 9.0.

Acid spills by industries lower the pH. This can kill organisms. Also, smelters and coal-burning power plants put sulfur dioxide into the air. It reacts with water vapour in the air to form sulfuric acid. When it rains or snows this acid gets into lakes and rivers. It is called **acid rain**. Its pH is usually below 4.0. It, too, can kill organisms.

Sewage usually raises the pH of water. Household sewage has many bases in it. Also, the decay of sewage by bacteria produces bases. Thus sewage can kill organisms by raising the pH.

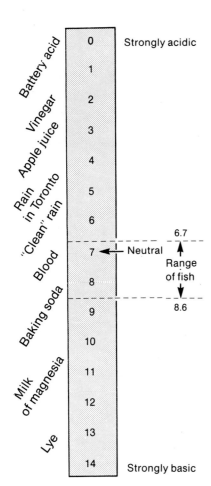

Fig. 20-4 The pH scale for acidity.

Section Review

1. How does carbon dioxide get into water?
2. What level of carbon dioxide harms organisms?
3. What is pH?
4. Explain the pH scale.
5. What pH range is good for fish?
6. What is acid rain? Why will it harm organisms?
7. How does sewage affect pH?

20.3 ACTIVITY Measuring the pH of Water Samples

You learned in Section 20.2 that pH is an important measure of water quality. In this activity you will measure the pH of several water samples. Then you will decide how the pH of each sample might affect animal life in the water.

Problem

How does the pH of the water vary with the nature of the water sample?

Materials

test tube
pH paper
water samples prepared or collected by students and teacher

Procedure

a. Copy Table 20-2 into your notebook.
b. Wash the test tube well.
c. Put 1-2 mL of distilled water (from the bottle labelled "Sample 1") in the test tube.
d. Find the pH of Sample 1 using the pH paper. Record your result.
e. Repeat steps (b) to (d) using tap water (Sample 2).
f. Repeat steps (b) to (d) for a weak acid (Sample 3) and a weak base (Sample 4).
g. Repeat steps (b) to (d) for the other samples provided. Your teacher may ask you to provide some samples. It will be interesting to test water from a local river or lake, from a swimming pool, and from an aquarium. You could also test water containing detergents, water containing decaying organic matter like fish food, dish water, rain water, and melted snow.
h. Read the part on pH in Section 20.2 again. Now complete the last column in your table.

Table 20-2   The pH of Water Samples

Sample number	Nature of sample	pH of sample	Effect on living things
1	distilled water		
2	tap water		
3	carbonic acid (a weak acid)		
4	limewater (a weak base)		
5			

Discussion

1. What is the pH of distilled water? Why?
2. Did tap water have a different pH than distilled water? Why?
3. Which has the higher pH, carbonic acid or limewater? Why?
4. Try to give a reason for the pH of each sample from number 5 on.

20.4 Nitrogen and Phosphorus

Fig. 20-5 A 4-9-15 fertilizer is 4% nitrogen, 9% phosphorus, and 15% potassium. All three nutrients are needed for growth.

Nitrogen and phosphorus are two elements that all organisms need. Therefore they are called **nutrients**. They make organisms grow (Fig. 20-5); but sometimes they cause too much growth and algal blooms result. In such cases, nitrogen and phosphorus are called **pollutants**.

Nitrogen

Nitrogen (N) occurs in two main forms in water. One form is ammonia (NH_3); the other is nitrate (NO_3).

When dead plants and animals decay, ammonia is formed. It is also formed when fecal matter and urea break down. Thus ammonia in water suggests sewage may be present. In urban areas the sewage is likely human. In rural areas it may come from beef feedlots.

Many fertilizers also contain ammonia. Therefore runoff from fields and lawns pollutes water with ammonia.

Ammonia does little harm directly. But it converts to nitrate. Then the nitrate causes plants and algae to grow. If the total nitrogen is over 0.30 $\mu g/g$, an algal bloom may result in a lake.

Phosphorus

Phosphorus (P) usually occurs in water as a phosphate (PO_4). Its sources are much the same as those for nitrogen: sewage, animal wastes, and decaying plants and animals. Fertilizers usually contain phosphate. But runoff from fields and lawns contains little phosphate. The phosphate particles stick to soil particles. Therefore, unless soil erosion occurs, phosphates do not wash from fields and lawns.

Phosphate causes plants and algae to grow. If the total phosphorus is just 0.015 $\mu g/g$, an algal bloom may result in a lake.

Section Review

1. What are nutrients?
2. When is a nutrient called a pollutant?
3. In what forms does nitrogen occur in water?
4. What are the sources of nitrogen in water?
5. How much nitrogen is needed for an algal bloom?
6. In what form does phosphorus occur in water?
7. How much phosphorus is needed for an algal bloom?

20.5 ACTIVITY Studying Suspended Solids in Water

You have probably seen a river or lake that looks dirty. This dirty appearance is due to tiny solid particles floating in the water called **suspended solids** [sus-PEN-ded]. Some solids like nitrates and phosphates are soluble. They dissolve in the water and cannot be seen. Suspended solids are insoluble; they do not dissolve. And, if you look closely, you can see them.

Some suspended solids are alive. They are tiny algae and other small organisms. Other suspended solids are non-living. They include sewage particles, silt, clay, and many industrial wastes.

In this activity you will separate suspended solids from water. Then you will look at them closely.

Problem

Can you separate the suspended solids from water?

Materials

funnel
filter paper
ring stand
iron ring

400 mL beaker (2)
water containing suspended solids (from a river after a storm)
hand lens

Procedure

a. Fold the filter paper as shown by your teacher. Then put it in the funnel.
b. Get a beaker full of the water.
c. Filter the water through the filter paper (Fig. 20-6).
d. Let the filter paper dry.
e. Examine the particles on the filter paper with a hand lens.

Discussion

1. What did the water look like before filtration?
2. What did the filtrate (liquid that went through the paper) look like?
3. Describe the appearance of the particles.
4. What solids might be in the filtrate? How could you prove this?
5. How do you think the suspended solids got into the water?
6. What could be done to keep the suspended solids out of the water?

Fig. 20-6 Filtration of a water sample

20.6 DEMONSTRATION Effect of Organic Matter on Water Quality

Organic [or-GAN-ik] matter is matter that contains carbon. Matter that is or was once part of a living organism is organic. Therefore, it contains carbon. It also contains nitrogen, phosphorus, and other elements. Many of these are nutrients. What do you think will happen if this kind of organic matter breaks down in water?

Problem
How does organic matter affect water quality?

Materials
small aquaria or pails of at least 5 L size (2)
several sprigs of an aquatic plant
100 W lamps (2)
organic matter (for example, fish food)
Hach water testing kits
thermometer (−10°C to 100°C)

Procedure
a. Fill both containers to within 2 cm of the top with water.
b. Add about 10 sprigs of aquatic plant to each container.
c. Place the containers side by side. Then mount a 100 W lamp over each one (Fig. 20-7).

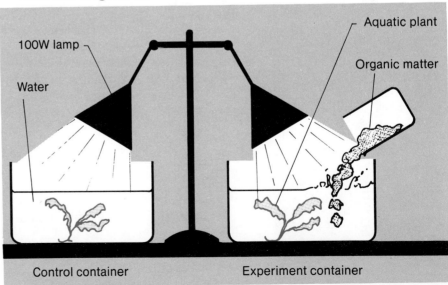

Fig. 20-7 How will organic matter affect water quality?

d. Dump a can of fish food (or other organic matter) into one container.

e. Wait at least 2 or 3 d. Then test the water in each container. If possible, test for oxygen, carbon dioxide, pH, ammonia, nitrate, and phosphate. Record your results in a table. Also record the water temperature.
f. Note any changes in the appearance of the water and plants.
g. If time permits, repeat the tests two weeks later. Also, note any changes in the appearance of the water and plants.

Discussion

1. No organic matter was added to one container. Why?
2. Explain any differences between the tests for the two containers.
3. Explain any differences in appearance between the two containers.
4. What difference did two weeks make? Why?

Main Ideas

1. Warm water holds less oxygen than cold water.
2. Fish need more oxygen at higher temperatures.
3. Sewage lowers the oxygen level in water. It also adds nutrients to the water.
4. Too much nutrient in water can cause algal blooms.
5. High levels of carbon dioxide can harm aquatic animals.
6. Very low and very high pH levels can harm aquatic animals.
7. Nitrogen and phosphorus are nutrients. They enter water from sewage, feedlots, and fertilizer. They can cause algal blooms.
8. Suspended solids are tiny particles of living things, sewage, silt, and clay.

Glossary

acid rain		rain that has a very low pH; caused mainly by industries
algal bloom	AL-gal BLOOM	the growth of algae at a rapid rate
nitrogen	NI-tro-jen	an element needed for growth of organisms
nutrients	NU-tree-ents	an element such as nitrogen that causes growth
organic	or-GAN-ik	matter containing the element carbon
phosphorus	FOS-for-us	an element needed for growth of organisms

sewage	SU-aj	waste from human sources
solubility	sol-you-BIL-i-tee	the mass of a substance that will dissolve in a certain mass of a liquid
suspended solids	sus-PEN-ded	tiny particles of living and non-living solids in water
thermal pollution	THUR-mal	pollution caused by heat

Study Questions

A. True or False

Decide whether each of the following sentences is true or false. If the sentence is false, rewrite it to make it true. (Do not write in this book.)
1. As water gets warmer, fish need more oxygen.
2. As water gets warmer, it holds more oxygen.
3. Photosynthesis is the main source of oxygen in a stream.
4. Nutrients like phosphorus can cause algal blooms.
5. Water with a pH of 9 is acidic.

B. Completion

Complete each of the following sentences with a word or phrase which will make the sentence correct. (Do not write in this book.)
1. Water should contain at least ▨▨▨ µg/g of oxygen.
2. Extra nutrients in a lake can cause an ▨▨▨.
3. Levels of carbon dioxide over ▨▨▨ can harm living things.
4. Acid rain is caused by ▨▨▨.
5. An algal bloom may occur if the total nitrogen in a lake is over ▨▨▨.

C. Multiple Choice

Each of the following statements is followed by four responses. Choose the correct response in each case. (Do not write in this book.)
1. Thermal pollution has warmed up a bay on a lake. The water temperature of the bay is about 30°C. Which one of the following statements is true?
   a) There will be trout in the bay.
   b) The bay will have an algal bloom.
   c) There could be carp in the bay.
   d) There will be no fish in the bay.
2. Suppose that some sewage escapes into a small lake. What is likely to happen first?
   a) The algae in the lake will die.
   b) The oxygen level will drop.

c) Nutrients will be used up.
   d) Everything in the lake will die.
3. A lake near a smelter has a pH of 3.5. This low pH is likely due to
   a) acid rain
   b) sewage
   c) too many fish in the water
   d) an algal bloom
4. A lake has a total nitrogen of 0.4 µg/g and a total phosphorus of 0.2 µg/g. What is likely to happen in this lake?
   a) All fish will die.
   b) All plants will die.
   c) The pH will go up.
   d) An algal bloom will develop.
5. Select two kinds of suspended solids.
   a) Clay and algae
   b) Nitrates and phosphates
   c) Silt and nitrates
   d) Clay and oxygen

D. Using Your Knowledge

1. What would happen in a lake if all the plants and algae died? Explain.
2. Thermal pollution often kills fish like trout. But it often helps produce more fish like carp. Why is this so?
3. Algae can be both good and bad for a lake. Why is this so?
4. Why should people not put too much fertilizer on their lawns?

E. Investigations

1. Design and conduct an experiment to prove that natural water contains *dissolved* solids.
2. Research the problem of soil erosion. What causes it? How serious is it? How can it be stopped?
3. Design and conduct an experiment to study the effect of an algal bloom on water quality.
4. Attend a tour of a sewage treatment plant. Write a report on your visit. It should explain the steps in sewage treatment. Take photographs if you can.
5. Design and conduct an experiment to study the effect of lawn fertilizer on water quality.
6. Obtain an oxygen Hach kit from your teacher. Then design and conduct an experiment to confirm the oxygen solubility data in Table 20-1, page 228.

21 How to be a Thinking Consumer

21.1 Problems Facing Consumers
21.2 Choosing Products the Scientific Way
21.3 Activity: Testing Detergents Using the Scientific Method

All of us want the best buy for our money. But how do you know when you are getting the best buy? Can you believe advertisements? Can you trust the labels on products?

You can use your knowledge of chemistry and its methods to help you make choices. You will see how that is done in this chapter. You will even get a chance to try this yourself.

21.1 Problems Facing Consumers

Fig. 21-1 Which detergent cleans best? Does it also cost more? Will it save me money because it works in cold water or because I need less per wash? Which one should I buy?

As consumers we have to make choices almost every day. Which cereal provides the most nutrient per serving? What type of margarine is best? What brand of soap cleans best (Fig. 21-1)? What brand of toothpaste prevents tooth decay the best? What colour television set gives the best picture? What car will last longest and burn the least fuel?

How do you make such choices? It certainly isn't easy! The advertisements on radio and television don't help. Everyone's product seems to be "the best". And newspaper advertisements aren't much better. Perhaps the labels on the products will help us. Let us consider some labels to see if they do help.

Labels on Cleaning Products

Label 1. The label on one cleaning product says "Picks up six times more dust." That sounds good, doesn't it? But the *thinking consumer* asks a question: Six times more than what? than all other products? than a dry dusty rag? By itself, the statement means little.

Label 2. Another cleaning product is recommended for sinks, stoves, refrigerators, and whitewall tires. Its label says: "Cuts tough dirt

and leaves a fresh pine fragrance". Again the *thinking consumer* asks questions: Just what is "tough dirt"? How much extra am I paying so my fridge and whitewall tires will smell like pine trees?
Label 3. The label on another cleaning product says: "Now stronger; chases dirt". But the *thinking consumer* should ask questions. Stronger than what? How much stronger? Can a chemical really "chase" dirt?
Label 4. "Lemon fresh" says another label. That's supposed to make you want the product. But you, the *thinking consumer*, should ask questions. How well does the product clean? Is a lemon smell really "fresh"? Do I want to pay extra so everything I clean will smell like a lemon?

No label said what was in the product. Therefore you cannot compare products. So how do you know what to buy?

Labels on Cereals

Not all labels are useless. Table 21-1 summarizes data from the labels on four cereals. In all cases a serving was defined as 28 g of cereal. Which cereal will give you the best breakfast? What else would you like to know about that cereal before you bought it?

Section Review

1. Do the labels on cleaners have qualitative or quantitative data on them?
2. Do the labels on cereals have qualitative or quantitative data on them?
3. Which seem to be more useful, qualitative or quantitative data? Why?

Table 21-1 Nutrient Value per Serving of Cereals

	Cereal 1	Cereal 2	Cereal 3	Cereal 4
Food energy	441 kJ	458 kJ	454 kJ	584 kJ
Protein	2.2 g	3.5 g	2.7 g	3.4 g
Fat	0.1 g	2.0 g	0.2 g	6.2 g

Total carbohydrates
    Sugars
    Starch and other carbohydrates
    Dietary Fibre
	24.4 g
	3.5 g
	20.5 g
	0.4 g
		20.2 g
		0.7 g
		18.5 g
		1.0 g
			23.7 g
			5.0 g
			16.9 g
			1.8 g
				17.3 g
				6.3 g
				11.0 g
				—

Vitamin B_1 (thiamine)	0.6 mg	—	0.6 mg	—
Vitamin B_2 (riboflavin)	1.0 mg	1.0 mg	1.0 mg	—
Vitamin, niacinamide	6.0 mg	6.0 mg	6.0 mg	—
Iron	5.5 mg	4.0 mg	4.0 mg	—

21.2 Choosing Products The Scientific Way

You learned about the scientific method in Chapter 2 and have been using it throughout this course. Now we want to make sure you can use it as a consumer. Then, hopefully, you won't get cheated when you buy things.

A Review of the Scientific Method

Most scientific studies use six steps:
1. Recognition of a problem;
2. Collecting information on the problem;
3. Making a hypothesis;
4. Doing an experiment;
5. Observing and recording results;
6. Making a conclusion.

Let us see how you can use this method to decide which paper towels to buy (Fig. 21-2).

Fig. 21-2 Which brand absorbs water the best? How can you find out?

Selecting Paper Towels

According to advertisements, several kinds of paper towels are "the best". Let's find out which one is really the best.
1. *The Problem:* What brand of paper towel absorbs water the best?
2. *Collecting Information:* You might talk to others who have used various brands. You could study the claims of the makers. And you could read studies like "Consumer Reports". If you find the information you need, your study is over. But suppose you don't. Then you must go on with the scientific method.

3. *Hypothesis:* The paper towel that picks up the most water per 100 g is the best.
4. *Experiment:* You could design and try this experiment.
   a) Weigh out 100 g of each type of towel.
   b) Put each type in water. Leave them there until they are saturated.
   c) Remove them and weigh them again.
5. *Observing and Recording Results:* You could use a table like Table 21-2 for your results.

Table 21-2 Studying Paper Towels

Brand	Mass of towel at start	Mass of towel plus water	Mass of water absorbed
A	100 g	150 g	50 g
B	100 g	170 g	70 g
C	100 g	142 g	42 g
D	100 g	185 g	85 g

6. *The Conclusion:* Since Brand D absorbed the most water per 100 g, it is the best.

Usually some questions come out of such a study. For example, is Brand D as cheap to use? In other words, will $1.00 worth of Brand D pick up as much water as $1.00 worth of other brands? Also, will Brand D pick up oils and other kitchen spills as well as the other brands? These questions suggest more experiments that you should do. You could spend all your life doing experiments!

The Consumers Association of Canada

Fig. 21-3 The wise consumer studies sources like these before buying.

Fortunately you do not have to do the experiments yourself. The Consumers Association of Canada (CAC) tests many products; its American affiliate tests even more. Between them they test almost everything you will ever want to buy: cereals, cameras, skates, bicycles, cars, vacuum cleaners, radios, and so on. The results of their tests are published in "Canadian Consumer" and "Consumer Reports" (Fig. 21-3). Each year books are published that summarize the main studies.

Our governments also test many products. For example, the federal government tests the fuel economy of all cars sold in Canada. The results are published in a booklet called the "Fuel Consumption Guide". Reports on cars are also published in "Popular Science" and other magazines.

The wise consumer never buys before studying sources such as these.

Section Review

1. Describe how to use the scientific method to find out what brand of paper towel absorbs water best.
2. The wise consumer uses the scientific method when buying things. But you may not have the time or equipment to do your own studies. What can you do instead?

21.3 ACTIVITY Testing Detergents Using the Scientific Method

You cannot test every product you are going to buy. You don't have the time or equipment. Besides, you don't need to. The CAC and other groups test them for you. However, you should understand how the testing is done. This activity gives you a chance to find out.

Obtain several brands of detergent that are used for washing clothes. Then use the scientific method to find out which detergent cleans best. The following suggestions will help you design and carry out your study.

1. The Problem

What is the problem to be studied? You want to find out which detergent cleans best. Write the problem in your notebook.

2. Collecting Information on the Problem

This time, don't read "Canadian Consumer" or "Consumer Reports". Read just the information on the cartons and in advertisements. Summarize your findings in your notebook.

3. Making a Hypothesis

Make a hypothesis using the information you gathered. For example, your hypothesis could say: Brand B has the greatest cleansing ability. But perhaps you should narrow the study down. Brand B might be best with just one fabric. Or it might be best at removing just one substance. Perhaps you should start with a hypothesis like this: Brand B has the greatest cleansing ability in removing oil from denim.

4. Doing an Experiment

Now, design your experiment. What detergents will you use? How much will you use? Will you use warm or cold water? What fabrics will you test? How will you mix the detergent, water, and fabric?

Be sure to pay attention to proper **controls**. That is, control all **variables** [VAI-ri-a-buls] but the one you are studying.

Write out your method. Discuss it with others. Then try your experiment.

5. Observing and Recording Results

Prepare a table for your results. Comparisons are easier to make if the data are in a table.

6. The Conclusion

Study the data carefully. Draw a conclusion. Does it agree with your hypothesis? How valid do you think your conclusion is? What further studies should be done? Is the detergent that cleans best always the best detergent? (Consider cost.)

Main Ideas

1. Labels on products do not always give enough information.
2. The scientific method can be used to compare consumer goods.
3. The Consumers Association of Canada (CAC) uses the scientific method to compare consumer goods.
4. You should always consult studies from the CAC and others before buying goods.

Glossary

consumer		person who buys a product
qualitative	KWOL-i-tay-tiv	a description that cannot be measured
quantitative	KWON-ti-tay-tiv	a description that can be measured
variable	VAI-ri-a-bul	a factor that can affect the results of an experiment

Study Questions

A. True or False

Decide whether each of the following sentences is true or false. If the sentence is false, rewrite it to make it true. (Do not write in this book.)
1. Advertisements are the best way to decide what to buy.
2. All labels on products are useless.
3. All cereals contain the same amount of vitamins in a 28 g serving.
4. Most scientific studies begin with a problem.
5. A hypothesis will always be true.

B. Completion

Complete each of the following sentences with a word or phrase which will make the sentence correct. (Do not write in this book.)
1. In Table 21-1, Cereal ▆▆▆ contains the most fat in a serving.
2. In Table 21-1, Cereal ▆▆▆ contains the most protein in a serving.
3. The labels on cleaning products give qualitative data. But the labels on cereal boxes give ▆▆▆ data.
4. Most consumer products are tested for us by ▆▆▆ .
5. The fuel economy of cars is tested by ▆▆▆ .

C. Multiple Choice

Each of the following statements is followed by four responses. Choose the correct response in each case. (Do not write in this book.)
1. Cleaner A has a label which says "Cleans best". Cleaner B says "Over twice as effective". Cleaner C says "New, improved". Which cleaner is likely the best?
   a) Cleaner A
   b) Cleaner B
   c) Cleaner C
   d) cannot tell from this information
2. Cleaner A leaves a "fresh pine smell". Cleaner B leaves a "fresh lemon smell". Cleaner C has no smell at all. Which cleaner is likely the best?
   a) Cleaner A
   b) Cleaner B
   c) Cleaner C
   d) cannot tell from this information
3. In Table 21-1, Cereal 4 gives the most food energy per serving. The extra energy comes mainly from
   a) fats b) carbohydrates c) proteins d) the lack of vitamins
4. The best place to read about tests on Canadian consumer products is
   a) government reports
   b) Popular Science
   c) Canadian Consumer
   d) test data from the makers of the products

5. The fuel economy of all cars sold in Canada is tested by
   a) The Consumers Association of Canada
   b) the Canadian government
   c) General Motors
   d) mechanics who work for Popular Science

RATING COTE	MANUFACTURER FABRICANT	MODEL MODÈLE	ENGINE SIZE CYLINDRÉE DU MOTEUR	CYLINDERS CYLINDRES	TRANSMISSION BOÎTE DE VITESSES
4.3	VOLKSWAGEN	RABBIT DIESEL	1.6	4	M4
4.9	VOLKSWAGEN	JETTA DIESEL	1.6	4	M5
	VOLKSWAGEN	RABBIT DIESEL	1.6	4	M5
5.1	CHEVROLET	CHEVETTE DIESEL	1.8	4	M5
	HONDA	CIVIC FE 2DR/PTES HB/SHA	1.3	4	M5
	PONTIAC	ACADIAN DIESEL	1.8	4	M5
5.3	TOYOTA	STARLET	1.3	4	M5
5.5	DATSUN	210	1.4	4	M5
5.7	HONDA	CIVIC 1500 GL 2DR/PTES	1.5	4	M5
	VOLKSWAGEN	JETTA DIESEL	1.6	4	A3
	VOLKSWAGEN	RABBIT DIESEL	1.6	4	A3
5.8	TOYOTA	TERCEL SP	1.5	4	M4
5.9	DODGE	COLT 3DR/PTES	1.4	4	M4
	PLYMOUTH	COLT 3DR/PTES	1.4	4	M4
6.0	HONDA	CIVIC 4DR/PTES	1.5	4	M5
	MAZDA	GLC	1.5	4	M5
6.1	CHEVROLET	CHEVETTE DIESEL	1.8	4	A3
	DODGE	COLT 3DR/PTES	1.4	4	M8
	DODGE	COLT 5DR/PTES	1.4	4	M8
	PLYMOUTH	COLT 3DR/PTES	1.4	4	M8
	PLYMOUTH	COLT 5DR/PTES	1.4	4	M8
	PONTIAC	ACADIAN DIESEL	1.8	4	A3
6.2	MAZDA	GLC	1.5	4	M4
	TOYOTA	TERCEL	1.5	4	M5
6.3	DATSUN	NISSAN STANZA	2.0	4	M5
	DATSUN	310	1.5	4	M5
	DATSUN	310	1.5	4	M4
	DODGE	COLT 3DR/PTES	1.6	4	M8
	PLYMOUTH	COLT 3DR/PTES	1.6	4	M8
	TOYOTA	TERCEL	1.5	4	M4
6.4	CHEVROLET	CHEVETTE	1.6	4	M5
	CHEVROLET	CHEVETTE	1.6	4	M4
	DODGE	COLT 5DR/PTES	1.6	4	M8
	DODGE	OMNI 024 MISER	1.7	4	M4
	HONDA	CIVIC 2DR/PTES HB/SHA	1.3	4	M5
	HONDA	CIVIC 4DR/PTES HB/SHA	1.3	4	M5
	PLYMOUTH	COLT 5DR/PTES	1.6	4	M8
	PLYMOUTH	HORIZON TC3 MISER	1.7	4	M4
	PONTIAC	ACADIAN	1.6	4	M5
	PONTIAC	ACADIAN	1.6	4	M4

Fig. 21-4 What do the top 20 cars have in common?

D. Using Your Knowledge

1. a) Figure 21-4 is a photograph of a page from the "Fuel Consumption Guide". It lists cars in order of ascending fuel consumption. That is, the first one burns the least fuel. Study the data carefully. Then make a list of things the first 20 cars have in common. Now make a generalization that describes the car a person should look for if he/she wants a car with low fuel consumption.
   b) What other factors should be considered before he/she buys the car?
2. The advertisement for one pain reliever says: "Contains twice as much pain reliever per pill as the most common brand". Make a list of questions you would like to ask the maker before you buy this pain reliever.
3. Why are off-road trucks given names like Bronco, Blazer, Renegade, and Ramcharger?
4. Why are some cars given names like Firebird, Cobra, and Eagle?
5. Do you think advertisements work? How do you know?

E. Investigations

1. Design and do an experiment to find out which brand of paper towel absorbs water the best. Compare at least three brands.
2. Visit a drugstore. Note the data on at least three brands of headache tablets that have the same ingredients. Note also the number of pills per bottle and the price. Then prepare a report that recommends the best buy.
3. Visit a grocery store. Pick out two brands of the same type of cereal such as cornflakes. Record the data you need to decide which of the two brands is the best buy. Write a report on your study.
4. Suppose you are shopping for a new car. You want a sub-compact. Its fuel economy rating must be 7.0 L/100 km or better. Prepare a data table you could use to compare three makes of cars. Visit three showrooms and collect the data.
5. Visit a library and research detergents. What are the active ingredients? How do they work? Do some detergents pollute more than others?
6. Design and do an experiment to test the pH of several shampoos. Find out the pH of a healthy scalp. Then give the name of the shampoo that will change the scalp's pH the least.
7. Design and do an experiment to test the pH of several brands of bottled water.

Mechanics and Machines

The earth exerts huge forces on matter. Sometimes these forces have devastating effects. Earthquakes shatter dams, bridges, and buildings. Volcanoes spew dust and rocks into the air. Hurricanes toss cars about like toys.

Mechanics is a branch of physics. It deals with the motion of objects. It also describes the action of forces. In this unit you will study forces. You will find out their effects on matter. You will learn how we harness forces. Machines help us harness forces. We use machines to make work easier.

CHAPTER 22
Force and Pressure

CHAPTER 23
Work, Energy and Power

CHAPTER 24
Machines

Fig. 22-0 Forces in nature

22 Force and Pressure

22.1 The Nature of Force
22.2 Activity: Measuring a Force
22.3 The Unit of Force and the Force of Gravity
22.4 Balanced and Unbalanced Forces
22.5 Activity: The Forces of Static and Kinetic Friction
22.6 Pressure: A Force Applied to An Area

What do an elastic band, a magnet, and gravity have in common? Why does a bicycle speed up when you push on the pedals? Why does it coast to a stop on a level road? Why do small heels push into carpet further than large ones? You will find answers to such questions through the study of forces and pressure.

22.1 The Nature of Force

Definition of a Force

Push a book away from you. Pull it toward you. You are exerting forces. *A force is a push or a pull.* There are many examples of pushes and pulls. The adult in Figure 22-1 is pushing a child on a swing. The people in Figure 22-2 are having a tug of war. They are pulling on a rope.

Fig. 22-2 Push and pull forces

Fig. 22-1 Push and pull forces

Effects of a Force

A force can change the shape of an object. Place the palm of your hand on top of a soft rubber ball. Push downward. The ball changes shape (Fig. 22-3,A).

A force can start or stop an object moving. Push sideways on a ball. The ball starts to move. Then push in a direction opposite to its motion. The ball stops. These actions are shown in Figure 22-3,B.

A force can change the speed of an object. Push a moving ball. It speeds up. Give a similar push in the opposite direction. The ball slows down (Fig. 22-3,C).

A force can also change the direction of a moving object. Push sideways on a moving ball. The ball changes direction. This is shown in Figure 22-3,D. These examples show that a force can change the shape and/or motion of an object.

Properties of a Force

All forces have two properties in common. Every force has a definite size. The size of a force is described using words like small or large. To be more precise scientists use a numeral and a unit. For example, they might describe a force as having a size of ten newtons. The numeral in this example is ten; the unit is the newton.

Each force also has a definite direction. Adjectives such as left, right, up, and down state directions. So do compass directions such as north, south, and southeast.

Types of Forces

There are many types of forces in nature. An athlete applies a muscular force to lift a weight. A magnet exerts a magnetic force to attract iron. A charged comb exerts an electric force to attract paper. A string exerts a tensile force to support a yo-yo. An apple is attracted to the earth by the force of gravity. The force of friction affects all moving objects. Figure 22-4 shows several types of forces.

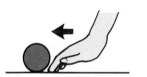
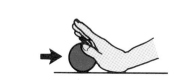
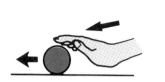
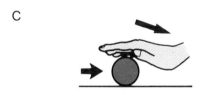
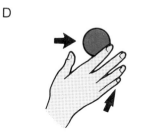

Fig. 22-3 The effects of force on an object

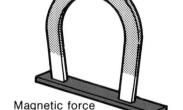
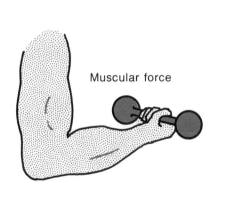

Magnetic force

Muscular force

Force of gravity

Fig. 22-4 Types of forces

Section 22.1 249

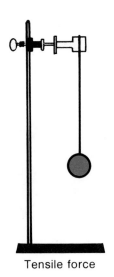
Tensile force

Electric force

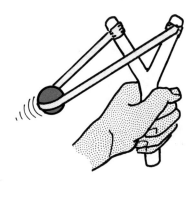
Elastic force

Section Review

1. Define force.
2. What effects can a force have on an object?
3. What two properties do all forces have in common?
4. What are two ways of stating the size of a force?
5. What are two ways of stating the direction of a force?
6. Name and give examples of 6 types of forces?

22.2 ACTIVITY Measuring a Force

In this activity you will find out how to measure force. You will measure the force of **gravity** on standard masses and find the force of gravity-mass ratio. Then you will calculate the force of gravity acting on your body.

Problem

How do we find the force of gravity-mass ratio?

Materials

newton spring scale
set of hooked masses
graph paper
kilogram bathroom scale

Procedure

a. Copy Table 22-1 into your notebook.
b. Hang a 0.10 kg mass on the lower end of the spring (Fig. 22-5).
c. Lower the mass gently until the spring supports it. Read the spring scale. This reading gives the force of gravity on the 0.1 kg mass.
d. Record the force of gravity in newtons acting on the mass.
e. Find the force of gravity acting on the other masses listed in the table by repeating steps (b), (c), and (d).

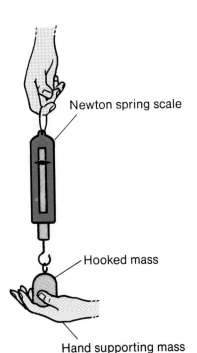
Fig. 22-5 Measuring a force

f. Divide the force of gravity by the mass in each case. Enter this in the table.
g. Measure and record your mass in kilograms using the bathroom scale.

Table 22-1   Measuring a Force

Mass (kg)	Force of gravity (N)	Force of gravity/mass (N/kg)
0.10		
0.20		
0.50		
1.00		
my body		

Discussion

1. What happens to the spring as larger masses are attached?
2. Compare the size of the force of gravity on an object with its mass.
3. Calculate the average value of the force of gravity-mass ratio. Include units.
4. Find the class average for the force of gravity-mass ratio.
5. Calculate the force of gravity acting on your body. To do this multiply your mass by the force of gravity-mass ratio.

22.3 The Unit of Force and the Force of Gravity

The Newton

The primary unit of force is the **newton** [NU-ton]. The symbol for the newton is N. One newton is roughly the force exerted by the earth on a 0.1 kg mass. Thus a 1 kg mass experiences a force of about 10 N.

Fig. 22-6 Humans can exert large forces for short times.

The newton is a small force. One newton is about equal to the force of gravity on a medium sized apple. Some humans can exert relatively large forces. A champion weight lifter is shown in Figure 22-6. He is exerting a force of about 500 N on the barbells. Imagine trying to lift 500 apples!

The Force of Gravity-Mass Ratio

The following table shows the forces of gravity acting on various masses.

Table 22-2   The Relationship Between Force of Gravity and Mass

Force of gravity (F) N	Mass (m) kg	$\dfrac{\text{Force of gravity}}{\text{Mass}} \left(\dfrac{F}{m}\right) \dfrac{N}{kg}$
1.0	0.10	10
5.0	0.50	10
10.0	1.00	10

The force of gravity increases with the mass. When the mass doubles, the force of gravity doubles. The force of gravity-mass ratio is constant for one location. But it does vary slightly from place to place. Can you suggest why?

The force of gravity-mass ratio on earth is about 10 N/kg. We can use this ratio to calculate the force of gravity for any mass on earth. First express the mass in kilograms. Then multiply the mass by the force of gravity-mass ratio. The equation $F = mg$ summarizes the relationship, where

F is the force of gravity in newtons (N)
m is the mass in kilograms (kg)
g is the force of gravity-mass ratio in newtons per kilogram (N/kg).

Sample Problem

Calculate the force of gravity the earth exerts on a 320 g mass.

Given
$m = 320$ g
$g = 10$ N/kg
Required
F
Analysis
mass must be in kilograms
1 kg = 1000 g

Solution
$m = 320$ g $= 0.32$ kg
$F = mg$
$F = 0.32$ kg x 10 N/kg
$F = 3.2$ N
Statement
The force of gravity on a 320 g mass is 3.2 N.

Section Review

1. What is a newton?
2. What symbol is used for the newton?
3. Give one example of a newton force.

4. How large is the force of gravity on a 1 kg mass?
5. a) How does the force of gravity change as mass increases?
   b) How does the force of gravity change as mass decreases?
6. a) What symbol is used for the force of gravity-mass ratio?
   b) What unit is used for the force of gravity-mass ratio?
7. What is the approximate value of the force of gravity-mass ratio on earth?
8. Is the force of gravity the same everywhere on earth? Discuss.
9. What equation relates F, m, and g?
10. Find the earth's force of gravity acting on each of the following masses: 1 kg; 4 kg; 0.5 kg; 100 g; 50 g; 250 g.

22.4 Balanced and Unbalanced Forces

Force Vectors

A force has both a size and a direction. It is represented on diagrams by an arrow. This arrow is called a **vector** [VEK-tor] (Fig. 22-7). The length of the arrow represents the size. The head shows the direction. To draw a force vector, first choose a scale. One unit of length equals a definite amount of force.

A vector is shown in Figure 22-8. It is drawn to represent a push on a toy car. The force is 30 N east. Each 1 cm of the vector represents 10 N of force. A vector to represent 30 N is 3 cm long. The head points east, showing that the force is east. The scale is shown with the diagram. The scale enables anyone to find the size of the force.

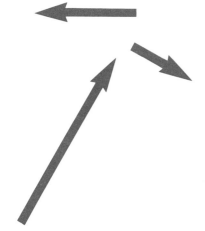

Fig. 22-7 A vector shows both the size and direction of a force.

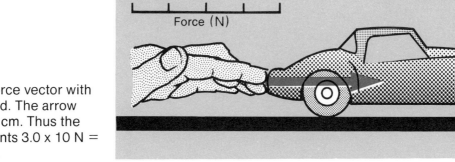

Fig. 22-8 A force vector with scale indicated. The arrow measures 3.0 cm. Thus the arrow represents 3.0 x 10 N = 30 N of force.

Balanced Forces: Objects at Rest

A mass suspended from a spring scale has two forces acting on it. One is the tension in the spring (the tensile force). The other is the force of gravity. These are shown using vectors in Figure 22-9. The **tensile** [TEN-sl] force acts vertically upward. The force of gravity acts vertically downward. The two forces are equal in size. But they are opposite in direction. Thus the forces are balanced. The mass stays at rest. It does not start to move because the forces are balanced.

Balanced Forces: Objects in Motion

Consider the forces acting on a car. The car is moving at a constant speed on a straight, level road (Fig. 22-10). There are two vertical forces acting on the car. The force of gravity pulls down. The road pushes up. The force of the road is called the **reaction force**. These two forces are balanced. Thus there is no motion of the car up or down. There are two horizontal forces. The moter exerts a force forward. The road and air exert a frictional force backward. The car is neither speeding up nor slowing down. It is moving at a constant speed. Thus the forces are balanced. An object moving at constant speed has balanced forces acting.

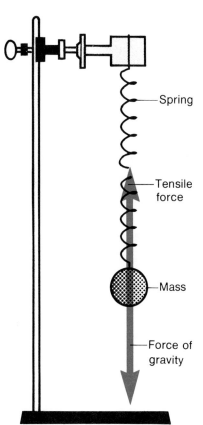

Fig. 22-9 A spring supporting a mass. Note the magnitude and direction of the forces.

Fig. 22-10 Forces acting on a car moving at constant speed in a straight line

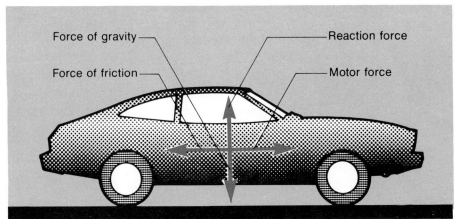

Unbalanced Forces: Objects in Motion

Suppose the driver of a car wishes to speed up. The driver presses on the gas pedal. The forward force exerted by the motor increases. It becomes greater than the force of friction (Fig. 22-11). Now the horizontal forces are unbalanced forward. The car increases in speed. If the driver eases up on the gas pedal, the force of the motor decreases. It becomes less than the force of friction (Fig. 22-12). The forces are unbalanced backward. The car slows down. An unbalanced force causes the speed of an object to increase or decrease.

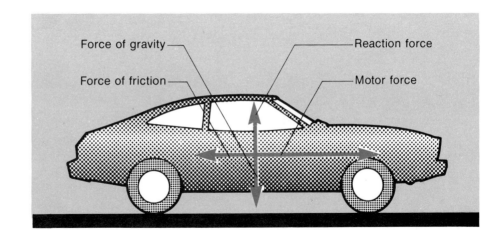

Fig. 22-11 An unbalanced force forward on the car causes it to speed up (accelerate).

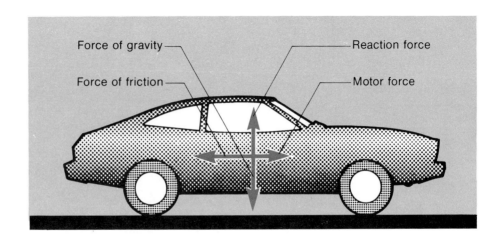

Fig. 22-12 An unbalanced force backward on the car causes it to slow down (decelerate).

Section Review

1. What is a vector?
2. When is a force vector used?
3. How is a force vector drawn?
4. What effect do balanced forces have on an object?
5. How do you know when the forces acting on an object are balanced?
6. What effect do unbalanced forces have on the speed of an object?

22.5 ACTIVITY The Forces of Static and Kinetic Friction

Friction [FRIK-shun] *is the force that opposes motion.* **Static friction** makes it difficult to start objects moving. **Kinetic friction** [ki-NET-ik] makes it difficult to keep objects moving. In this activity you will compare static and kinetic friction. You will find the effect of surface area on friction. You will see if friction is larger with a larger load.

Problem

What are the properties of friction?

Materials

newton spring scale
set of masses
string

smoothly sanded wooden block
smoothly sanded wooden board

Procedure

a. Copy Table 22-3 into your notebook.
b. Attach the newton spring scale to the block as shown in Figure 22-13.

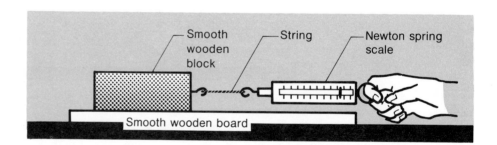

Fig. 22-13 The force of friction between a wooden block and a wooden board.

c. Place the block on top of the board with the largest surface down.
d. Pull slowly on the spring scale. Just before the block begins to move, read and record the force. This is the largest force of static friction.
e. Repeat step (d) three times. Record the average value for static friction.
f. Pull slowly on the spring scale. Make the block slide at constant speed. Record the force needed to keep the object moving. This is the force of kinetic friction.
g. Repeat step (f) three times. Record the average value for kinetic friction.
h. Place a 1 kg load on the upper surface of the block. This increases the load.

i. Measure and record the force of kinetic friction with this load. Take three readings and average them.
j. Remove the 1 kg mass. Face one of the smaller surfaces of the block down.
k. Repeat step (f) with the smaller surface in contact.

Table 22-3 Studying Friction

Trial	Force of static friction (N)	Force of kinetic friction (N)		
		Large surface No load	Large surface Load of 1 kg	Small surface No load
1				
2				
3				
Average				

Discussion

1. Which is larger, the force of static or kinetic friction?
2. What happens to the force of friction as the load increases?
3. Does the area of the surface in contact affect the force of friction? Discuss.

22.6 Pressure: A Force Applied to an Area

What is Pressure?

A clothes iron exerts pressure to remove wrinkles. Air exerts pressure to inflate tires. Water exerts pressure on submarines. All solids, liquids, and gases exert pressure on surfaces. **Pressure** [PRESH-ur] *is defined as the force acting on a unit area of a surface.*

What Affects Pressure?

Two factors affect pressure. One is the size of the force applied to a surface. The other is the surface area which the force acts on. Press the point of a pencil gently against your hand. Observe how much it pushes in on your skin. Feel the pressure. Increase the force by pressing harder. You feel more pressure. Increasing the force applied to an area increases the pressure (Fig. 22-14).

Fig. 22-14 The larger the force applied to an area, the larger the pressure.

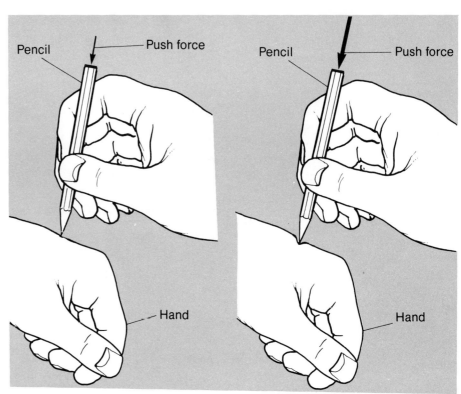

Press the blunt end of the pencil gently against your palm. Feel the pressure. Press the point of the pencil against your palm with the same force. You feel more pressure. The same force on a smaller area increases the pressure (Fig. 22-15).

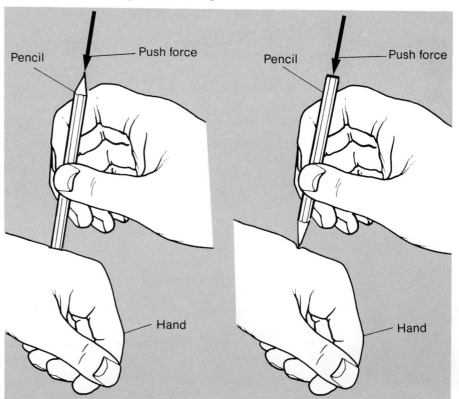

Fig. 22-15 The larger the area to which a force is applied the smaller the pressure.

How is Pressure Measured?

Two measurements determine pressure. One is the applied force in newtons. The other is the area in square metres. To calculate pressure, divide the force in newtons by the area in square metres. **Pressure = Force/Area.**

If p stands for pressure, F for force and A for area, the equation for pressure is: $p = F/A$.

The SI Unit of Pressure

The primary unit of pressure is the **newton per square metre (N/m^2)**. The newton per square metre is given a special name, **pascal (Pa)**. This is in honour of the French scientist Blaise Pascal (1623-1662).

One pascal (Pa) is the pressure when one newton (N) of force is applied to an area of one square metre (m^2).
$$1 \text{ Pa} = 1 \text{ N}/m^2$$
The pascal is a very small pressure. A dollar bill lying flat on a table exerts a pressure of about 1 Pa. A more useful unit is the **kilopascal (kPa)**. A kilopascal is one thousand times larger than a pascal.

Sample Problem

A block of nickel 15 cm long and 20 cm wide rests on a table. The block has a mass of 27 kg. Calculate the pressure exerted by the block on the table. Force of gravity-mass ratio = 10 N/kg.

Given
m = 27 kg
l = 15 cm = 0.15 m
w = 20 cm = 0.20 m
g = 10 N/kg
Required
p
Analysis
Since $p = \dfrac{F}{A}$, we have to find F and A. Thus:
Find force
Find area
Divide force by area

Solution
$F = mg$
    = 27 kg x 10 N/kg
    = 270 N down
$A = l \times w$
    = 0.15 m x 0.20 m
    = 0.03 m^2
$p = F/A$
    = $\dfrac{270 \text{ N}}{0.03 \text{ m}^2}$
    = 9000 N/m^2
    = 9000 Pa
    = 9 kPa down

Statement
The pressure exerted by the block on the table is 9 kPa down.

Some Example Pressures

Standard air pressure is about 100 kPa. Think of 100 000 dollar bills stacked on a table. This pressure acts on every surface of your body.

Section 22.6

A tire pressure gauge reads the difference between the inside and outside pressure of a tire. Suppose the air pressure inside a bicycle tire is 500 kPa. The tire gauge will read only 400 kPa. Can you see why?

Section Review

1. Define pressure.
2. What two factors affect pressure? How?
3. What is the SI unit of pressure?
4. What two things must be measured to determine pressure? What is the equation for pressure?
5. What is standard air pressure in kPa?
6. A brick 20 cm long by 10 cm wide rests on a table. It has a mass of 3 kg. Calculate the pressure in kPa it exerts. g = 10 N/kg. Remember that the area must be in square metres and the force in newtons.

Main Ideas

1. A force is a push or a pull. There are many types of forces.
2. A force can change the shape and/or motion of an object.
3. Every force has a definite size and direction.
4. Balanced forces do not change the shape and/or motion of an object.
5. Friction is the force that opposes motion.
6. Pressure is the force acting on a unit area of surface.

Glossary

force		a push or a pull
friction	FRIK-shun	a force opposing motion
gravity		the force of attraction between any two masses
newton	NU-ton	a unit of force
pascal	pas-KAL	primary unit of pressure
pressure	PRESH-ur	a force acting on a unit area of surface
vector	VEK-tor	an arrow drawn to scale to represent both the direction and size of a quantity

Study Questions

A. True or False

Decide whether each of the following sentences is true or false. If the

sentence is false, rewrite it to make it true. (Do not write in this book.)
1. Each force has a definite size and direction.
2. The force of gravity-mass ratio is the same everywhere.
3. Balanced forces act on an object moving at constant speed.
4. Static friction is smaller than kinetic friction.
5. The larger the area on which a force acts, the larger the pressure.

B. Completion

Complete each of the following sentences with a word or phrase which will make the sentence correct. (Do not write in this book.)
1. The _____ of a force vector represents the size of the force.
2. One _____ is about equal to the force of gravity on a 0.1 kg mass.
3. Friction is the force that _____ motion.
4. _____ friction is larger than _____ friction.
5. Pressure increases when the _____ increases.

C. Multiple Choice

Each of the following statements or questions is followed by four responses. Choose the correct response in each case. (Do not write in this book.)
1. Which of the following forces of gravity on earth is about equal to a force of 1 N? The force the earth exerts on a
   a) medium sized apple
   b) 10 g mass
   c) 1 kg mass
   d) 10 kg mass
2. The force of gravity on an elevator is 5000 N. The tensile force exerted upward by the cable is 5000 N. Which of the following statements is *not* correct? The elevator may be
   a) stopped at one of the floors
   b) increasing in speed
   c) moving down with a constant speed
   d) moving up with a constant speed
3. Which of the following will *not* reduce the force of kinetic friction?
   a) polish the surfaces
   b) reduce the size of the load
   c) apply a thin film of oil
   d) decrease the surface area
4. An astronaut collected data on a planet. The data are in Table 22-4.

Table 22-4

Force of gravity (N)	Mass (kg)
1.0	0.5
2.0	1.0
3.0	1.5
4.0	2.0

What is g in N/kg according to these data?
a) 1.6   b) 2.0   c) 4.0   d) 10
5. What is the equation for finding pressure?
a) $\rho = A/F$ b) $\rho = F/A$ c) $\rho = F + A$ d) $\rho = FA$

D. Using Your Knowledge

1. Determine the mass the earth attracts with each of the following forces:
a) 4 N   b) 20 N   c) 300 N   d) 1.5 kN   e) 10 kN   f) 8 MN
2. Explain how a spring scale force measurer uses the principle of balanced forces.
3. A small car has a mass of 900 kg. It is travelling along a straight level road at a constant speed. The force of friction acting on the car is 2000 N. Draw a sketch of the car. Draw force vectors to show the forces acting on the car. Use a scale of 1 cm to represent 2000 N.
4. a) Should a driver in a drag race spin the car tires during takeoff? Why?
   b) Why is traction improved if a bag of sand is placed over the driving wheels of a car?
5. How do snowshoes for adults differ from those for children? Why?

E. Investigations

1. Engineers are doing research to reduce air resistance in cars. Find out what features are being changed. Find out how the changes are affecting gasoline consumption. Submit a report to your teacher.
2. How do you think the pressure on a liquid changes with depth? How do you think it changes with volume? Design and do an experiment to test your predictions. Be careful to control all variables.
3. Do an experiment with a loaded bicycle. Determine the pressure the tires exert on the road in kilopascals. First measure the force of gravity using a bathroom scale. Then make tire prints for the loaded bicycle on centimetre graph paper using ink. Find the area. Calculate the pressure in kilopascals.

23 Work, Energy and Power

23.1 Work
23.2 Activity: Measuring Work to Lift Loads
23.3 Energy
23.4 Power
23.5 Activity: Measuring Leg Power

A farmer strains against a stone but it won't move. Is any work being done? What do work and energy have in common? What is power? How can I measure my leg power? How can my body power be improved? You will find answers to questions like these in this chapter.

23.1 Work

When Work is Done

Work is done when a force moves an object. **Work** *is defined as a force acting through a distance.* Three conditions are needed before work is done.
1. A force is applied to an object.
2. The object moves through a distance.
3. The force and the distance are in the same direction.

Figure 23-1 shows three examples involving work. In example A, a person lifts a box from the floor to a table. The person exerts a force up. The box moves a distance up. The force and the distance are in the same direction.

In example B a person shoves a box across the floor. The push is forward. The box moves forward through a distance.

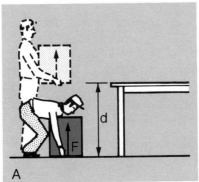
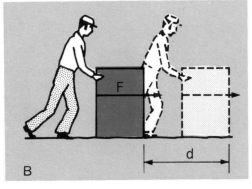

Fig. 23-1 Work is being done in these examples.

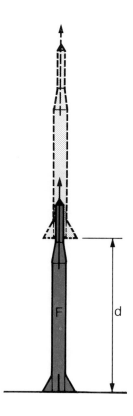

In example C a rocket is being launched. The burning fuel provides the force up. The rocket moves a distance up. In these examples all three conditions are met.

Figure 23-2 shows three examples where work is not done. In example A, a person holds a box. The person exerts a constant force up against the force of gravity. This force balances the force of gravity. The box stays at rest. Since there is no motion, no work is done.

In example B, the box moves horizontally. There is no friction so no horizontal force is needed. The person exerts a constant force up to support the box. But the force and the distance moved are at right angles. Therefore no work is done.

In example C, a capsule coasts far out in space with its engines off. There is very little gravity far out in space. Also, space is very nearly a vacuum. Thus there is very little friction. Because of this, the space capsule keeps moving at a constant speed. Since there is no applied force, there is no work.

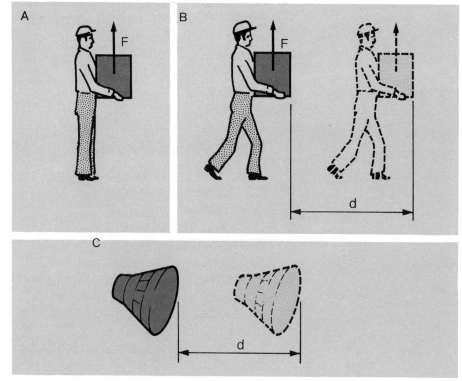

Fig. 23-2 No work is being done in these examples.

How Work is Determined

Work involves a force and a distance. To determine work, we need two measurements. One is the size of the applied force. The other is the distance the object moves. Work is calculated by multiplying the force times the distance.

Work = force x distance

If W is work, F is force, and d is distance, then: $W = Fd$

Remember that the force and the distance must be in the same direction.

Sample Problem

A force forward of 1 N is applied to an object. It moves forward through a distance of 1 m. What is the work done?

Given
$F = 1$ N
$d = 1$ m
Required
W

Analysis
The units are consistent.

Solution
$W = Fd$
$= 1$ N x 1 m
$= 1$ N•m

Statement
The work done is 1 N•m

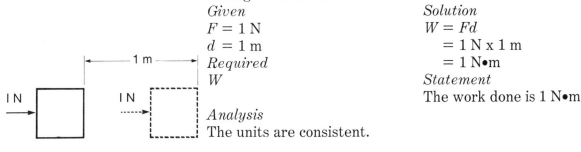

Fig. 23-0

The SI Unit of Work

The primary unit of force is the newton. The primary unit of distance is the metre. We multiply these units to calculate work. Thus the primary unit of work is the **newton metre (N•m)**. Scientists have given the newton metre a special name. They call it the **joule (J)**. This is in honour of James Prescott Joule (1818-1885). He made many early measurements to find the work done.

The joule (J) is defined in terms of the newton and the metre. *One joule (J) is the work done when a force of one newton (1 N) acts through a distance of one metre (1 m).*

$$1 \text{ J} = 1 \text{ N•m}$$

You may wonder how large one joule is. Lift a medium sized apple from the floor to chest height. You have done about one joule of work (Fig. 23-3).

Fig. 23-3 One joule of work is done to lift the apple to chest height.

Section Review

1. Define work.
2. What three conditions are needed before work is done?
3. What two measurements are needed to find work?
4. What is the word equation to calculate work?
5. What is the equation in symbols for work?
6. What is the unit of work?
7. Define the joule.
8. What is the symbol for the joule?

23.2 ACTIVITY Measuring Work to Lift Loads

In this activity you will lift masses from one level to another. You will measure the force used. You will also measure the distance the objects move. You will compare the work done using your senses. Then you will calculate the work done.

Section 23.2 265

Problem

How can we find the work done to life masses?

Materials

a set of hooked masses   2 metre sticks   a newton spring scale

Procedure

a. Copy Table 23-1 into your notebook.
b. Attach a 1.0 kg mass to the spring scale (Fig. 23-4).
c. Raise the mass slowly from the floor to a height of 0.5 m.
d. Record the force in newtons needed to lift the mass.
e. Repeat steps (c) and (d), but lift the mass 1.0 m and then 1.5 m.
f. Use your senses to compare the work done. Then calculate the work.
g. Attach a 0.5 kg mass to the spring scale. Record the force. Lift the mass 1.0 m.
h. Repeat step (g) using the 0.2 kg mass and then the 0.1 kg mass.
i. Use your senses to compare the work done. Then calculate the work.

Fig. 23-4 Finding the work done to lift a load

Table 23-1 Working to Lift Masses

Mass of object (kg)	Force applied (N)	Distance moved (m)	Work done (N•m = J)
1.0		0.5	
1.0		1.0	
1.0		1.5	
0.5		1.0	
0.2		1.0	
0.1		1.0	

Discussion

1. Compare the work done when the same load is lifted different distances.
2. How does distance affect work?
3. Compare the work when different loads are lifted the same distance.
4. How does force affect work?
5. What two measurements are needed to determine work?
6. What is done with these two quantities to calculate work?
7. How well were you able to sense the differences in the work done?

23.3 Energy

What is Energy?

Energy *is the ability to do work.* It can exert a force. It can make things move. It can produce heat. The moving air in a tornado can pick up heavy objects and carry them great distances. The energy in a raised hammer can exert a force on a nail and drive it into wood. It also makes the head of the nail hot. The energy in a stretched spring can do work. Dynamite has lots of energy. When it explodes it exerts large forces to shatter rock and move it about (Fig. 23-5).

Fig. 23-5 Dynamite has energy.

Two Basic Kinds of Energy

The energy a moving object has because of its motion is called **kinetic energy**. Moving air and water can do work; so can moving solids such as a pile driver (Fig. 23-6). Sometimes energy is not doing work. It is being stored. A raised object has stored energy. A stretched spring has stored energy. *The energy an object has because of its position or shape is called* **potential energy**.

An example of potential energy of position is the energy in water at the top of a dam. When released it can do work turning turbines. A stretched rubber band has potential energy of shape. If the band is part of a slingshot it can do work. When released it exerts a force on a pebble. The pebble is flung forward.

The SI Unit of Energy

Energy is the ability to do work. Thus it has the same unit as work. The unit of both work and energy is the **joule (J)**. The joule is a very small amount of energy. Two larger units are used. The **kilojoule (kJ)** equals one thousand joules. The **megajoule (MJ)** equals one million joules.

1 kJ = 1000 J   1 MJ = 1000 kJ = 1 000 000 J

Fig. 23-6 A pile driver can do work.

Measuring Energy

Energy is found by measuring work. When work is done, energy is used. In fact, more energy is used than the work done. This is because not all energy does useful work. Friction is always present. Thus some energy is used to overcome friction. This becomes waste heat. In fact, about one-quarter of the energy in gasoline used to move cars does useful work. The rest becomes waste heat due to friction. Where would you expect to find forces of friction in a moving car? What is done to make these forces smaller so that less energy will be wasted?

Sample Problem

A window is located 500 m up the CN tower. A window washer has a mass of 80 kg. Find the least energy used to lift the window washer to the job. The force of gravity-mass ratio g is 10 N/kg.

Given	Solution
$m = 80$ kg	$F = mg$
$d = 500$ m	$\quad = 80$ kg x 10 N/kg
$g = 10$ N/kg	$\quad = 800$ N down
Required	The force needed is 800 N up
E	(to lift the load)
Analysis	$W = Fd$
Energy needed = work done	$\quad = 800$ N x 500 m
To find work, find force and distance	$\quad = 400\,000$ N•m
	$\quad = 400\,000$ J = 400 kJ

The size of the force up = force of gravity down
The distance is given.

Statement
400 kJ of energy are used to lift the worker. In addition, energy is used against friction. This becomes waste heat.

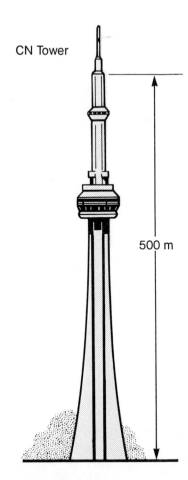

CN Tower

500 m

Fig. 23-0-0

Section Review

1. Define energy.
2. What three things can energy do?
3. Define kinetic energy and give an example.
4. Define potential energy. Give an example of shape potential energy. Give an example of position potential energy.
5. What is the symbol for energy?
6. What is the SI unit of energy?
7. Compare the SI unit of energy and work.
8. What are two larger units of energy?
9. How is energy measured?
10. Why is the energy used always more than the work done?
11. Find the energy needed to raise a 20 kg box to a height of 40 m.

23.4 Power

What is Power?

Work done does not depend on the time taken. This is also true for energy. Climb a flight of stairs in 10 s. Climb the same flight of stairs in 20 s. You do the same work. You use the same energy. But the *rate* of using energy and doing work is larger when you climb the stairs in 10 s. **Power** *is defined as the rate of doing work or using energy.*

What Factors Affect Power?

Two factors affect power. One is the amount of work done in a given time. Suppose two people are lifting boxes. Person A lifts 10 boxes in 5 min. Person B lifts 20 boxes in 5 min. Person B does twice the work. The time taken is the same. Person B has twice the power.

The other factor is the time taken to do a given amount of work. Again two people are lifting boxes. Person A lifts 10 boxes in 10 min. Person B lifts 10 boxes in 5 min. Both do the same work. But Person B does it twice as fast. Person B has twice the power. The more work you do in a given time, the greater your power. The less time it takes to do the work, the greater the power.

How is Power Measured?

Two measurements are needed to find power. One is the work done in joules. The other is the time taken in seconds. To calculate power, divide the work done by the time taken. **Power = $\frac{\text{Work}}{\text{Time}}$**. If P stands for power, W for work, and t for time, the equation for power is:
$$P = \frac{W}{t}.$$

The SI Unit of Power

Power is work divided by time. The primary unit of power is the **joule per second (J/s).** The joule per second is given the special name **watt**. This is in honour of James Watt (1736-1819). Watt was a Scottish engineer who made steam power practical in 1769. Steam power is still in use today (Fig. 23-7).

One **watt (W)** *is the power when one joule (J) of work is done in one second.* $\qquad$ 1 W = 1 J/s

Lift a medium sized apple from the floor to chest height. Do it in one second. You have used a power of one watt.

Another unit of power in SI is the **kilowatt**. It is one thousand watts. $\qquad$ 1 kW = 1000 W = 1000 J/s

Fig. 23-7 James Watt made the steam engine practical.

Sample Problem

A champion weight lifter at the Montreal Olympics lifted a mass of 255 kg. The mass was raised a height of 2.0 m. It took 5 s. The force of gravity to mass ratio g is 10 N/kg. Calculate the athlete's power output.

Given
$m = 255$ kg
$d = 2.0$ m
$t = 5$ s
$g = 10$ N/kg

Required
P

Analysis
To find power, find work.
To find work, find force.
The size of the force up = force of gravity down.

Solution
$F = mg$
$\quad = 255 \text{ kg} \times 10 \text{ N/kg}$
$\quad = 2550 \text{ N}$
The force needed is 2550 N up
$W = Fd$
$\quad = 2550 \text{ N} \times 2 \text{ m}$
$\quad = 5100 \text{ N} \cdot \text{m}$
$\quad = 5100 \text{ J}$
$P = W/t$
$\quad = \dfrac{5000 \text{ J}}{5 \text{ s}}$
$\quad = 1020 \text{ W}$

Statement
The power output of the athlete was at least 1020 W. This is greater than the power output of ten 100 W light bulbs burning brightly. Actually the power was even larger since some energy was also used to lift the athlete's large arms.

People are not capable of such a power output for very long. Our average power output is only about 75 W. The power output when we sleep is less than a 25 W light bulb! Remember, power is the *rate* of using energy or doing work.

Section Review

1. Define power.
2. What two factors affect power? How?
3. What is the SI unit of power?
4. How is power measured? What is the equation for power?
5. Find the least power to raise a 40 kg box. The box is raised a height of 3 m. It takes 2 s.

23.5   ACTIVITY Measuring Your Leg Power

A top athlete has a power of about 1700 W. A person running has a power of about 1000 W. A person walking has a power of 230 W. In this activity you will find your leg power in watts.

Problem

What is your leg power in watts?

Materials

roll of string and a bob   metre stick   stop watch

Procedure

a. Find a flight of stairs to run up.
b. Use the string, bob, and metre stick to measure the height of the stairs.
c. Determine your mass in kilograms.
d. Determine your force of gravity in newtons.
e. Measure the fastest time you can climb the stairs. CAUTION: Do not over exert yourself.
f. Calculate your leg power in watts.
g. Calculate the average power of the class. Your teacher will help you.

Discussion

1. What is your maximum leg power in watts?
2. Compare your power with a top athlete.
3. Compare your power with the power of a person i) running ii) walking.
4. Who develops the most power in your class? Can you suggest why?
5. What is the average leg power in the class?

Main Ideas

1. Work is a force applied through a distance.
2. Energy is the ability to do work.
3. Power is the rate of doing work or using energy.
4. Power is calculated by dividing work by time.

Glossary

joule	JOOL	the work done when a force of 1 N is applied through 1 m distance
kilojoule	KIH-lo-jool	one thousand joules
kilowatt	KIH-lo-wat	one thousand watts
power		the rate of doing work or using energy
watt		joule per second
work		a force applied through a distance

Study Questions

A. True or False

Decide whether each of the following sentences is true or false. If the sentence is false, rewrite it to make it true. (Do not write in this book.)
1. A person carrying a box along the level is doing work.
2. The unit of work is the newton.
3. Energy has the same definition as work.
4. Power is measured by dividing force by time.
5. The power output is twice as large when twice the work is done in the same time.

B. Completion

Complete each of the following sentences with a word or phrase which will make the sentence correct. (Do not write in this book.)
1. Work is a force applied through a ▬▬▬ .
2. Energy is the ability to ▬▬▬ .
3. Energy is measured in terms of the ▬▬▬ .
4. The less time to do work, the ▬▬▬ the power
5. One thousand joules is equal to one ▬▬▬ .

C. Multiple Choice

Each of the following statements or questions is followed by four responses. Choose the correct response in each case. (Do not write in this book.)
1. Work is calculated by
   - **a)** multiplying force by time
   - **b)** multiplying force by distance
   - **c)** dividing force by time
   - **d)** dividing force by distance
2. A force of 10 N forward is applied to an object. The object moves 0.50 m forward. How much work is done?
   - **a)** 0
   - **b)** 0.05 J
   - **c)** 5.0 J
   - **d)** 20 J
3. The unit of energy is the same as the unit of
   - **a)** force
   - **b)** distance
   - **c)** work
   - **d)** power
4. Which of the following is a correct description of power? Power is
   - **a)** work x time
   - **b)** work x distance
   - **c)** work/time
   - **d)** work/distance
5. What is the unit for power?
   - **a)** joule
   - **b)** joule second
   - **c)** newton metre
   - **d)** watt

D. Using Your Knowledge

1. A satellite orbits the earth. The earth exerts a force on it. Is the earth doing work on the satellite? Explain your answer.

2. A force of 350 N is required to push a piano. The piano has a mass of 400 kg. The piano is pushed 5.0 m. How much work is done?
3. A skier has a mass of 70 kg. The ski hill is 300 m high. A ski tow runs from the bottom to the top. What is the least energy needed to raise the skier from the bottom to the top?
4. A crane lifts a 1000 kg load through a distance of 30 m in 40 s. Calculate:
   a) the force exerted by the crane.
   b) the work done by the crane.
   c) the power of the crane in watts and kilowatts.
5. A car engine exerts a force of 4000 N. The car speeds up over a distance of 100 m. It does this in 8.0 s. Calculate the power of the engine in watts and kilowatts.

E. Investigations

1. Predict which is larger, the power of your arms or legs. Design a way to measure the power of your arms. (Hint: use a bathroom scale calibrated in newtons or kilograms.) Compare this with the power of your legs.
2. a) Predict whether your right arm or left arm is more powerful. Design a way of finding the power of each arm.
   b) One might predict that a right-handed person has a more powerful right arm than left arm. Does this prediction agree with your results?
3. Do an experiment with an exercise bicycle to determine your leg power. Put the wheel under tension. Measure the average pedal force to turn the wheel. Measure the circumference traced out by the pedal. Pedal the cycle for a known number of turns in a known time.

24 Machines

24.1 Kinds and Uses of Machines
24.2 Activity: How an Inclined Plane Helps Us
24.3 The Lever
24.4 Activity: How First Class Levers Help Us
24.5 Activity: How Single Pulleys Help Us
24.6 Mechanical Advantage and Efficiency

In Chapter 23 you studied work. In this chapter you find out how to make work easier. Does a machine decrease the work we do? What are the different machines? Is there a way to make machines more efficient? What is efficiency? The study of this chapter provides the answers to such questions.

24.1 Kinds and Uses of Machines

*A **machine** is any device that helps us to do work.* All machines are classified as either simple or compound. A **simple machine** does work with one movement. The bottle opener in Figure 24-1 is an example. **Compound machines** consist of two or more simple machines. The C clamp in Figure 24-2 is an example. It consists of a lever and a screw. How many motions does it combine? The study of simple machines will help you understand compound machines.

Fig. 24-1 A bottle opener is a simple machine.

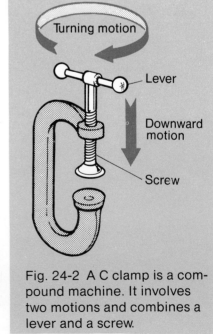

Fig. 24-2 A C clamp is a compound machine. It involves two motions and combines a lever and a screw.

The Six Simple Machines

The two basic kinds of simple machines are the inclined plane and the lever. The **inclined plane** is a slanted surface. It is used to raise objects from one level to another (Fig. 24-3). Stairs, hilly roads, and barn ramps are examples of inclined planes. The **wedge** developed from the inclined plane. It is thicker at one end than the other. Examples of wedges are a knife, axe, can opener, and log splitter. The **screw** is another adaptation of the inclined plane. It is a spiral inclined plane such as a barn jack and wood screw.

The **lever** is a bar that can turn about a support (Fig. 24-4). The support is called the fulcrum. Hammers, bottle openers, and fishing poles are levers. The **pulley** is based upon the lever. It is a wheel that turns on an axle. A sailboat has many pulleys. The **wheel and axle** also came from the lever. The wheel turns with the axle. The front wheel of a tricycle is an example. Figure 24-5 shows examples of the six simple machines. Can you figure out to which class each belongs?

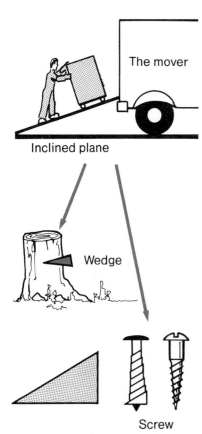

Fig. 24-3 The wedge and the screw came from the inclined plane.

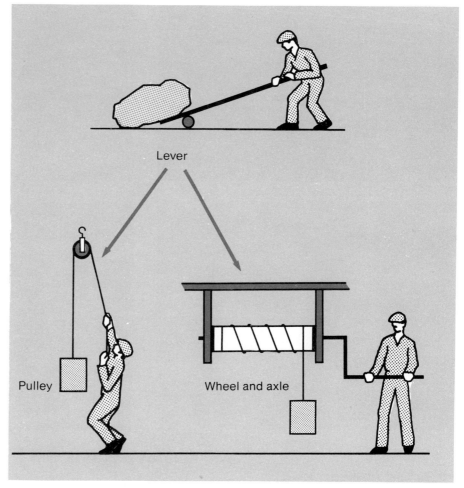

Fig. 24-4 The pulley and the wheel and axle came from the lever.

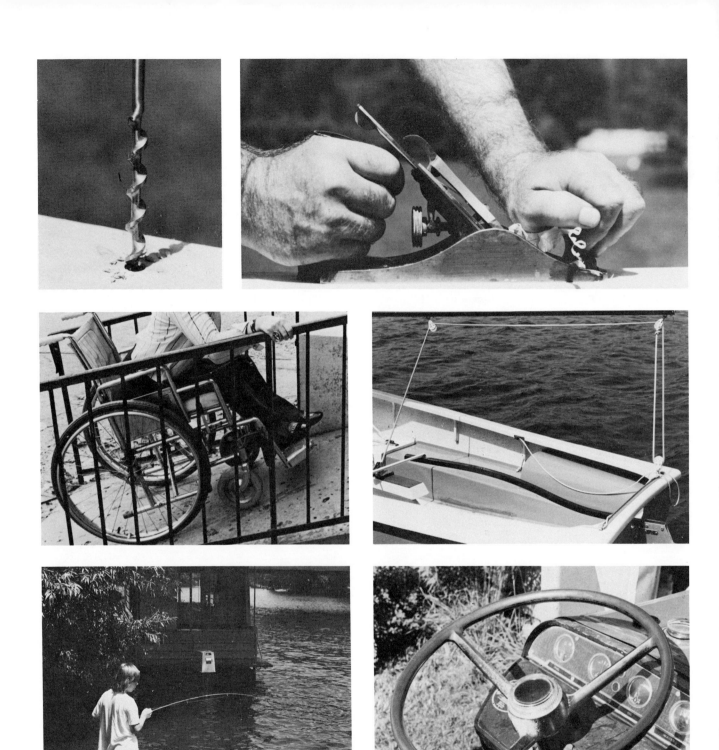

Fig. 24-5 Common examples of the six simple machines

How Machines Help Us

Machines help us to do work in many ways. They change the kind of force. For example, a hydro-electric generator changes the force of gravity to an electric force (Fig. 24-6,A). Machines can change the location of a force. For example, a typewriter transfers forces from the keys through the symbols to the paper (Fig. 24-6,B). Machines such as a car jack can change the size of a force. A small muscular force becomes a large lifting force (Fig. 24-6,C). Machines can change the direction of a force. For example, the pulley at the top of the flagpole changes the direction of a force (Fig. 24-6,D). A machine can change the speed of a force. A tricycle changes the speed of a force. The child applies a slow force to the pedals, the rim of the wheel goes faster than the pedals and the child moves quickly along the sidewalk (Fig. 24-6,E).

Fig. 24-6 How machines help us

Section Review

1. What is a machine?
2. List five ways machines change forces.
3. What is the difference between a simple and a compound machine?
4. Sketch and describe the two basic simple machines.
5. Sketch and describe the two simple machines that come from the
   a) lever   b) inclined plane

24.2   ACTIVITY   How an Inclined Plane Helps Us

In this activity you will raise an object from one level to another with and without an inclined plane. Will the force needed be larger with or without the incline? How will the work compare?

Problem

Will an inclined plane make it easier for us to raise an object?

Materials

1 kg bag of sand or brick   inclined plane   books
newton spring scale   metre stick

Procedure

a. Copy Table 24-1 into your notebook.
b. Set up the inclined plane as shown in Figure 24-7. Incline the plane at an angle of about 30°.

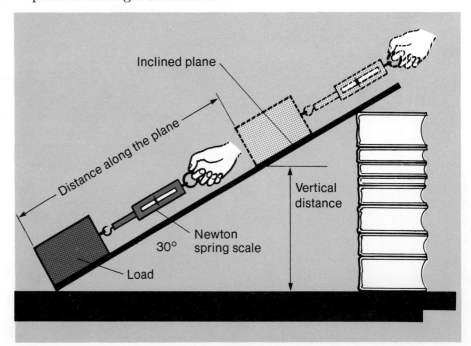

Fig. 24-7 How an inclined plane helps us

c. Mark off a distance of about 0.5 m along the plane to a point x as shown. Record this distance.
d. Measure the vertical distance from the table top to the point x. Record your measurement.
e. Use the spring scale to pull the load slowly, at a constant speed, up the inclined plane to the point x. Record the average force.
f. Use the spring scale to lift the load vertically to point x. Record the force.
g. Repeat steps (d) to (f) using a smaller angle than 30°.
h. Repeat steps (d) to (f) using a larger angle than 30°.
i. Calculate and record the work to move the mass to the point x. Do this for each trial with and without the inclined plane.

Table 24-1 Using an Inclined Plane

Trial Angle	Lifting the load using the inclined plane			Lifting the load without the inclined plane		
	Force (N)	Distance (m)	Work (J)	Force (N)	Distance (m)	Work (J)
Approx. 30°						
Smaller angle						
Larger angle						

Discussion

1. Which is larger, the lifting or the dragging force?
2. How did the dragging force change as the steepness of the incline increased? decreased?
3. Which is greater, the vertical or the ramp distances to the point x?
4. Was more work done with or without the inclined plane?
5. Use the results of this activity to explain why people use inclined planes.

24.3 The Lever

Parts of a Lever

A lever consists of a **bar** and a support (Fig. 24-8). The support is called the **fulcrum** [FUL-krem]. The bar turns about the fulcrum. The force you apply to the bar is called the **effort force**. The **effort arm** is the distance from the effort force to the fulcrum. The lever moves an object called the **load**. The load applies a **load force** to the lever. The **load arm** is the distance from the fulcrum to the load.

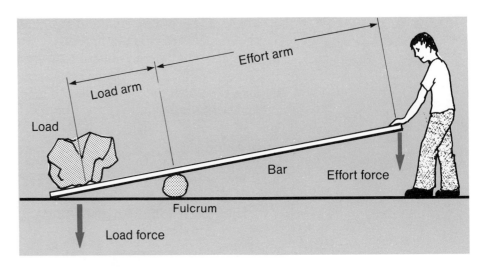

Fig. 24-8 The parts of a lever

The Three Classes of Levers

In the **first class lever**, the fulcrum is located between the load force and the effort force (Fig. 24-9). The load force and effort force both act in the same direction on the lever. A beam balance and a pump handle are first class levers.

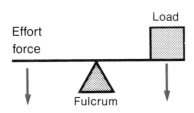

Fig. 24-9 A first-class lever with examples

In the **second class lever**, the load force is between the fulcrum and the effort force (Fig. 24-10). The effort force and the load force act in opposite directions on the lever. A wheelbarrow and a nutcracker are second class levers.

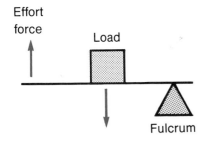

Fig. 24-10 A second-class lever with examples

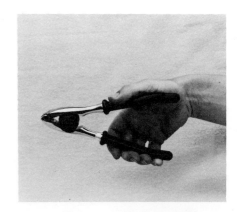

In the **third class lever**, the effort force is between the load force and the fulcrum (Fig. 24-11). The effort force and the load force act in opposite directions on the lever. Tweezers, a curling broom, and a hockey stick are third class levers.

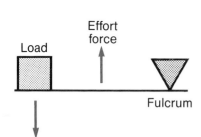

Fig. 24-11 A third-class lever with examples

Section Review

1. **a)** Draw and label the parts of a lever. Show the forces acting.
   b) State the meaning of effort force, load force, effort arm, and load arm.
2. How are first and second class levers different?

3. **a)** How are second and third class levers the same?
   b) How are they different?
   c) Identify the fulcrum, load force, and effort force for the hockey stick in Figure 24-11.

24.4   ACTIVITY   How First Class Levers Help Us

In this activity you will lift an object using a first class lever. You will vary the load arm and measure the effort force. If the load arm is longer than the effort arm, which will be larger, the load force or the effort force? How will the work compare?

Problem

Can a first class lever help us?

Materials

newton spring scale
0.5 kg load
C clamp
knife-edge holder

ring stand, clamp, and rod
3 metre sticks
unbent paper clip

Procedure

a. Copy Table 24-2 into your notebook.
b. Attach the knife-edge holder (fulcrum) at the 50 cm mark on the metre stick.
c. Suspend the metre stick from the ring stand (Fig. 24-12).
d. Use a paper clip to hang the 0.5 kg load at the 15 cm mark. The load arm will be 35 cm long.
e. Use a paper clip to attach the newton spring scale at the 85 cm mark. The effort arm will be 35 cm long.
f. Pull downward on the lever using the spring scale. Raise the load 5 cm.
g. Measure and record the effort force and the vertical distance it moved.
h. Move the load closer to the fulcrum (to the 30 cm mark). Repeat steps (f) and (g).
i. Move the load farther from the fulcrum (to the 0 cm mark). Repeat steps (f) and (g).
j. Calculate and record the work done on the load and by the effort.

282   Chapter 24

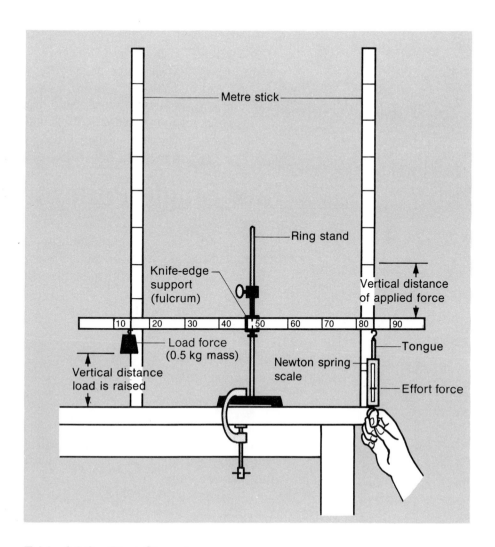

Fig. 24-12 How a first class lever helps us.

Table 24-2 First Class Lever

Load arm (m)	Load force (N)	Vertical distance load moved (m)	Work done on load (J)	Effort arm (m)	Effort force (N)	Vertical distance effort force moved (m)	Work done by effort force (J)
0.35	5	0.05		0.35			
0.20	5	0.05		0.35			
0.50	5	0.05		0.35			

Discussion

1. When the load arm equals the effort arm, how does the load force compare to the effort force?
2. When the load arm is less than the effort arm, which is greater
   a) the load force or the effort force?
   b) the vertical distance moved by the load force or the distance moved by the effort force?

Section 24.4

3. When the load arm is larger than the effort arm, which is greater
   a) the load force or the effort force?
   b) the vertical distance moved by the load force or the distance moved by the effort force?
4. Which is greater, the work done on the load or the work done by the effort force in each case? Or, are they the same?
5. Use the results of this activity to explain why people use levers.

24.5 ACTIVITY How Single Pulleys Help Us

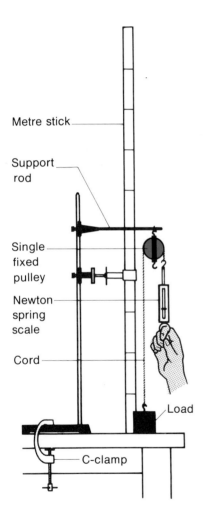

Fig. 24-13 How a single fixed pulley helps us

A pulley consists of a wheel that turns on an axle. A rope, chain, or belt turns the wheel. In this activity you will study a fixed pulley and a movable pulley. You will measure the effort force, the load force, and the distance the forces move. Then you will compare the work done by the effort with the work done on the load.

Problem

How can single pulleys help us?

Materials

newton spring scale   cord   single pulley
1 kg mass for load   C clamp   metre stick
ring stand, clamps, rod

Procedure

a. Copy Table 24-3 into your notebook.
b. Lift the load slowly and at a constant speed without a pulley. Use the newton spring scale. Record the force. This is the load force.
c. Set up the apparatus as shown in Figure 24-13.
d. Measure and record the force you apply to raise the load a short distance using the single fixed pulley. This is the effort force.
e. Raise the load through a vertical distance of 0.25 m. Record the distance the effort force moves.
f. Set up the apparatus as shown in Figure 24-14.
g. Repeat steps (d) and (e) with the single movable pulley.

Table 24-3 Simple Pulleys

Kind of single pulley	Load force (N)	Vertical distance load moves (m)	Work done on load (J)	Effort force (N)	Vertical distance effort moves (m)	Work done by effort (J)
Fixed	10	0.25				
Movable	10	0.25				

Discussion

1. For both the fixed pulley and movable pulley, compare:
   a) the size of the load and effort force;
   b) the direction of the load and effort force;
   c) the distance moved by the load and effort;
   d) the work done on the load with the work done by the effort.
2. a) What useful purpose is served by the fixed pulley? Given an example.
   b) What useful purpose is served by the movable pulley? Give an example.
   c) Which is more useful? Why?

24.6 Mechanical Advantage and Efficiency

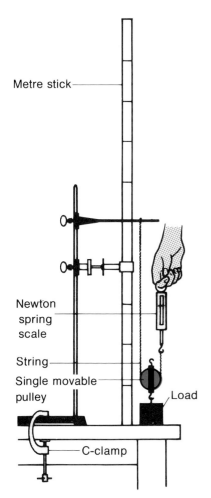

Fig. 24-14 How a single movable pulley helps us

Mechanical Advantage

Some machines enable us to lift a large load with a small effort. An inclined plane is an example. We say the machine multiplies the effort force. **Mechanical advantage** [muh-CAN-i-cal ad-VAN-tij] *is the number that tells us how many times a machine multiplies an effort force.* Suppose the mechanical advantage is 7. That means the machine multiplies our effort force seven times. Therefore, we can lift a load seven times larger than we could without the machine.

Measuring Mechanical Advantage

Two measurements are needed to find the mechanical advantage. One is the effort force in newtons. The other is the load force in newtons. To calculate mechanical advantage, divide the load force by the effort force. This gives a number without units.

$$\text{Mechanical Advantage} = \frac{\text{Load Force}}{\text{Effort Force}}$$

In the first class lever, the effort force can be less than, equal to, or greater than the load force. This depends on where the fulcrum is placed. Thus the mechanical advantage can be greater than, equal to, or less than one. In the second class lever, the load force is always larger than the effort force. Therefore the mechanical advantage is greater than one. In the third class lever, the load force is always less than the effort force. Therefore the mechanical advantage is less than one.

Sample Problem

Figure 24-15 shows a simple block and tackle. It consists of a single fixed pulley and a single movable pulley. The machine is lifting a load of 24 N. The effort force is 10 N. Calculate the mechanical advantage of the machine.

Given
Load Force = 24 N
Effort Force = 10N
Required
Mechanical Advantage
Analysis
Find the ratio of the forces

Solution
$$\text{Mechanical Advantage} = \frac{\text{Load Force}}{\text{Effort Force}}$$
$$= \frac{24\,\text{N}}{10\,\text{N}} = 2.4$$

Statement
The mechanical advantage of the block and tackle is 2.4. The machine will lift a load 2.4 times as large as the effort force.

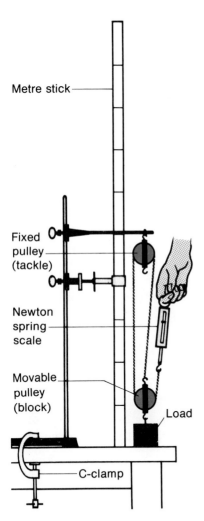

Fig. 24-15 A simple block and tackle

Efficiency

The purpose of a machine is to help us do useful work. All machines need energy to do work. The energy is used to do work on the machine. The work done on the machine is the effort force times the distance the effort moves. This is called the **work input**.

Work input = effort force x distance effort moves.

The machine does work on the load. This is called the **work output**. The work output is the load force times the distance the load moves.

Work output = load force x distance load moves.

The ratio of the work output to the work input is the **efficiency** [ee-FISH-un-see] of the machine.

$$\text{Efficiency} = \frac{\text{Work Output}}{\text{Work Input}}$$

Efficiency is usually expressed as a percentage. To do this multiply the ratio by 100.

Sample Problem

The simple block and tackle in Figure 24-15 lifts the load a distance of 0.50 m. To do this, the effort moves a distance of 1.5 m. Calculate the efficiency of the machine.

Given
Load Force = 24 N
Distance Load Moved = 0.5 m
Effort Force = 10 N
Distance Effort Moved = 1.5 m
Required
Efficiency
Analysis
Find: work output, work input, and efficiency

Solution
Work Output = Load Force x Distance Load Moves
= 24 N x 0.5 m
= 12 N•m = 12 J
Work Input = Effort Force x Distance Effort Moves
= 10 N x 1.5 m

$$\text{Efficiency} = \frac{\text{Work Output}}{\text{Work Input}}$$

$$= 15 \text{ N·m} = 15 \text{ J}$$

$$= \frac{12 \text{ J}}{15 \text{ J}} = 0.80$$

Statement
The efficiency of the machine is 0.80 or 80%.

The Limits of Machines

In an ideal machine the work output would equal the work input. The efficiency would be 100%. But no machine is ideal. Friction is present in every machine. Some effort force is always needed to overcome friction. Thus the work output in a real machine is always less than the work input. More work is needed with the machine than without it. The efficiency is always less than 100% (See Table 24-4). Modern machines attempt to reduce friction using ball bearings and lubricants, but friction is never entirely removed. The work to overcome friction becomes waste heat.

Table 24-4 Efficiencies of Some Devices

Device in good condition	Maximum efficiency
Large electric motor	92%
Wind generator	60%
Diesel engine	36%
Steam turbine	35%
Gasoline engine	26%
Jet engine	25%
Steam locomotive	9%

Main Ideas

1. Machines can change the kind, size, speed, location or direction of forces.
2. The six simple machines are: inclined plane, wedge, screw, lever, pulley, and wheel and axle.
3. There are three classes of levers depending on the position of the fulcrum, the effort and the load.
4. The mechanical advantage of a machine is the number of times it multiplies the effort force.
5. The efficiency of a machine is the ratio of the work output to the work input usually expressed as a percentage.
6. The efficiency of all machines is less than 100% because of friction.

Glossary

machine		device that can change the kind, speed, direction, or size of a force
mechanical advantage	muh-CAN-i-cal ad-VAN-tij	the ratio of the load force to the effort force
efficiency	ee-FISH-un-see	the ratio of the work output to the work input

Study Questions

A. True or False

Decide whether each of the following sentences is true or false. If the sentence is false, rewrite it to make it true. (Do not write in this book.)

1. The inclined plane developed from the wedge.
2. It takes more work to raise a load with an inclined plane than without it.
3. The lever is one of the two basic simple machines.
4. A first class lever has an efficiency of 100%.
5. A single fixed pulley changes the size of a force.

B. Completion

Complete each of the following sentences with a word or phrase which will make the sentence correct. (Do not write in this book.)

1. The wheel and axle and the ▩▩▩▩ developed from the lever.
2. In the first class lever, the ▩▩▩▩ force is larger than the ▩▩▩▩ when the effort arm is longer than the load arm.
3. A single fixed pulley changes the ▩▩▩▩ of a force.
4. If the mechanical advantage of a machine is 1, the effort force is ▩▩▩▩ the load force.
5. The efficiency of a machine is the ratio of the ▩▩▩▩ to the ▩▩▩▩ expressed as a percentage.

C. Multiple Choice

Each of the following statements or questions is followed by four responses. Choose the correct response in each case. (Do not write in this book.)

1. The pulley developed from the
   a) screw b) wheel c) axle d) lever
2. The inclined plane helps us by decreasing the
   a) effort force
   b) load force
   c) distance the load moves
   d) work done on the load

3. In the second class lever the
   a) fulcrum is between the load force and the effort force
   b) load force is between the fulcrum and the effort force
   c) effort force is between the fulcrum and the load force
   d) effort force is larger than the load force
4. A tweezer is an example of
   a) first class lever
   b) second class lever
   c) third class lever
   d) wedge

D. Using Your Knowledge

1. Look at the diagram of the wedge and screw in Figure 24-3. Explain why they are considered to be modifications of the inclined plane.
2. a) When is the mechanical advantage of a first class lever equal to 1? greater than 1? less than 1?
   b) Name three common second class levers. Explain why each is a second class lever.
   c) How do third class levers make our work easier?
3. The skeleton of the human body is a system of levers. Figure 24-16 shows two such levers.
   a) Compare the distance the muscles shorten or lengthen with the distance the ends of the limbs move.
   b) Is the mechanical advantage greater than, equal to, or less than 1 for these levers?
   c) Is the human skeleton built for speed or strength? Explain.
4. Sketch a pulley system using one stationary and one movable pulley and some cord. Make it give a mechanical advantage without friction of 2.
5. An effort force of 30 N moves a distance of 2.0 m and moves a load of 90 N a distance of 0.5 m.
   a) Determine the mechanical advantage.
   b) Determine the efficiency.

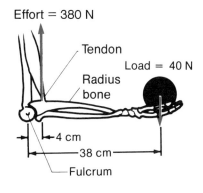

A. The elbow is a third-class lever.

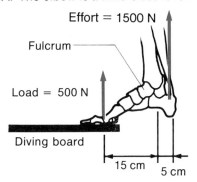

B. The foot is a first-class lever.

Fig. 24-16 Two levers in the skeleton of the human body (simplified).

E. Investigations

1. Examine the data for Activity 24.4. Try to find a relationship between the load force, the load arm, the effort force, and the effort arm. This relationship is called the **Law of the Lever**.
2. Examine a block and tackle. Predict the relationship between the mechanical advantage without friction and the number of cords supporting the movable tackle. Design and do an experiment to test your prediction.
3. Examine a multi-speed bicycle. Predict whether the mechanical advantage is greater than, equal to, or less than 1. Does the mechanical advantage change with the gears? Design and do an experiment to test your prediction.

Vibrations and Waves: Sound

CHAPTER 25
The Nature of Sound

CHAPTER 26
Transmission of Sound

CHAPTER 27
Characteristics of Sound

The study of sound is really a study of vibrations and waves. In your work with vibrations, you will learn what causes sound. You will work with wave motion and learn to measure the different characteristics of sound. Sound is very important in our lives. This unit should help you understand how sound affects all of us.

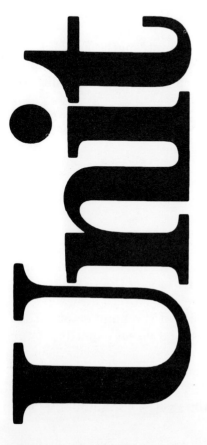

Fig. 25-0 Musical sounds give enjoyment to most people.

25 The Nature of Sound

25.1 What is Sound?
25.2 Activity: The Source of Sound
25.3 Sound and Vibrations
25.4 Transverse Vibrations
25.5 Activity: The Study of Longitudinal Vibrations
25.6 The Range of Sound

In this chapter you will find out what sound is. You will find out how it is produced and how it moves from place to place. Learn the terms in this chapter well. You will use them in the next two chapters.

25.1 What is Sound?

Telephone

Alarm clock

Starter's gun

Imagine a world of total silence. You could never hear laughter, music, or people speaking. You could not hear traffic, fire alarms, birds, or even the sound of your own voice! Many people cannot stand more than twenty minutes of total silence. Can you?

Sound is Communication

Without sound, you could never hear music on the radio. You could not use a telephone. You could not understand your favourite television shows. The next time you watch television, turn off the sound. You will soon see how much you rely upon hearing (Fig. 25-1). More than 70% of the sounds we make in English are not visible by lip reading.

Babies learn to talk by mimicking the sounds they hear. But deaf children never hear these sounds, they only see your lips move. Could you learn to speak a new language this way? Suppose you could only lip read the words taught in language class. How would you ever learn how to pronounce them, especially if you could not hear yourself speak! Speech is lost without sound.

Smoke detector

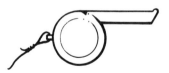
Referee's whistle

Ambulance Siren

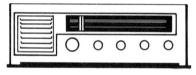
Radio

Fig. 25-1 Many sounds are important in our daily lives.

You react to the world through sound. You wake to an alarm. You answer doorbells and telephones. You hear the kettle boil. You respond to a car horn. Sound makes you aware of the activity of others. If someone sneaks up behind you, you are startled.

Think of all the jobs involving sound. Your doctor listens to your heartbeat. Ambulances use sirens to alert traffic. Policemen are summoned by radio. Firemen respond to alarms. The mechanic listens to car engines. Dancers, musicians and airline pilots all need sound — the list is endless. Most people use sound in their work.

Sound is Energy

What is sound? Sound is a form of energy that stimulates our eardrums. Loud noises can shake buildings. Noise can rattle windows. Some sounds can shatter glass. Others hurt our ears. Some sounds cause intense pain. Yet, others we cannot even hear.

What causes sound? Energy cannot be created or destroyed. It simply changes from one form to another. Therefore sound energy must come from another form of energy. Activity 25.2 should give us some answers.

Section Review

1. Why is speech without sound so difficult?
2. State five ways in which sound affects your daily life.
3. Name five different jobs which need sound.
4. What is sound?
5. Name five different sources of sound.

25.2 ACTIVITY The Source of Sound

The purpose of this activity is to find out what causes sound. You will test different sources of sound. See what they have in common. Study your observations. Then try to form a theory on what causes sound.

Problem

What causes sound?

Materials

ruler
elastic band
tuning fork
rubber mallet

pith ball on a thread
beaker of water
drum (optional)
sand grains (optional)

Procedure

a. Copy Table 25-1 into your notebook.
b. Place your thumb and forefinger on your throat just below your "Adam's apple" (Fig. 25-2). What do you observe when you are totally silent? Now hum softly. What do you notice? Hum as loudly as you can. Record your observations.
c. Stretch an elastic band between your partner's two hands. Observe the elastic when it is still. Pluck it gently. Record what you see and hear. Pluck the elastic harder. Record any changes you notice.
d. Hold a ruler firmly on the desk. Let half of the ruler stick out over the edge. Tap the free end lightly with your hand. Now tap harder. Describe what you see and hear. Try different positions of the ruler. Record any changes you observe.
e. Place your ear close to a quiet tuning fork. What do you observe? Strike the tuning fork gently with a rubber mallet. Record what you see and hear. Strike the tuning fork harder. Record any changes. Strike the tuning fork again. Place your fingers on the tips of the prongs. Describe what happens.
f. Touch a pith ball to the quiet tuning fork. Observe. Now strike the tuning fork and repeat this test. Record your observations.
g. Touch the surface of the water in the beaker with the prongs of the quiet tuning fork. Observe. Repeat this after you strike the tuning fork gently. Now strike the tuning fork harder. Repeat the test. Record your observations.
h. Place some sand on the flat surface of a drum. Describe the motion of the sand before and after you strike the drum. Record your observations.

Table 25-1 Testing Sources of Sound

Sound source	No motion	Motion
Voice		
Elastic band		
Ruler		
Tuning fork		
Tuning fork and pith ball		
Tuning fork and water		
Drum and sand grains (optional)		

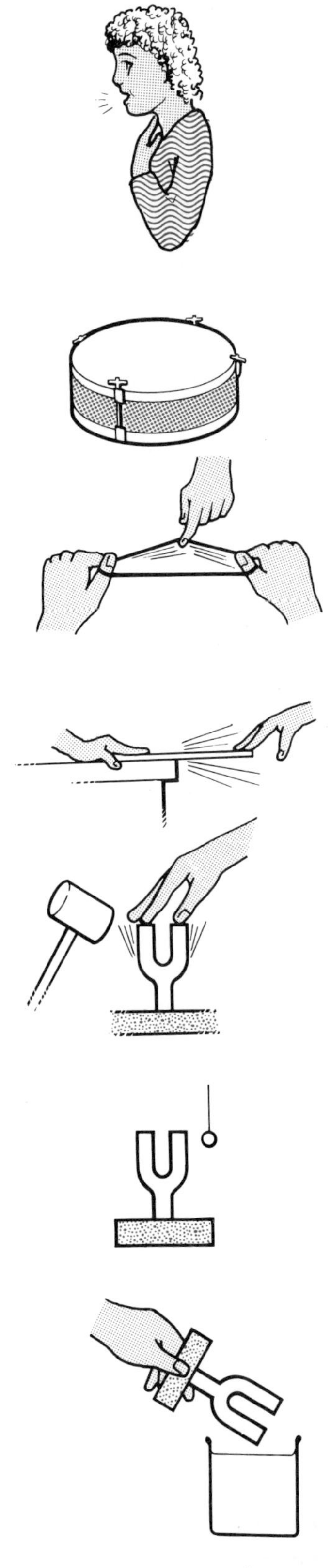

Fig. 25-2 Testing sources of sound

Discussion

1. Did all of the materials tested produce sound?
2. What condition was necessary to produce sound? Which observations can you use as scientific evidence for your answer?

3. What happened when you touched the humming tuning fork with your fingers? Why?
4. What did the tests using the tuning fork with the pith ball and the water show?
5. How can you make a sound louder? What conditions are you changing? Which observations support your answer?
6. You have now tested several sources of sound. Make a theory that describes how sound is produced.
7. Now test your theory. Try to explain how sound is produced by one of these: a guitar; a piano; your voice.

25.3 Sound and Vibrations

Activity 25.2 showed that sound is produced by objects moving rapidly back and forth. This kind of motion is called **vibration** [vy-BRAY-shun]. The elastic, ruler, and tuning fork vibrate rapidly to produce sound. When the vibrations stop, the sound stops.

Sound and Energy

You used mechanical energy to cause rapid vibrations. These produce sound energy. Greater mechanical energy causes stronger vibrations. These, in turn, produce greater sound energy. This creates *louder* sound.

Rapid vibrations are often hard to see. The wings of a mosquito vibrate so fast they look blurry. But you can hear the wings buzzing if the mosquito comes close to your ears. You could tell the tuning fork prongs were moving. You watched the pith ball and the water. The vibration of a drum can also be seen. You can watch sand grains move on a flat surface.

Types of Sound

Wind vibrates the branches and leaves of trees. This creates rustling sounds (Fig. 25-3). Waves crash against each other. The vibrating water and air above cause the roar of the ocean. Crickets make sound vibrations too. They rub their wings against their bodies. A bell clapper vibrates the metal bell. This vibrates the surrounding air. Air molecules vibrate when they are blown into hollow tubes. This is how wind instruments, such as trumpets, produce sound (Fig. 25-4).

What causes the sound of thunder or explosions? Heat expands air. This can make air molecules vibrate rapidly enough to produce sound. Heat from lightning expands the air suddenly. The rapidly vibrating air molecules cause thunder. When a volcano erupts, hot lava quickly expands the air in the crater. The molecules of air and

Fig. 25-4 Wind instruments produce sound when a column of air is vibrated. How many of these instruments can you name?

Section 25.3 295

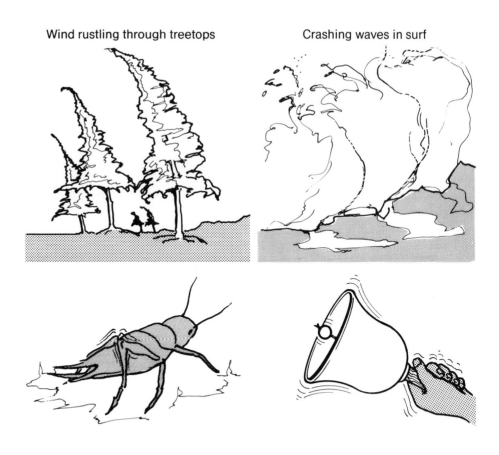

Fig. 25-3 Rapidly vibrating matter produces sounds.

other hot gases from the volcano vibrate rapidly. The sound produced is a loud explosion. The heat from burning dynamite causes a similar sound.

Your Voice

How do you make sound? Air moves through your windpipe. Then it vibrates the two vocal cords in your voice box (Fig. 25-5). You can feel these vibrations. Your mouth parts change the sound to form speech. The sound made by a vibrating elastic band changes as it stretches. You change your sounds when you stretch or relax your vocal cords. Your voice results from the vibrations of your vocal cords and the air in your windpipe.

Section Review

1. What is vibration?
2. How is sound produced?
3. How can you make a sound louder?
4. Rapid vibrations are often difficult to see. How can you show that the prongs of a tuning fork are really moving?
5. What is vibrating to produce each of these sounds: the rustling of wind through the trees; the crashing of ocean waves; the ringing of a bell; the notes of a trumpet?

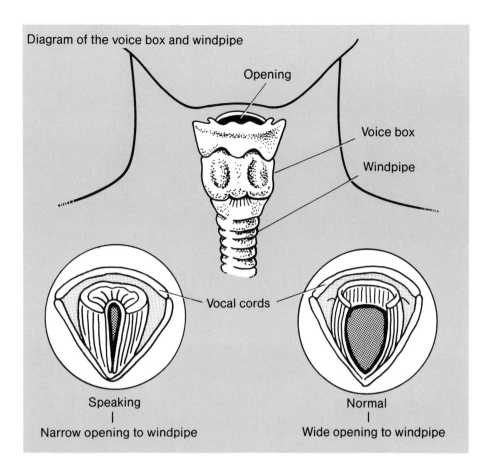

Fig. 25-5 The opening to your windpipe is normally wide. When you speak, this opening narrows.

6. How does heat affect air?
7. What causes thunder?
8. What produces the sound of your voice?

25.4 Transverse Vibrations

Before we learn more about sound, we must study vibrations in detail. Most objects producing sound vibrate so quickly that we cannot study their motion. But anyone who has ever pushed a swing has created slow transverse vibrations. **Transverse vibrations** [trans-VURS] *occur when an object moves back and forth at right angles to its length.* Figure 25-6 shows three examples.

Amplitude and Cycle

You may have studied a pendulum in Activity 2.2 (page 14). You can also use a pendulum to study transverse vibrations. To swing the pendulum, the bob must be moved or displaced from its **rest position** (Fig. 25-7). The horizontal displacement of the bob is called the

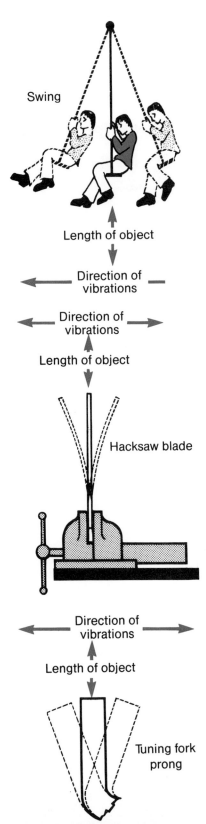

Fig. 25-6 Transverse vibrations occur at right angles to the length of the object.

amplitude [AM-pli-tood] of vibration. Suppose the bob is released from position A and makes one complete back and forth motion (A to B and back to A). We say the pendulum has made one vibration or **cycle**.

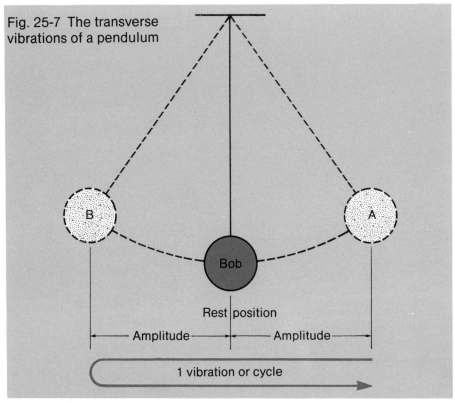

Fig. 25-7 The transverse vibrations of a pendulum

Period and Frequency

The time taken for one cycle is the **period** (T) of vibration. The number of cycles per second is the **frequency** [FREE-kwen-see] (f). The unit used to measure frequency is the **hertz (Hz)**. This name was chosen to honour the German scientist Heinrich Hertz (1857-1894). He discovered radio waves.

$$1 \text{ hertz (1 Hz)} = 1 \text{ cycle per second}$$

Cycles are counted and not measured. Therefore 1 Hz = 1/s. (The unit hertz is simply a reciprocal second.)

This example will help you understand these terms. Suppose the pendulum makes 1 complete cycle in 0.5 s. This is the period of vibration. The number of cycles in 1 s is the frequency of vibration. This pendulum completes 2 cycles in 1 s. Therefore, the frequency of vibration is 2 Hz.

$$\text{frequency} = \frac{1}{\text{period}} \text{ or } f = \frac{1}{T}$$

If you know the period of vibration, you can calculate the frequency. You can also rearrange this equation:

$$\text{period} = \frac{1}{\text{frequency}} \text{ or } T = \frac{1}{f}$$

Now suppose the frequency (*f*) is increased to 10 Hz. The pendulum completes ten cycles in one second. The period (*T*), or time taken for one cycle, can be calculated using $T = \frac{1}{f}$. Therefore, $T(s) = \frac{1}{f(Hz)} = \frac{1}{10 \text{ Hz}} = \frac{1}{10 \text{ 1/s}} = 0.1$ s. The period of the pendulum has decreased to 0.1 s. Therefore, as the frequency increases, the period decreases.

Test Your Understanding

You may have already studied the relationship between the displacement and period of a pendulum. If not, you should go back and do Activity 2.2. You can also test your understanding of transverse waves with the following activity:

Place your elbow on the desk with your forearm vertical. Your fingers should be pointing upright (Fig. 25-8). This is your rest position. Now vibrate your forearm transversely with

a) an amplitude of 10 cm
b) an amplitude of 20 cm
c) a frequency of 1 Hz
d) a frequency of 3 Hz
e) a period of 1 s
f) a period of 0.5 s

Your partner can check your accuracy.

Fig. 25-8 Test your understanding of transverse vibrations.

Section Review

1. What are transverse vibrations?
2. State three examples of transverse vibrations.
3. Define the following terms: amplitude of vibration; period; frequency; hertz.
4. A pendulum completes two cycles in one second. What is the frequency in hertz?
5. State the relationship between period and frequency.

25.5 ACTIVITY The Study of Longitudinal Vibrations

Some objects vibrate with longitudinal vibrations. **Longitudinal vibrations** [lon-ji-TOO-dinul] *occur along the length of the vibrating object* (Fig. 25-9). You will study two examples of longitudinal vibrations in this activity.

Problem

What are longitudinal vibrations?

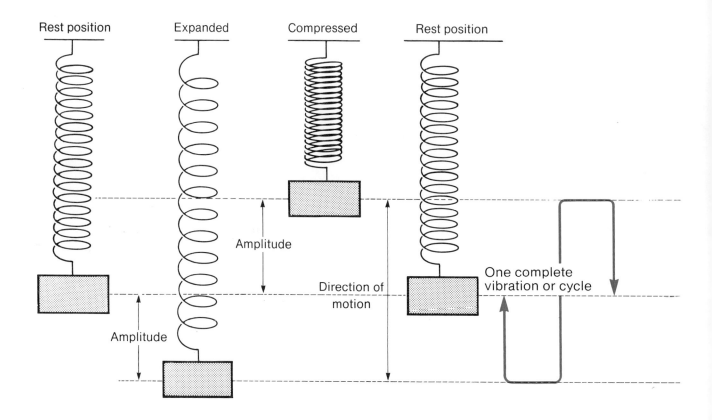

Fig. 25-9 Longitudinal vibrations occur along the length of the spring.

Materials

ring stand	test tube clamp	ruler
iron rod	rubber mallet	timer
pith ball on a string	coil spring	splint
ring clamp	mass	tape

Procedure A Longitudinal Vibrations in a Rod

a. Set up the apparatus as shown in Figure 25-10. (Wrap a paper towel around the iron rod. Then it will fit tightly in the clamp.)
b. Strike the far end of the iron rod *gently* with the rubber mallet. Record your observations.
c. Now strike the rod harder. Record your observations.
d. Strike the rod again. Touch it lightly with your fingertips. What do you observe?

Procedure B Longitudinal Vibrations in a Coil

a. Set up the apparatus as shown in Figure 25-11.
b. Copy Table 25-2 into your notebook.
c. Record the rest position of the mass in your table.
d. Pull the mass down below rest position. Observe the new position of the splint marker on the ruler. Release the mass. Note the distance the marker moves above rest position. Record the amplitude of this vibration in your table. Use this amplitude while you

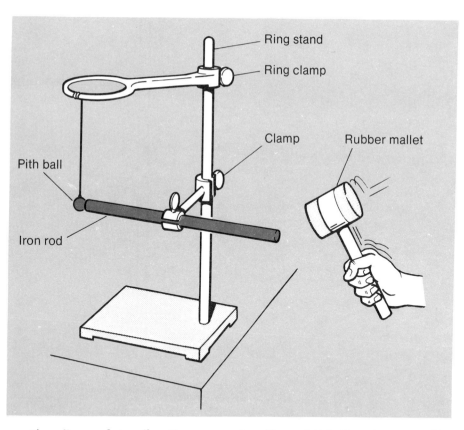

Fig. 25-10 Studying longitudinal vibrations

time 5 complete vibrations or cycles. Record this time in your table.
e. Repeat this measurement for 2 more trials.
f. Determine the average time for the 3 trials. Calculate the period of the vibrations.
g. Calculate the frequency of these vibrations.
h. Now repeat this procedure using 2 other amplitudes.

Table 25-2

Rest position: ___ cm

Trial	Amplitude = ___ cm	Amplitude = ___ cm	Amplitude = ___ cm
	Time for 5 vibrations (s)	Time for 5 vibrations (s)	Time for 5 vibrations (s)
#1			
#2			
#3			
Average (s)			
Period (Time for 1 vibration)			
Frequency (Hz) (cycles per second)			

Section 25.5

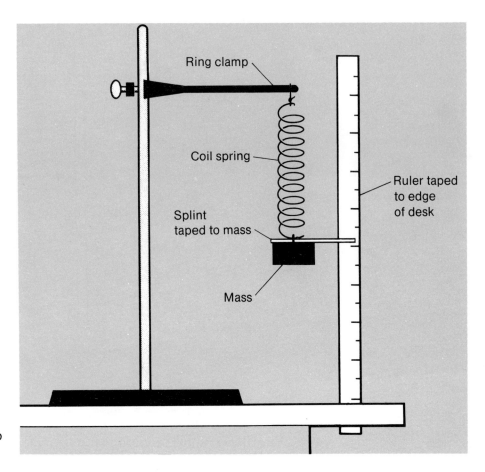

Fig. 25-11 Apparatus used to study longitudinal vibrations

Discussion A

1. What do you observe when you strike the rod gently? Why?
2. What happens when you strike the rod harder? Explain.
3. What did you feel when you touched the rod?
4. Explain which type of vibration is occurring.

Discussion B

1. Why are the vibrations of the coil spring called longitudinal vibrations?
2. Did an increase in amplitude increase or decrease the frequency of the vibrations?
3. Compare your results to the class results.

25.6 The Range of Sound

The pendulum let you study transverse vibrations in slow motion. But only very rapid vibrations produce sound. How fast would your arm have to vibrate before you could hear any sound?

Audible Sound

An **audible sound** [AW-di-bul] *is one which you can hear.* Humans can hear sound from objects vibrating with frequencies ranging from about 16 Hz to 20 000 Hz (20 kHz). Old age or damage to your hearing can reduce your audible range.

Many animals have a greater range of hearing than humans (Fig. 25-12). They can detect sounds which we cannot hear. For instance, dogs can hear sounds with frequencies as high as 35 kHz.

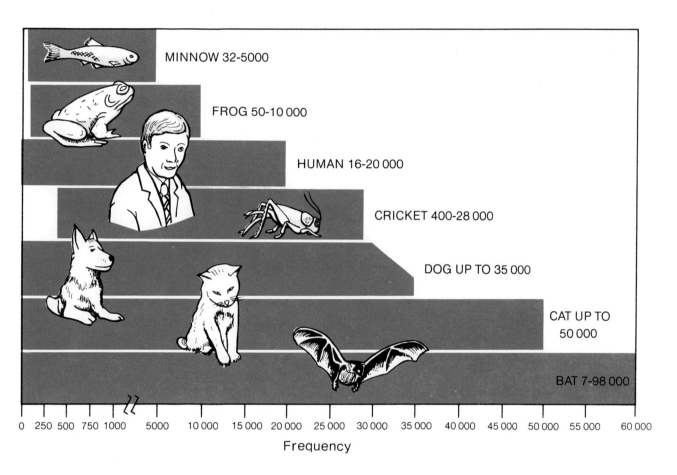

Fig. 25-12 The frequency of a sound depends upon the number of cycles the vibrating source makes in one second. (1 cycle per second = 1 Hz.) Humans can rarely hear sounds above 20 000 Hz (20 kHz) or below 16 Hz.

Ultrasonic Vibrations

A dog whistle vibrates above 20 kHz. Therefore, you cannot hear it, but a dog can. This sound is **ultrasonic** [ul-tra-SON-ik]. (*Ultra* means

Section 25.6 303

beyond. *Sonic* means sound.) Ultrasonic sounds are beyond the audible range for humans.

Bats can fly in total darkness. They use ultrasonics to guide themselves with their ears. Bats emit high-pitched shrieks, yet you cannot hear them. Their sounds have a frequency of about 50 kHz. These sounds strike nearby objects. Then they bounce back as echoes (Fig. 25-13). Bats use the ultrasonic echoes to avoid hitting these objects. Dolphins use the same system underwater.

Sonar

Scientists now use ultrasonics to bounce echoes off the ocean floor. They measure the time needed for sound to reach the bottom and echo back to the ship. They can then calculate the depth. This method is called **sonar** [so-nar]. The name stands for "**SO**und **N**avigation **A**nd **R**anging". Sonar uses sound frequencies of about 50 kHz. It can detect submarines and submerged icebergs (Fig. 25-14).

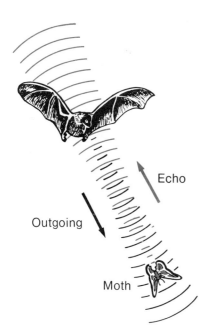

Fig. 25-13 Bats fly in the dark. They use ultrasonic echoes to locate prey and nearby objects.

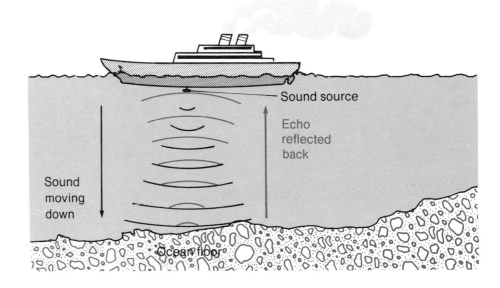

Fig. 25-14 Sonar uses echoes to measure water depths. It can also locate schools of fish.

The Growth of Ultrasonics

Canadian scientists were pioneers in the study of ultrasonics. These high frequency sounds are becoming more important every day. The rapid sound vibrations can cook food. They can clean dishes and clothes. Ultrasonic waves sterilize and homogenize milk. They are used in silent burglar alarms. Doctors can use ultrasonic beams for surgery and for photographing unborn babies. Dentists even use them for drilling teeth. An ultrasonic dental drill operates with almost no heat or sound. Many important industries also use ultrasonics.

Infrasonic Vibrations

Some frequencies are too low to be audible. (*Infra* means below. *Sonic* means sound.) Sound vibrations below 16 Hz are in the **infrasonic** [in-fru-SON-ik] range. Low frequency vibrations can break up hard rock. Infrasonic sound has been used to drill deep oil wells.

Section Review

1. What is audible sound?
2. State the frequency range of audible sound.
3. What is ultrasonic sound?
4. How do bats and dolphins navigate?
5. How is sonar used?
6. What is infrasonic sound? How is it used?

Main Ideas

1. Sound is a form of energy produced by rapidly vibrating objects.
2. Transverse vibrations occur when an object vibrates at right angles to its length.
3. Longitudinal vibrations occur along the length of the vibrating object.
4. The amplitude is the maximum displacement of a vibrating object from the rest position.
5. The period (T) is the time required for one vibration or cycle.
6. The frequency (f) is the number of cycles per second.
7. Frequency is measured in hertz. One cycle per second = one hertz (1 Hz).
8. Most humans can hear vibrations ranging from 16 Hz to 20 000 Hz (20 kHz).

Glossary

amplitude	AM-pli-tood	the distance the bob moves from rest position
audible	AW-di-bul	heard by the human ear
frequency	FREE-kwen-see	the number of vibrations per second
infrasonic	in-fru-SON-ik	below the range of audible sound
longitudinal	lon-ji-TOO-dinul	lengthwise
sonar	SO-nar	a sounding device using echoes
transverse	trans-VURS	crosswise
ultrasonic	ul-tra-SON-ik	beyond the range of audible sound
vibration	vy-BRAY-shun	a back and forth motion

Study Questions

A. True or False

Decide whether each of the following sentences is true or false. If the sentence is false, rewrite it to make it true. (Do not write in this book.)
1. Sound is caused by any vibrating object.
2. Transverse vibrations occur at right angles to the length of the object.
3. Frequency is the time required for one complete vibration.
4. Only dogs and humans can hear ultrasonic sound.
5. Amplitude measures the displacement from rest position.

B. Completion

Complete each of the following sentences with a word or phrase which will make the sentence correct. (Do not write in this book.)
1. Sound is _____ caused by _____ .
2. A vibration is a _____ motion.
3. Your voice is produced by vibrations of _____ and _____ .
4. The period of a vibration is _____ .
5. Most humans can hear sounds with frequencies ranging from _____ to _____ .

C. Multiple Choice

Each of the following statements is followed by four responses. Choose the correct response in each case. (Do not write in this book.)
1. Louder sounds can be made by
   a) more rapid vibrations
   b) stronger vibrations
   c) ultrasonic vibrations
   d) vibrations of lower frequency
2. A hertz is the unit used to measure
   a) amplitude   b) period   c) frequency   d) displacement
3. Increasing the frequency of a vibration makes a sound
   a) higher   b) louder   c) lower   d) infrasonic
4. If the period of a vibration gets longer, the frequency
   a) increases
   b) decreases
   c) does not change
   d) becomes ultrasonic
5. A low sound audible to the human ear might have a frequency of
   a) 21 kHz   b) 16 Hz   c) 16 000 Hz   d) 25 Hz

D. Using Your Knowledge

1. Explain the source of sound in each of the following:
   a) a door slams
   b) chalk scrapes on the chalkboard
   c) car tires squeal on pavement
   d) a firecracker explodes

2. Select five different jobs which rely greatly on sound. Now try to name five jobs where deaf people might have an advantage.
3. A chimney can be cleaned using ultrasonic vibrations. Yet no sound can be heard inside the house. Explain.
4. You can hear rain on a roof. Yet you don't hear a snowfall. Why?
5. If you remove the pebble from a referee's whistle, you cannot hear it blow? Why?

E. Investigations

1. Consult the local association for the deaf. Find out how deaf people function in a hearing world. How do they manage without alarm clocks or telephones? Can they dance without music? How do they express feelings without speech? Is a deaf person allowed to drive a car?
2. Musical instruments are classified as stringed, wind, or percussion. Every instrument has a vibrating part and a way of changing the sound. Select five different instruments. Classify each one. Explain how each one works. Consult the library or the music room for information.
3. Consult your library to learn more about ultrasonics. How many different fields use this high frequency sound?
4. Select one of the following and find out how it works:
    a) a telephone   **b)** a record   **c)** a recording tape

26 Transmission of Sound

26.1 Sound and Matter
26.2 How Sound Travels
26.3 Activity: The Study of Wave Energy
26.4 The Speed of Sound
26.5 Sound Detectors

In the last chapter you learned how sound is formed. You also studied the kinds of vibrations. In this chapter you will use that knowledge. You will find out how sound moves from place to place. And you will find out how your ear detects sound.

26.1 Sound and Matter

All sound comes from rapidly vibrating objects. It travels to our ears through different types of matter, or media. Can sound travel through a vacuum? Could a person hear sound in outer space?

Testing a Vacuum

You can test sound in a **vacuum** [VAK-yoom]. Use the apparatus shown in Figure 26-1. An electric bell is suspended inside the bell jar. The bell is connected to a battery. The ringing bell can be heard clearly. Then the air inside the jar is removed using the vacuum pump. As air leaves the jar, the ringing sound gets fainter. Soon you can barely hear the sound. But the bell is still vibrating. There is almost a complete vacuum in the jar. If you tilt the jar, the bell touches the glass. The ringing grows loud and clear. Now straighten the jar. Pump the air back in. The faint sound of the bell keeps getting louder. Finally all the air is returned. The sound of the bell is normal again.

This experiment shows that sound does *not* travel through a vacuum. Air, glass, or some other medium is needed. A **medium** is a form of matter.

Types of Media

Most of the sound you hear travels through air. But sound can travel through any form of matter. Many solids transmit sound easily. Sound can travel through a wall or a closed door. Any apartment dweller knows that! Indians used sound to track moving buffalo

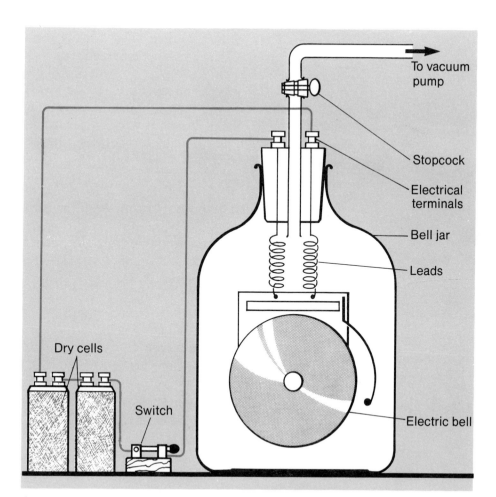

Fig. 26-1 Testing sound transmission in a vacuum

herds. They pressed their ear to the ground. They could hear distant hoof beats. The sound travelled through the earth better than through the air. It also travels faster. In fact, some solids transmit sound about fourteen times faster than air! Similarly, early travellers could detect an approaching train. They would press their ear to the track. Vibrations from the train moved much faster and louder through the rail than through the air.

Have a classmate stand at one end of your desk. Then place your ear against the other end of the desk. Close your eyes. Now tell your partner to scratch the end of the desk. Can you hear it? Now stand up. Keep your eyes closed. Tell your partner to scratch the desk again. Can you hear it?

Sound also travels through liquids. Water is a good medium. Swimmers underwater can hear distant motorboats. Sound travels almost five times faster in water than in air. Sounds made underwater are very loud to a diver. Test for sound in the bathtub. Place one ear in the water. Then scratch the bottom of the tub.

Sound Absorption

Some surfaces tend to reflect sound. Others absorb sound. It depends on the type of material. Hard materials vibrate more than soft materials. Surfaces such as metal, plaster, and wood are hard.

They reflect sounds. This makes sounds louder. Surfaces such as carpets, drapes, and cushioned furniture are soft. They absorb sound (Fig. 26-2).

Carpets in an apartment can muffle the sound of footsteps or moving furniture. But sounds are very loud in an empty room or apartment. Materials which absorb sound are used in soundproofing. We will discuss this further in Chapter 27.

Section Review

1. What happens to the sound as air is pumped out of a bell jar? Why?
2. How does sound energy differ from light energy?
3. If you tip the evacuated jar, the vibrating bell touches the glass. What happens to the sound? Why?
4. What is a medium?
5. Swimmers underwater can often hear motorboats they could not hear at the surface. Why?
6. Why do hard surfaces reflect sound?
7. What kind of materials are used for soundproofing?

Fig. 26-2 Which surfaces in this room absorb the sounds of the children at play?

26.2 How Sound Travels

Rapidly vibrating objects produce sound energy. This energy is transmitted through a medium. But how does sound travel from the source to your ears?

Transverse Waves

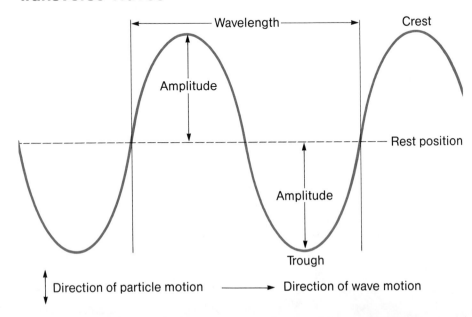

Fig. 26-3 Transverse wave motion. The particles of the medium move at right angles to the direction of the wave.

Drop a pebble into a small pan of water. It makes a pattern of waves. These are the ripples on the surface. Watch the waves travel. The water bobs up and down at right angles to the direction of the waves. This is called **transverse wave motion**.

Greater energy produces bigger waves. They have a greater **amplitude**. The high part of each wave is called the **crest**. The low part is called the **trough** [TROFF]. One crest and one trough form one **wavelength** (Fig. 26-3). After the wave energy passes, the water is still.

Float a cork in the centre of the pan. Drop in another pebble. Watch the motion of the cork. The wave energy makes it bob up and down. But the cork does not move along with the wave. Only the energy of the waves is travelling.

Sound Waves

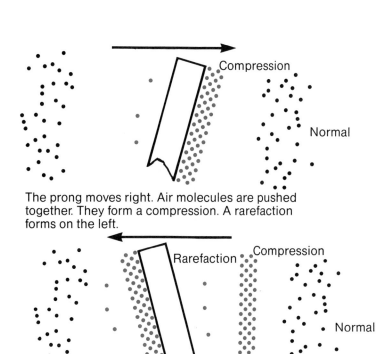

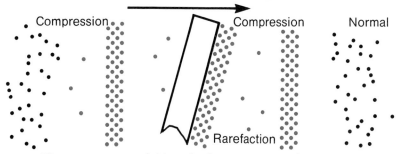

Fig. 26-4 The production of sound waves by a tuning fork

Sound energy also travels in waves. But sound waves are different from water waves. Picture the prong of a vibrating tuning fork. Imagine what happens to the surrounding air molecules (Fig. 26-4). The prong moves to the right. The air molecules there are pushed together. This region of crowded molecules is called a **compression** [cum-PRES-shun].

This starts a chain reaction. The energy from the tuning fork causes the compressed air molecules to vibrate rapidly. They collide with nearby molecules. Thus, the energy is passed along. This produces a moving compression.

Then the prong moves back to the left. The molecules spread out. This forms a region called a **rarefaction** [rar-a-FAK-shun]. Then the prong swings back again. This forms a new compression.

The effect is similar to someone shoving at the back of a lineup. Each person in turn gets pushed into the person ahead. But nobody moves far from his or her original position. Everyone has a chance to straighten up. Then the person at the back starts shoving again.

Each complete vibration of the prong forms one compression along with one rarefaction. A series of vibrations cause a series of compressions and rarefactions. These move outwards from the prong (Fig. 26-5). These are sound waves. *Each **sound wave** consists of one compression followed by a rarefaction. A series of waves is called a **wave train**.*

Longitudinal Waves

The vibrating molecules simply transfer the energy to other molecules. As they push together and spread apart, they move back and forth. This motion is parallel to the direction of the moving wave. This is **longitudinal wave motion** [lon-ji-TOO-dinul]. In a longitudinal wave, the medium particles vibrate parallel to the direction of the moving wave.

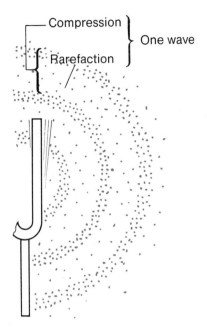

Fig. 26-5 Sound waves spread out from a tuning fork.

Fig. 26-6 A longitudinal sound wave can be represented by a transverse wave pattern.

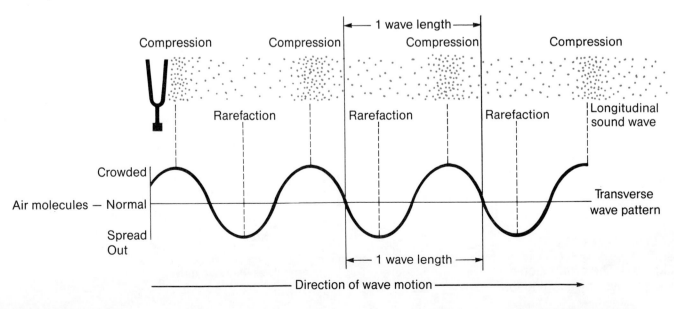

Figure 26-6 compares a longitudinal wave with a transverse wave. Sound waves are longitudinal. but transverse wave patterns are much easier to draw. So we will represent many sound waves using transverse wave patterns. You can do the same.

Features of Sound Waves

You should know the following facts about sound waves:
1. Every sound wave has a vibrating energy source.
2. Greater energy produces larger waves. Larger waves have a greater amplitude.
3. Sound waves move and carry energy.
4. Sound waves must travel in a medium.
5. When a sound wave passes through a medium, the particles are disturbed. But their movement is limited. The particles of the medium are not carried along with the wave.
6. A wavelength consists of one compression and one rarefaction.
7. The frequency of sound waves depends upon how rapidly the source vibrates.
8. The speed at which sound waves travel depends mostly on the medium.
9. Sound waves spread out. They decrease in amplitude as they move away from the source.

Section Review

1. How can you increase the amplitude of a wave?
2. Describe a crest, a trough, and a transverse wave.
3. How do sound waves differ from water waves?
4. Describe a compression, a rarefaction, and a longitudinal wave.
5. What determines the frequency of sound waves?
6. Why does sound become fainter as you move farther from the source?

26.3   ACTIVITY   The Study of Wave Energy

In this activity you will observe the transfer of energy through a medium. You will also compare the motion of transverse and longitudinal waves.

Problem

How does energy move through a medium?

Materials

5 marbles or spheres
ruler with groove
small sheet of stiff cardboard

candle with holder
slinky or loose coil
paper clip or coloured marker

Procedure A Transfer of Energy by Marbles

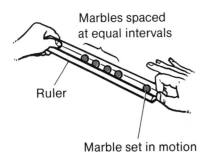

Fig. 26-7 Procedure A: Studying the transfer of energy

a. Place 4 marbles at equal intervals along a ruler groove. The marbles should be close, but not touching (Fig. 26-7).
b. Place your hand at one end of the ruler.
c. Strike the first marble at the other end with a fifth rolling marble.
d. Describe the resulting motion.
e. What happens if you place the marbles closer together?

Procedure B Transfer of Energy Through Air

a. Set up the candle in a holder on your desk. Light it.
b. Hold the stiff cardboard upright about 20 cm from the flame.
c. Tap the back of the cardboard with your other hand.
d. Describe what happens to the flame. What happens if you strike the cardboard harder?

Procedure C Transverse Wave Motion

a. Choose one coil near the centre of a slinky. Attach a paper clip or coloured marker to it.
b. Put the slinky flat along the floor. You hold one end. Your partner holds the other. Gently stretch the slinky. Do not stretch too tightly. You will damage the coils.
c. Your partner holds one end still. You form a wave by moving your hand from side to side in front of you (Fig. 26-8). Remember to keep the slinky flat on the floor. Observe the motion of the coils.

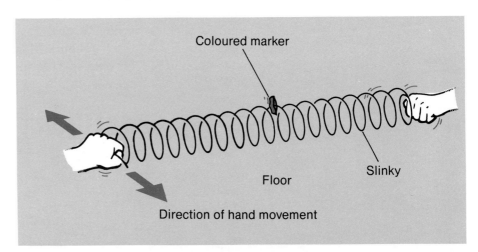

Fig. 26-8 Procedure C: Form a wave pattern. Move your hand back and forth at right angles to the slinky.

d. Now vibrate your end by pulling it back and forth to form a series of waves. Keep the other end still. Study the motion of the marker on the centre coil. Which direction is it moving? Which direction are the waves moving?
e. Try to form a smaller wave. Then form a larger wave. What factor are you changing?
f. Form faster waves. What factor are you changing?

Procedure D Longitudinal Wave Motion

a. Now form a different type of wave. Hold the far end of the slinky still. Pinch several of the coils at your end together (Fig. 26-9). Then release them. Repeat this to form a regular wave pattern.
b. Describe the motion of the coil. Which direction is the wave moving?
c. Watch the marker on the coil. Which direction is it moving?
d. Form faster waves. What factor are you changing?

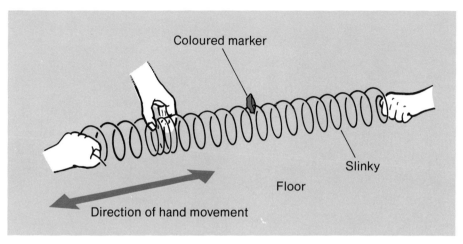

Fig. 26-9 Procedure D: Form a wave pattern. Pinch several slinky coils together. Then release them.

Discussion A

1. How far did each of the marbles move?
2. How was the energy of the first marble transferred to the last marble?
3. How could you increase the energy given to the last marble?
4. How could you transfer the energy faster?
5. How does this test resemble sound waves travelling through air?

Discussion B

1. What happens to the candle flame when you tap the cardboard? Why?
2. How do waves of energy travel through air?
3. What happens when you tap the cardboard harder? Why?

Discussion C

1. Draw a straight line to represent the slinky at rest. Now draw the pattern formed by the moving coils.
2. Label one wavelength.
3. Label the maximum amplitude of one wave.
4. Show the direction in which the coil marker moves. Show the wave direction.
5. Which type of wave motion did you form? Explain your choice.
6. How can you make faster waves?
7. How can you change the amplitude of the waves you form?

Section 26.3

Discussion D

1. Describe the motion of the moving coils.
2. Which way does the coil marker move? Which direction do the waves move?
3. Which type of wave motion did you form? Explain your choice.
4. Compare the two motions in (Fig. 26-10). Which one resembles sound waves? Why?

Fig. 26-10 Which of these wave patterns resembles sound waves? Why?

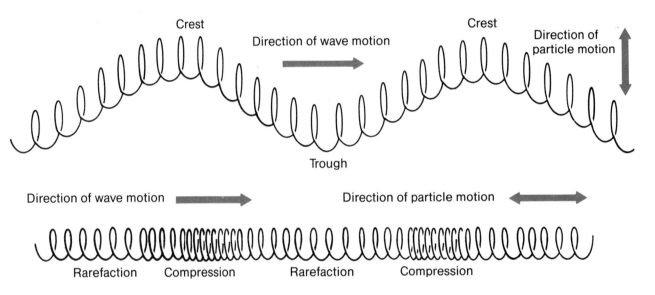

26.4 The Speed of Sound

Sound usually seems to move instantly. You hear words while a person's lips are moving. But you often see a flash of lightning before you hear the crash of thunder it produces. Or watch someone in the distance strike a ball. The ball is in the air before you hear a sound. Light moves much faster than sound. In fact, light travels nearly a million times faster than sound!

Radar uses microwaves. These travel the same speed as light. A radar signal to the moon is reflected back in about 2.5 s. Suppose sound could travel to the moon. It would take nearly 26 d to return!

The Effect of Media

How fast is sound? That depends on the medium. Look at Table 26-1.

Table 26-1 The Speed of Sound in Different Media

Solid at 20°C	Speed m/s	Liquid	Speed m/s	Gas at 20°C	Speed m/s
Iron	5130	Water at 20°C	1461	Hydrogen	1270
Aluminum	5104	Seawater at 9°C	1450	Air	344
Glass	5000	Water at 9°C	1435	Oxygen	317
Oak	3850			Carbon dioxide	258
Copper	3560			Air at 0°C	332
Brass	3500				

Sound travels faster in solids than in liquids. It moves faster in liquids than in gases. Note that the speed also changes with temperature. As the temperature increases, the speed increases. Sound travels faster in warm air than in cold. You may have noticed how voices carry on a warm summer night.

Supersonic Speeds

Jet propulsion has increased the speed of aircraft. Some planes and many missiles travel faster than sound. These speeds are called **supersonic**.

The **Mach Number** compares the supersonic speed to the speed of sound in air. A plane flying at Mach 1 has a speed of 1207 km/h. This is the speed of sound in air at 5°C. If a plane goes twice as fast as sound, its Mach Number is 2.

The Sound Barrier

A slow airplane sets up sound waves. These move away from the plane in all directions. But suppose the plane speeds up to Mach 1. It catches up with its own sound wave. This giant wave is called the **sound barrier**. It causes great stress on the plane. To keep moving, the plane must break this sound barrier (Fig. 26-11). An explosive sound called a **sonic boom** results. Sonic booms shatter windows. Supersonic jets have to fly at high altitudes so that the sonic boom cannot damage buildings (Fig. 26-12).

Fig. 26-11 Breaking the sound barrier. Supersonic jets must penetrate their own giant sound waves.

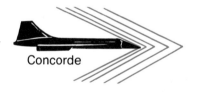

Concentrated sound waves

Fig. 26-12 Supersonic jets must be far from cities when they pass through the sound barrier. The cruising altitude is very high. This reduces ground disturbance.

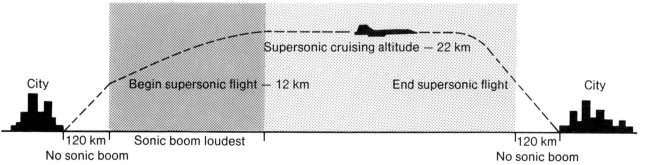

Echoes

Many hard surfaces reflect sound waves. They go back to where they started. This returning sound is called an **echo**. An echo must take at least 0.1 s to return. Otherwise, you cannot tell the reflected sound from the original.

Reverberations

Some large auditoriums have many reflecting surfaces. You hear the speaker at the front. But you also hear echoes of the words. The sounds get jumbled together. You cannot hear the words properly. This is called **reverberation** [ree-ver-buh-RAY-shun] (Fig. 26-13). Such a room has poor **acoustics** [ah-KOOS-tiks]. You cannot hear sounds clearly.

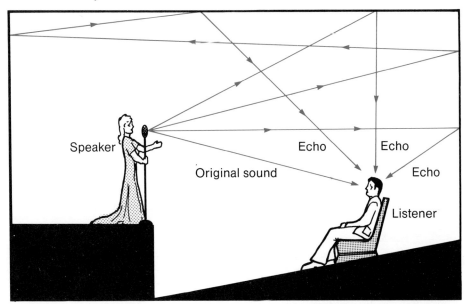

Fig. 26-13 Reverberation causes confusing sound. Many reflecting surfaces make echoes. These interfere with the original sound.

Fig. 26-14 This cheerleader is using a megaphone. It directs all of the sound waves towards the crowd. Otherwise the sound would spread out in all directions.

318 Chapter 26

Sometimes architects reflect sound on purpose. They make halls with curved walls and ceilings. But they reflect all sounds straight to the listener. Even a tiny whisper can be heard across the hall. The capitol building in Washington D.C. has a famous "Whispering Gallery".

Sound spreads out in all directions from a source. A megaphone reflects all the sound waves toward the listeners (Fig. 26-14). This concentrates the sound and helps reduce reverberations.

Section Review

1. Compare the speed of sound to the speed of light.
2. Compare the speed of sound in solids, liquids, and gases.
3. How does temperature affect the speed of sound?
4. What is supersonic speed?
5. Why must supersonic jets fly at high altitudes?
6. How do echoes form?
7. Why are reverberations undesirable in an auditorium?

26.5 Sound Detectors

Every sound has a rapidly vibrating source. Sound waves travel through a medium. But how is sound detected? How does a microphone work? What happens when sound waves reach our ears? How do we identify different sounds?

Changing Energy

Electrical energy can generate sound waves and sound waves can generate electrical energy. To detect a sound wave, some of its vibrational energy must be changed. It must form a different kind of energy signal we can measure. It can be an electrical signal through wires in a microphone. Or it can be an electrical signal through nerves to our brain. Although ears and microphones do not look the same, they work in a similar way.

The Microphone

Sound waves travel through the air to the microphone. They strike a very thin membrane. It vibrates. This membrane is part of an electrical circuit. The vibrations produce an electrical signal. Good microphones are sensitive. They can detect a wide range of sound frequencies. They can pick up faint as well as loud sounds.

The Human Ear

The human ear has three main parts: the outer ear, the middle ear, and the inner ear (Fig. 26-15). The outer ear is shaped to collect sound waves. The **ear canal** leads these waves to the **eardrum**. The eardrum is a very sensitive membrane. It is only 0.008 cm thick! Sound waves make the eardrum vibrate. Sometimes an eardrum breaks. Then the ear is deaf.

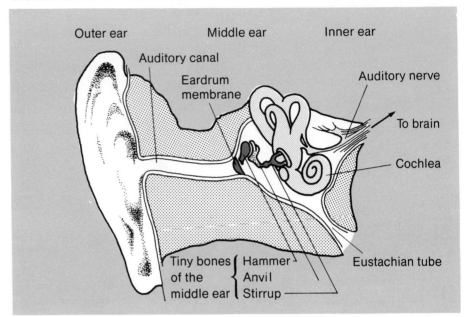

Fig. 26-15 The human ear. Sound waves enter the auditory canal. They vibrate the ear drum. These vibrations are amplified and passed to the inner ear. The auditory nerve carries the signals to the brain.

There are three tiny bones in the **middle ear**. They pick up sound vibrations from the ear drum. Then the vibrations pass to the inner ear.

The **inner ear** has a spiral-shaped hearing organ called the **cochlea** [KOK-lee-uh]. It contains a fluid called **lymph** [LIMF]. The inner ear also has thousands of tiny **nerve hairs**. When sound vibrations reach the inner ear, the fluid vibrates. This vibrates the nerve hairs.

Recognizing Sounds

Some of these nerve hairs can detect low frequency sound. Others can detect high frequency sound. Loud sounds have high energy vibrations. They make the fluid vibrate harder. Then the nerve hairs vibrate harder. Soft sounds have low energy vibrations. The fluid and nerve hairs do not vibrate as hard. The nerve hairs send out electrical signals. They pass through the **auditory nerve** [AW-di-tow-ree] to the brain. The brain can tell whether a sound is loud, soft, high, or low. It learns to recognize words.

Ears are very efficient. They are almost as sensitive to frequency as the best microphones. They work entirely on internal power. Ears can detect much fainter sounds than microphones can.

Shock Waves

Some sound waves have huge energy. We do not need microphones or ears to detect them. Sound waves from explosions or earthquakes can destroy buildings. Sound waves from supersonic jets can shatter windows. These super energy waves are called **shock waves**. They travel faster than sound waves and can transmit energy long distances. Shock waves are only formed when the sound source moves faster than the speed of sound in air.

Section Review

1. How are sound waves related to electrical energy?
2. Which part of a microphone do sound waves affect first?
3. Describe the function of the outer ear, the eardrum, the middle ear, and the nerve hairs of the inner ear.
4. How does the ear detect different frequencies?
5. How do loud sounds affect the inner ear?
6. How are sounds transmitted to the brain?

Main Ideas

1. Sound waves cannot travel in a vacuum.
2. Sound travels fastest in solids, slower in liquids, and slowest in gases.
3. The speed of sound increases with temperature.
4. Sound energy travels in the form of longitudinal waves.
5. Hard surfaces reflect sounds as echoes. Soft surfaces absorb sounds.
6. Sound detectors change sound wave energy into electrical signals.

Glossary

acoustics	ah-KOOS-tiks	sound quality
auditory	AW-di-tow-ree	dealing with sound and hearing
cochlea	KOK-lee-uh	the hearing organ of the inner ear
compression	cum-PRES-shun	a region of particles crowded together
lymph	LIMF	fluid of the inner ear
rarefaction	rare-a-FAK-shun	a region of widely separated particles
reverberation	ree-ver-buh-RAY-shun	a series of echoes in an enclosed space
trough	TROFF	the lowest point of a transverse wave
vacuum	VAK-yoom	the total absence of matter

Study Questions

A. True or False

Decide whether each of the following sentences is true or false. If the sentence is false, rewrite it to make it true. (Do not write in this book.)
1. Sound can travel in outer space.
2. Sound moves faster in water than in wood.
3. Sound waves are longitudinal.
4. Sound travels faster in warm air than in cold.
5. The eardrum is a thick membrane.

B. Completion

Complete each of the following sentences with a word or phrase which will make the sentence correct. (Do not write in this book.)
1. A longitudinal wave consists of a ▓▓▓ followed by a ▓▓▓.
2. The amplitude of a wave depends upon ▓▓▓.
3. ▓▓▓ reflect sound while ▓▓▓ absorb sound.
4. The frequency of sound waves depends upon ▓▓▓ the source ▓▓▓.
5. Shock waves form when a sound source moves ▓▓▓.

C. Multiple Choice

Each of the following statements or questions is followed by four responses. Choose the correct response in each case. (Do not write in this book.)
1. The speed of sound in water is
   a) faster than in copper
   b) slower than in iron
   c) the same speed as in glass
   d) slower than in air
2. A jet flies at Mach 3. Compared to the speed of sound in air, the plane is moving
   a) too fast to see
   b) 1/3 as fast
   c) too slow to break the sound barrier
   d) 3 times faster
3. Reverberations are caused when
   a) reflecting surfaces cause confusing echoes
   b) sound-proofing materials reflect sound
   c) a megaphone is used
   d) an echo is heard 0.1 s after the original sound
4. A speaker with a megaphone is heard more clearly because
   a) the sound energy of the source is increased
   b) the sound waves can spread out in all directions
   c) the sound waves are directed at the listeners
   d) the megaphone produces many echoes
5. Greater sound energy forms larger waves. Larger waves have
   a) a higher frequency
   b) a greater amplitude
   c) more compressions and rarefactions
   d) slower speed

D. Using Your Knowledge

1. An astronaut sees two small meteorites collide near his spaceship. Can he hear the crash? Explain.
2. The time official of a long distance race stands at the finish line. When should he begin timing? When he hears the starter's pistol? Or when he sees the pistol's flash? Explain.
3. Thunder in the mountains often sounds much louder than in the city. Why?
4. Suppose you are designing a quiet study room. What would you do to reduce sound reflection?

E. Investigations

1. Find out what kinds of materials are used for soundproofing. You may need to visit the local building supply centre. Examine some acoustical title. How does it absorb sound? Examine the library or the theatre arts room. How have the designers tried to reduce sound reflections? You may enjoy loud rock music. How could you soundproof your room?
2. Design an experiment to measure the speed of sound. How accurate is your measurement? Consult the library. Compare your experiment with the original efforts to measure the speed of sound.
3. Do some research on the latest jets. At what Mach Number do they fly? How high must they fly to avoid building damage from sonic booms? How does this high altitude flight affect the ozone layer in the upper atmosphere? Why are many scientists worried?
4. Consult the library to find out what causes deafness. Can modern surgery correct any forms of deafness? How does a hearing aid work? Can a hearing aid help all deaf people?

27 Characteristics of Sound

27.1 Sound Intensity
27.2 Activity: Factors Which Affect Pitch
27.3 Sound and Pitch
27.4 Sound Quality
27.5 Noise Pollution

Sound has three characteristics. They are intensity, pitch, and quality. When any one of these changes, the character (nature) of the sound changes. In this chapter you will find out how these three characteristics affect sound. You will also study noise pollution.

27.1 Sound Intensity

Sounds can be recognized by three different characteristics — loudness, pitch, and quality. **Sound intensity** *is the loudness of the sound*. It depends upon the vibrational energy of the source. Strong vibrations produce loud sounds. Weak vibrations produce faint sounds.

Measuring Sound Waves

Our ears are marvelous instruments. But they are not precise. They cannot accurately measure the loudness of sound. Before they can detect any change at all, the sound intensity must increase 26%. So scientists use an instrument called an **oscilloscope** [uh-SIL-o-skope]. It has a tube similar to a TV picture tube. A **microphone** detects sound waves. Their energy is changed into an electrical current. This produces waves on a graph on the oscilloscope screen (Fig. 27-1). This wave pattern is a "picture" of the sound waves. Now you can watch the sound waves and measure their characteristics.

Remember that the oscilloscope pattern represents longitudinal sound waves. The graph shows how the air molecules move back and forth to make compressions and rarefactions. The graph pattern is read from left to right. Where the line rises, the molecules are being

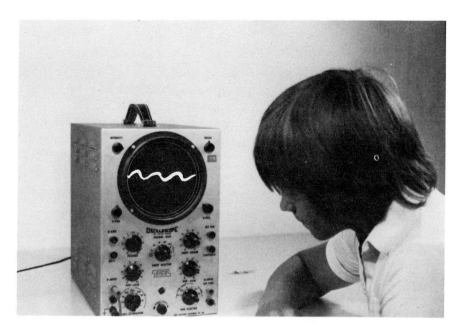

Fig. 27-1 This student is measuring sound intensity with an oscilloscope. He is "watching" sound waves on the screen.

pushed together by the vibrating object. This forms a compression. Where the line falls, the molecules are spreading out again. This forms a rarefaction. Each time the source vibrates, a sound wave forms. Each compression and rarefaction appears as one cycle on the graph pattern. The height or amplitude shows how hard the molecules are being pushed (Fig. 27-2). This energy determines the loudness or sound intensity.

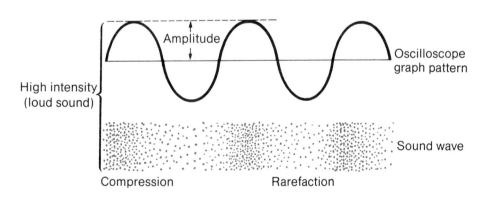

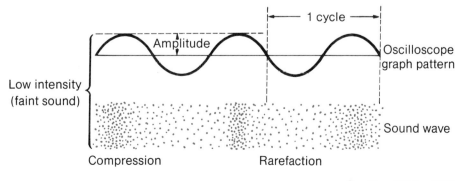

Fig. 27-2 Strong vibrations cause dense compressions. The sound is loud. Weaker vibrations do not compress the molecules as much. The sound is less intense. The amplitude of the graph indicates the sound intensity. How do the frequencies of these sound waves compare?

Section 27.1 325

Other Factors

The intensity of sound also depends upon the distance from the source. Sound is energy. It must be transmitted through a medium. As it passes from particle to particle, energy is lost. So loudness decreases as you move away from the source. If you move twice as far from a sound source, the loudness decreases by 75%. If you move four times as far, the loudness decreases by 84%.

Some types of matter transmit sound energy more easily than others. So the nature and temperature of the medium also affect loudness. For example, sounds are much louder underwater.

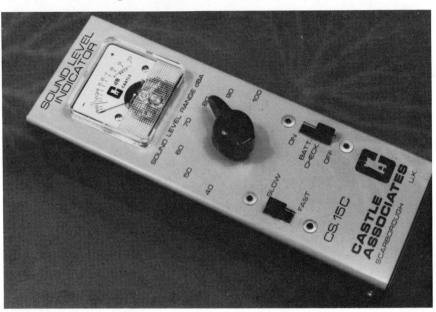

Fig. 27-3 This sound-level meter measures sound intensity in decibel units.

The Decibel Scale

You can also use a sound-level meter to measure loudness. The sound energy forms an electric current. This moves a pointer on the scale (Fig. 27-3). Louder sounds have greater energy. This moves the point farther. The meter scale measures in units called **decibels (dB)** [DES-i-bels].

When one sound is ten times louder than another, the difference in intensity is 1 **bel**. The unit honours Alexander Graham Bell. He invented the telephone and worked with the deaf. The bel is a large unit. It is divided into ten smaller units called decibels (dB). Each decibel is 0.1 bel.

The decibel scale starts at zero. This is about the faintest sound detected by the human ear. It is called the **threshold of hearing**. Hearing ability usually decreases with age. Thus, older people may have a higher threshold of hearing.

A sound level of 120 dB is the highest intensity the ear can endure without pain. This level is called the **threshold of pain**. Higher sound levels can injure your ears. A short exposure at 150 dB can cause hearing loss. Sound energy at 155 dB can burn your skin!

Remember, a sound of 20 dB is not twice as loud as one at 10 dB. A difference of 10 dB equals 1 bel. So the sound is ten times louder. 120 dB is one thousand million times more intense than the faintest sound! Table 27-1 shows how sound intensity increases on the decibel scale.

Table 27-1 Relative Intensities of Sounds

Sound level (dB)	Relative intensity	Example
0	1	Acute threshold of hearing — distant rustle of leaves
10	10	Average threshold of hearing — quiet whisper at 1 m
20	100	Average whisper at 1 m
30	1000	Watch ticking
40	10 000	Normal talking at 1 m
50	100 000	Interior of a quiet running car
60	1 000 000	Background music
70	10 000 000	Office with typewriters
80	100 000 000	Heavy traffic
—	—	Beginning of permanent hearing damage if prolonged
90	1 000 000 000	Power lawnmower
100	10 000 000 000	Accelerating motorcycle
120	100 000 000 000	Rock band Threshold of pain
130	1 000 000 000 000	Artillary fire
140	10 000 000 000 000	Jet taking off

Section Review

1. Describe three factors that affect sound intensity.
2. Name two instruments which can accurately measure sound intensity.
3. Which part of an oscilloscope pattern measures sound intensity?
4. What is a decibel?
5. What is the threshold of hearing?
6. What is the threshold of pain?

27.2   ACTIVITY   Factors Which Affect Pitch

A second characteristic of sound is pitch. *The* **pitch** *of a sound makes it high or low.* In this activity you will study the factors which affect pitch. You can also test your understanding of intensity.

Problem

What factors affect the pitch of a sound?

Materials

small wooden board
5 test tubes in a holder
a set of 4 guitar strings
8 eye screws
water
or a set of 4 elastic bands (2 thin bands of different length and 2 thick bands of different length)

Procedure A A Stringed Instrument

a. Make a stringed instrument as shown in Figure 27-4. The strings should all be tightened to the same degree. They must not touch the board. If they do, they will not vibrate properly.

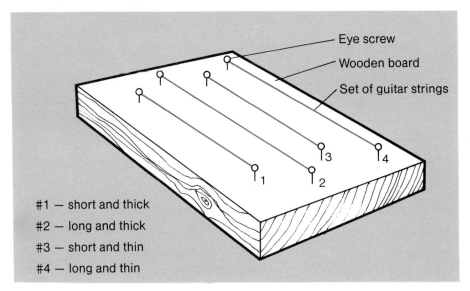

Fig. 27-4 Testing pitch in a stringed instrument. Stretch and number each string using the sequence above.

#1 — short and thick
#2 — long and thick
#3 — short and thin
#4 — long and thin

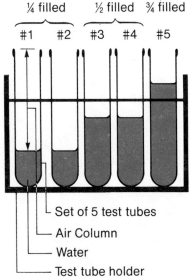

Fig. 27-5 Testing pitch in a wind instrument. Fill and number each test tube using the sequence above.

b. Test the effect of string length on pitch. Pluck #1 and #2. Compare pitch. Then pluck #3 and #4. Compare pitch. Record your observations.

c. Test the effect of string thickness. Pluck #1 and #3. Compare pitch. Then pluck #2 and #4. Compare pitch. Record your observations.

d. Test the effect of string tension. Continue to pluck #1 as you tighten it. How does the pitch change? Test each of the other strings in turn. Record your observations.

e. Test the effect of energy. Pluck #1 gently. Then pluck it harder. Does the pitch change? Test each of the other strings. Record your observations.

Procedure B A Wind Instrument

a. Make a wind instrument as shown in Figure 27-5. Blow across the top of each test tube to make a sound. Record your observations for each of the following steps.

b. Compare the pitch of #1 and #2.
c. Compare the pitch of #3 and #4.
d. Compare the pitch of #1, #3, and #5.
e. Add some water to #2. Compare #1 and #2.
f. Remove some water from #3. Compare #3 and #4.
g. Adjust the water level in #4 until #4 has the same pitch as #5.
h. Blow gently across #1. Now blow harder. Does the pitch change?

Discussion A

1. How does the length of string affect pitch?
2. How does string thickness affect pitch?
3. How does string tension affect pitch?
4. How does the energy of vibration affect pitch? How does the energy of vibration affect loudness?

Discussion B

1. How does the length of a column of air affect pitch?
2. What happens to the air column when you add water to a test tube? What happens to the pitch?
3. What happens to the air column when you remove water from a test tube? What happens to the pitch?
4. How do you obtain the same pitch from two different test tubes?
5. What happens to the pitch when you blow harder? What happens to the sound intensity? Why?

27.3 Sound and Pitch

*The **pitch** of a sound depends upon the rate of vibration (frequency) of the source.* Pitch describes how sound waves of different frequencies affect our ears. A high frequency produces a high pitched sound. A low frequency produces a low pitched sound.

Stringed Instruments

Middle C on the piano has a frequency of 256 Hz. High C has a frequency of 512 Hz. Its frequency is twice as great. These notes are one **octave** [OK-tiv] apart. You can change the pitch of a stringed instrument three different ways.

First, you can vary the string length. A long string vibrates slower. Its pitch is lower than that of a short string. Guitarists can increase the pitch of any string. They slide their finger up the neck of the guitar. This shortens the effective length of the string (Fig. 27-6).

Second, you can vary the string thickness. Thin strings vibrate faster. They have a higher pitch. A deep bass voice has a frequency of about 60 Hz. A high soprano voice has a frequency of about 1200 Hz. A man's vocal cords are one-third longer and thicker. A child's

Fig. 27-6 This guitarist can shorten the effective length of any string. The finger positions on the neck change the pitch of the strings.

voice is high pitched. The vocal cords are short and thin. When a child grows older, the vocal cords grow longer and thicker. Then the voice has a deeper pitch. When a person gets **laryngitis** [la-rin-JY-tis], the vocal cords thicken. They vibrate more slowly. The voice deepens.

Finally, you can vary the string tension. Tighter strings vibrate faster and have a higher pitch. You can hear the change of pitch when a guitar is tuned. The tuning key tightens or loosens the vibrating string. Muscles can change the tension in your vocal cords which alters the pitch of your voice.

Pitch is not affected by the energy of vibration. Pluck a string harder. The sound gets louder. But the pitch remains the same. The air molecules vibrate harder, but not faster.

Wind Instruments

The length of a vibrating string determines its pitch. The length of a vibrating air column determines the pitch of a wind instrument. Shorten the column. The pitch gets higher. Lengthen the column. The pitch gets lower. Instruments like the flute and clarinet use keys which change the effective air column length. This changes the pitch. Note that blowing harder changes the loudness but it does not change the pitch.

Measuring Pitch

Each time a sound source vibrates, a sound wave forms. Each compression and rarefaction appear as one cycle on an oscilloscope graph. As the number of vibrations per second increases, the frequency of the sound waves increases. The number of compressions and rarefactions increases. The number of cycles on the graph increases.

Compare the oscilloscope graphs of two sounds of different pitch (Fig. 27-7). The low pitch sound shows only a few cycles. The high pitch sound shows many cycles.

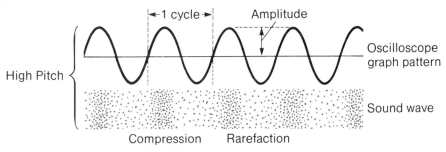

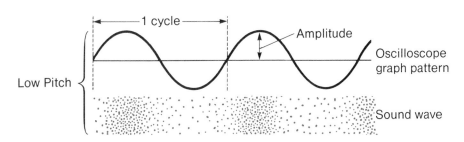

Fig. 27-7 Rapid vibrations produce sounds of high frequency or high pitch. The frequency is measured by the number of cycles occurring per second. The amplitude determines the loudness of the sound. Which sound is louder?

Section Review

1. What is pitch?
2. What factor does pitch depend upon?
3. State three ways of changing the pitch of a stringed instrument.
4. Why do men's voices have a lower pitch than women's voices?
5. Why does a child have a high pitched voice?
6. How do you change the pitch of a wind instrument?

27.4 Sound Quality

A rock band is performing. Some say it is noise not music. How do you distinguish between them? A note is played on the piano. Then a trumpet plays the same note. The loudness and pitch are the same. Yet you can tell them apart. And how do you recognize different voices? You use a characteristic of sound called **quality**.

Fundamentals and Overtones

Quality depends on how a source vibrates. Consider a guitar string. When the whole string vibrates, the tone heard is the **fundamental frequency** [fun-duh-MEN-tal]. But a string can also vibrate in parts (Fig. 27-8). Suppose the string vibrates in two shorter parts. Each part vibrates twice as fast as the whole string. The frequency doubles. The pitch is higher. This tone is called the **first overtone**. If the string vibrates in three shorter parts, the frequency triples. The pitch is even higher. This tone is the **second overtone**.

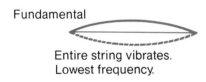

Third overtone
String vibrates in 4 parts.
Frequency quadruples.

Second overtone
String vibrates in 3 parts.
Frequency triples.

First overtone
String vibrates in 2 parts.
Frequency doubles.

Fundamental
Entire string vibrates.
Lowest frequency.

Fig. 27-8 A string vibrates in parts. It produces tones of higher pitch called overtones. The blending of the fundamental with its overtones creates the quality of a tone.

A string can vibrate as a whole and in parts at the same time. You hear a mixture of fundamental and different overtones. *The number and kinds of overtones determine the* quality *of a sound* (Fig. 27-9). Sometimes there are only a few weak, higher overtones. Then the note is mellow. Sometimes there are many strong higher overtones. Then the note is harsh and metallic. Wind instruments also produce overtones. A saxaphone can produce up to fifteen different overtones.

Your Voice

Your vocal cords are vibrating membranes. They produce overtones too. Your voice quality depends upon the blending of these overtones. The sound is reinforced by vibrating air in your mouth, throat, and nose cavities. This is why a head cold affects the sound of your voice.

Studying Sound Quality

Musical instruments vibrate in very complex ways. You can distinguish musical tones of the same intensity and pitch. The different sound qualities result from different overtones. Each different instrument produces a different oscilloscope pattern (Fig. 27-10). Notice the pattern of the tuning fork. It shows a pure tone. A tuning fork does not produce overtones.

Music is the regular repetition of vibrations. Noise is caused by irregular vibrations. Musical instruments produce oscilloscope graphs which are smooth and regular. Noises produce oscilloscope graphs which are jagged and irregular (Fig. 27-11). Test a recording of your favourite rock group. Is their sound music or noise?

Section Review

1. How is fundamental frequency produced?
2. How are overtones produced?
3. What determines the quality of a musical tone?
4. What determines the quality of your voice?
5. How does the production of noise and music differ?
6. How do the oscilloscope patterns of noise and music differ?

27.5 Noise Pollution

You can identify noise on an oscilloscope graph. But in fact, **noise** *is any unwanted sound*. And noise pollution is far more serious than you may think.

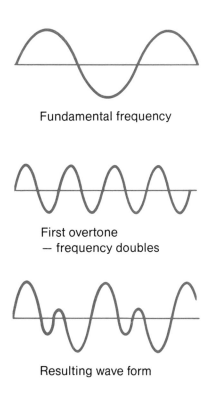

Fig. 27-9 A fundamental wave and its first overtone blend. The resulting wave has a richer tone quality.

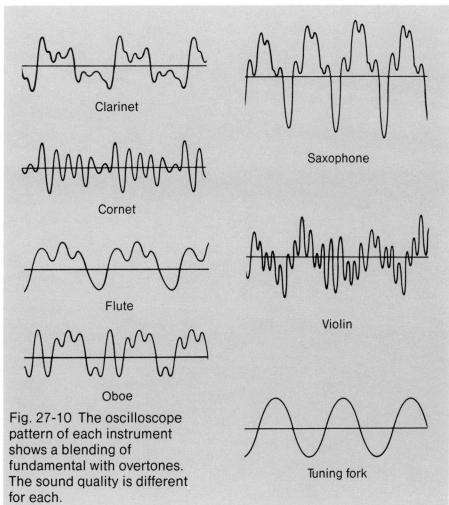

Fig. 27-10 The oscilloscope pattern of each instrument shows a blending of fundamental with overtones. The sound quality is different for each.

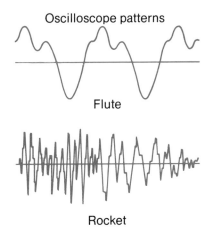

Fig. 27-11 Music versus noise. A musical tone produces a smooth, regular pattern. Noise produces an irregular, jagged pattern.

The Effects of Sound

Sound is energy. It can affect your body like a blow or a burn. Sound levels above 100 dB are deafening. Levels above 120 dB cause pain. 140 dB will rupture your eardrums. 165 dB can kill animals such as cats. 180 dB of sound energy will kill you! Figure 27-12 shows how lower sound levels affect you.

City noise levels are usually greater than 80 dB. This level is considered a threshold. Continuous noise above this level can injure your ears. The damage depends upon three things: the level of noise; how long you are exposed; and your hearing sensitivity.

Loud noises put stress on the nerve hairs of your inner ear. They stop working properly. They usually recover after rest in a quiet place, but not if they are overworked. Then, hearing loss is permanent.

Exposure Time

The louder the sound, the less time you should be exposed. Consider a farmer running a 92 dB tractor. A working day longer than 6 h may

Fig. 27-12 The effects of noise levels

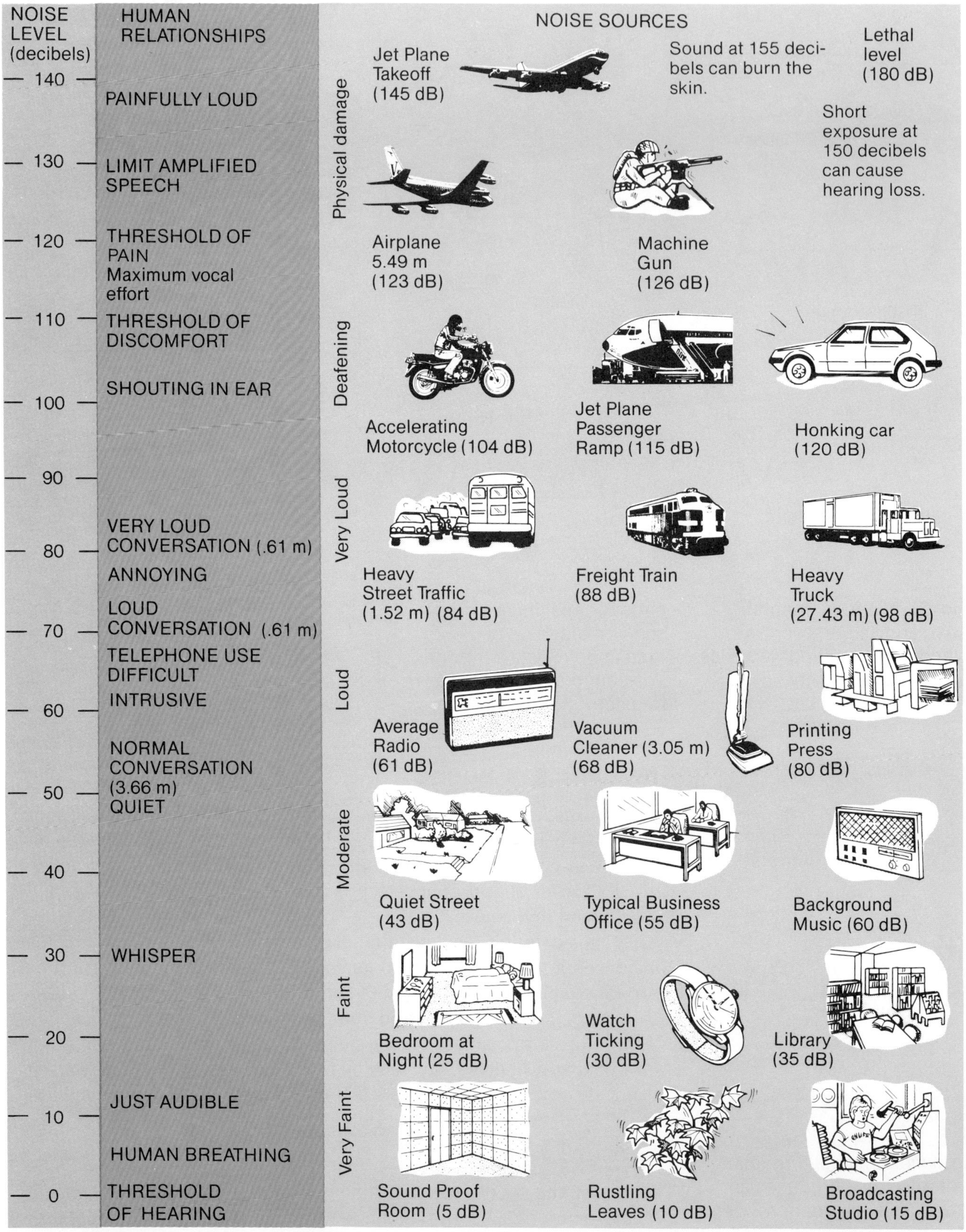

NOISE LEVEL (decibels)	HUMAN RELATIONSHIPS		NOISE SOURCES		
140	PAINFULLY LOUD	Physical damage	Jet Plane Takeoff (145 dB)	Sound at 155 decibels can burn the skin.	Lethal level (180 dB)
130	LIMIT AMPLIFIED SPEECH				Short exposure at 150 decibels can cause hearing loss.
120	THRESHOLD OF PAIN — Maximum vocal effort		Airplane 5.49 m (123 dB)	Machine Gun (126 dB)	
110	THRESHOLD OF DISCOMFORT	Deafening			
100	SHOUTING IN EAR		Accelerating Motorcycle (104 dB)	Jet Plane Passenger Ramp (115 dB)	Honking car (120 dB)
90		Very Loud			
80	VERY LOUD CONVERSATION (.61 m) — ANNOYING		Heavy Street Traffic (1.52 m) (84 dB)	Freight Train (88 dB)	Heavy Truck (27.43 m) (98 dB)
70	LOUD CONVERSATION (.61 m) — TELEPHONE USE DIFFICULT	Loud			
60	INTRUSIVE		Average Radio (61 dB)	Vacuum Cleaner (3.05 m) (68 dB)	Printing Press (80 dB)
50	NORMAL CONVERSATION (3.66 m) — QUIET	Moderate			
40			Quiet Street (43 dB)	Typical Business Office (55 dB)	Background Music (60 dB)
30	WHISPER	Faint			
20			Bedroom at Night (25 dB)	Watch Ticking (30 dB)	Library (35 dB)
10	JUST AUDIBLE — HUMAN BREATHING	Very Faint			
0	THRESHOLD OF HEARING		Sound Proof Room (5 dB)	Rustling Leaves (10 dB)	Broadcasting Studio (15 dB)

cause hearing loss. Firearms can produce noise of 130 dB or more. Many soldiers in training damage their hearing. People who work constantly around loud noises often lose their hearing. People who enjoy very loud music also damage their ears. Permanent hearing loss may take years or only months. By the time you notice a problem, the damage is done. Your ears may be exposed to loud noises. How often during the day does this happen?

Other Effects

Loud noise affects you in other ways too. At levels above 70 dB your blood pressure and pulse rate increase. Your nervous tension increases. You become angry and annoyed more easily. Your concentration is poor. You are not even safe when you sleep! Your nervous system still responds. Noise through the night can make people more prone to heart problems.

Noise and Concentration

Any sound which bothers your activity is noise. Noise affects some kinds of activities more than others. The level of noise is also important.

Noise levels above 70 dB make conversation difficult. Workers who must talk with one another often tend to become annoyed. Physical tasks do not seem to be affected as much as mental tasks.

Many people enjoy background music when they study or solve problems. But remember that noise levels above 70 dB affect your blood pressure and concentration. At noise levels between 90 dB and 110 dB you must concentrate much harder. And you cannot concentrate as long. Your errors increase. So the next time you sit down to study or work, consider the noise level. The music you enjoy could be making study more difficult.

Reducing Noise Pollution

There are many ways of reducing noise. In busy traffic areas houses can be built with solid walls facing the street. Barrier walls are used to deflect sound. Double windows in buildings help deaden street noise. Sound insulation is used in the walls between many apartments. Cars use mufflers. Noisy machinery can be isolated or placed in a soundproof area. Machines can be designed to run more quietly.

Where noise cannot be reduced, people should wear ear protectors. Ground controllers guide jets to loading platforms. The noise level exceeds 130 dB, yet their ear protectors shut out almost all of the sound.

Preventing noise is surely easier than trying to control it. There are many laws involving noise pollution. But they are difficult to enforce. Simple consideration of others would solve many of our noise problems. Turn our stereos down. Make sure cars have proper mufflers. Don't squeal car tires. There is much we can do.

Section Review

1. How does music affect study?
2. 80 dB is considered to be an important threshold. Why?
3. What factors affect damage to hearing?
4. Describe how loud noise affects you.
5. List five different ways of reducing noise.

Main Ideas

1. Oscilloscope patterns can be used to measure three characteristics of sound: intensity, pitch, and quality (Fig. 27-13).

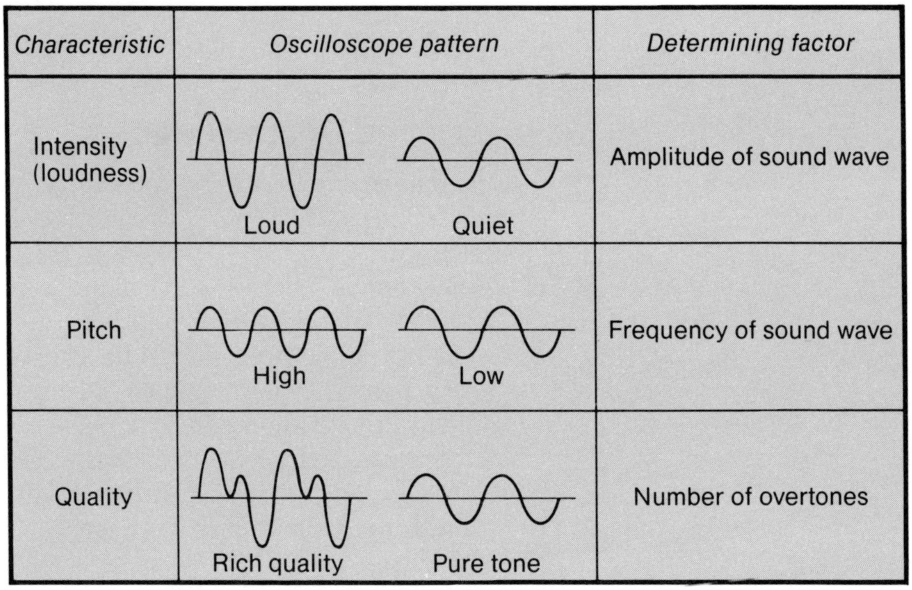

Characteristic	Oscilloscope pattern		Determining factor
Intensity (loudness)	Loud	Quiet	Amplitude of sound wave
Pitch	High	Low	Frequency of sound wave
Quality	Rich quality	Pure tone	Number of overtones

Fig. 27-13 Oscilloscope patterns can show the three characteristics of sound — intensity, pitch, and quality.

2. The intensity of a sound depends upon the vibrational energy of the source. It is measured by the amplitude of the wave pattern.
3. The pitch of a sound depends upon the frequency of vibration.
4. The quality of a sound depends upon the number of overtones blending with the fundamental.
5. Noise pollution seriously affects hearing and general health.

Glossary

decibel	DES-i-bel	the unit used to measure sound intensity
fundamental	fun-duh-MEN-tal	the tone frequency produced when the whole source vibrates
laryngitis	la-rin-JY-tis	a thickening of the vocal cords

microphone		an instrument used to detect sound waves
octave	OK-tiv	a separation of 2 overtones
oscilloscope	uh-SIL-o-skope	an instrument used to display sound waves

Study Questions

A. True or False

Decide whether each of the following sentences is true or false. If the sentence is false, rewrite it to make it true. (Do not write in this book.)

1. Sound intensity depends upon vibrational energy.
2. Middle C sung by a woman has a different pitch from Middle C sung by a man.
3. Sound levels above 80 dB can damage hearing.
4. Noise produces a smooth oscilloscope pattern.
5. Overtones have the same frequency as the fundamental.

B. Completion

Complete each of the following sentences with a word or phrase which will make the sentence correct. (Do not write in this book.)

1. The tone produced by a string vibrating in one part is called the ▓▓▓▓▓ .
2. The C just above middle C on a piano has ▓▓▓▓▓ the frequency of middle C.
3. Children have high pitched voices because their vocal cords are ▓▓▓▓▓ and ▓▓▓▓▓ .
4. As you move farther away from a sound source, the intensity ▓▓▓▓▓ .
5. A noise level of 140 dB can break your ▓▓▓▓▓ .

C. Multiple Choice

Each of the following statements or questions is followed by four responses. Choose the correct response in each case. (Do not write in this book.)

1. Noise is produced
    a) when matter vibrates slowly
    b) by regular vibrations
    c) by irregular vibrations
    d) when matter vibrates strongly
2. The pitch of a sound depends upon the
    a) overtones
    b) frequency
    c) amplitude
    d) energy of vibration

3. If you shorten a vibrating string, the pitch of the sound produced
   a) increases
   b) decreases
   c) gets louder
   d) does not change
4. If you tighten a vibrating string, the pitch of the sound produced
   a) increases
   b) decreases
   c) gets louder
   d) does not change
5. Decibels measure
   a) pitch
   b) frequency
   c) quality
   d) loudness

D. Using Your Knowledge

1. Why is an oscilloscope used to measure sound characteristics? Why is an oscilloscope better than your ear?
2. Describe the changes you would hear in the sound if the oscilloscope patterns changed as shown in Figure 27-14.

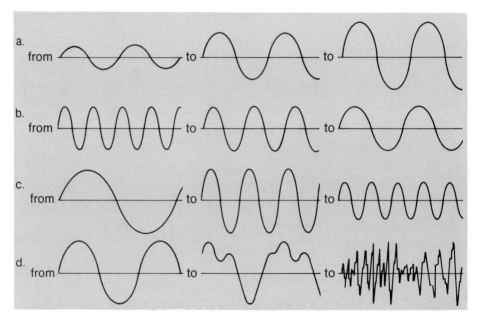

Fig. 27-14 Changes in oscilloscope patterns

3. Long exposure to a noise level of 90 dB could be as damaging as a short exposure to 120 dB. Why?
4. How do you tune a guitar? What factors are you changing? Why?

E. Investigations

1. Test your voice quality. Set up an oscilloscope to measure sound waves. Compare the graph patterns formed by different voices. Can you identify a speaker by the pattern? Try singing in unison. Then try harmony. How does the quality of your singing voice compare with your spoken voice?
2. Test several musical instruments with an oscilloscope. Does the skill of the musician affect the sound quality?
3. Test your hearing sensitivity. Set up an audio-oscillator with an amplifier and speaker. A good quality speaker should produce constant loudness. Listen to the sound. Vary the frequency from 0

to 20 kHz. Can you detect sound through the entire frequency range? Is there a range of frequencies you hear more clearly?
4. Obtain a sound-level meter. Test different areas of your school. Which areas have the highest noise levels? Which areas are best for quiet study? Is your school cafeteria a relaxing place to eat lunch?
5. Most people can tolerate a noise level of 70 dB. This is the average sound level of a busy street. Higher noise levels cause irritation to residents. Select the noisiest neighbourhoods in your town or city. Measure the noise levels with a sound meter. What steps have been taken to reduce sound effects? Suggest further improvements.

Heat Energy

Lightning flashes across the sky. It traces a jagged path from the clouds to the earth. There it strikes a tinder dry forest. The forest bursts into flames. Soon many square kilometres of forest are destroyed. This is an example of energy out of control.

In this unit you will study heat energy. You will first review how heat travels. You will then learn some of the effects it has on matter. Heat increases the temperature of a substance. It also causes melting and vapourization, two of the changes of state. You will learn the differences between heat energy and temperature and how to measure both. This unit will give you a better understanding of heat energy. Hopefully it will enable you to control heat energy and use it wisely.

CHAPTER 28
Heat Transfer, Expansion, and a Theory

CHAPTER 29
Quantity of Heat

CHAPTER 30
Heat and Changes of State

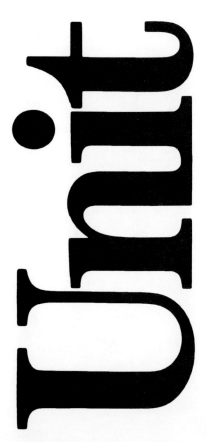

Fig. 28-0 Heat energy started this fire. The fire will produce still more heat energy.

28 Heat Transfer, Expansion, and a Theory

28.1 How Heat Travels in Solids, Liquids, and Gases
28.2 Activity: Comparing Insulating Materials
28.3 Activity: The Transfer of Radiant Energy
28.4 Thermal Expansion and Measuring Temperature
28.5 The Kinetic Molecular Theory of Heat

Why is the surface of a lake warmer than the bottom in summer? Why is a flat-plate solar collector the colour it is? What are the explanations for how heat travels and its effects on matter? How is temperature measured? You will find answers to these and other questions in this chapter.

28.1 How Heat Travels in Solids, Liquids, and Gases

What is Heat?

Heat is a form of energy that we can feel. We know that heat is energy because it can do work. Heat can exert a force. It can make things move. Heat water in a pan on the stove. Watch it bubble and boil. Notice how the water moves vigourously about.

There are many other forms of energy. Some of these are electrical, chemical, and nuclear energy. All these forms of energy can be converted to heat. In fact, the end form of all forms of energy is waste heat. For example, chemical energy is stored in natural gas. This gas is burned in furnaces to heat homes. But all this heat eventually ends up in the environment. It is spread very widely throughout the environment. Thus it cannot be collected and used economically. That is why there is so much concern about wasting fossil fuels.

Conduction

Heat moves through solids by **conduction** [kon-DUK-shun]. Figure

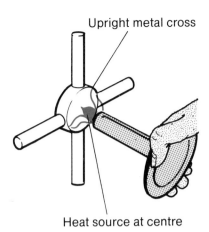

Fig. 28-1 Heat travels by conduction through solids.

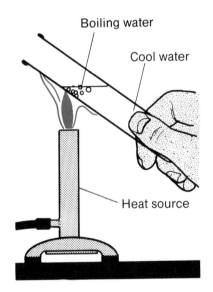

Fig. 28-2 Water is a poor conductor of heat.

28-1 shows an upright metal cross being heated at the centre. Heat moves outward from the centre along the arms. Heat moves from the hot part to the cool parts. The heat moves but the solid doesn't move along with it. The heat moves up, down, and sideways. Heat moves by conduction in all directions. A material which permits heat to move by conduction is called a **conductor** [kon-DUK-tor]. Different materials have different abilities to conduct heat. Copper and silver are two of the best conductors. Aluminum is also a good conductor.

Some materials restrict the flow of heat. These are called **insulators** [IN-suhl-ators]. Cork, glass, and most plastics are insulators. Water is also a poor conductor. To show this hold a test tube in your hand by the base. Hold it at an angle of 45°. Heat it with a hot flame near the surface as shown in Figure 28-2. If water is a poor conductor what do you expect to happen?

Convection

Water heated near the surface of an inclined test tube boils at the surface. The water at the bottom remains cool. If the test tube is heated at the bottom all the water becomes hot. This is because heat travels in a liquid or gas by convection. Liquids and gases are fluids, *A fluid is anything that flows.* **Convection** [kon-VEK-shun] *is defined as the transfer of heat by the movement of a fluid.* Both hot liquids and hot gases rise. In so doing they transfer heat upward. Heat is not transferred in all directions as with conduction. When a layer of hot water is below cold water it rises. As the cold water sinks, it pushes the hot water upward. This action continues as long as the water at the bottom is warmer. Convection does not occur if the bottom layer is colder. That is why the surface of a lake stays warmer than the bottom in the summer.

A hot fluid rises because it is lighter than the same volume of cool fluid. In order for this to happen it must expand when heated. This makes it less dense than the surrounding fluid. The less dense fluid rises for the same reason that a hot air balloon rises. Thus hot water rises and cold water sinks. Hot air rises and cold air sinks.

Section Review

1. How do we know that heat is a form of energy?
2. Name five sources of heat energy.
3. What is the end form of all forms of energy?
4. Define conduction.
5. Describe and give an example of a good conductor.
6. Define and give an example of an insulator.
7. Define convection.
8. What are two differences between convection and conduction?
9. What conditions are needed before convection can take place?

28.2 ACTIVITY Comparing Insulating Materials

Insulating materials differ in their ability to limit heat transfer. Manufacturers quote this as the R value. In this activity you will compare the insulating abilities of different materials.

Problem

Which is the best insulator: cellulose fibre, fibreglass, styrofoam, or a large air space?

Materials

4 identical large tin cans
4 identical small tin cans
4 cardboard covers
4 thermometers
hot water source
corks or rubber stoppers
cellulose fibre insulation
fibreglass insulation
styrofoam insulation

CAUTION: Hot water is dangerous. Fibreglass strands can be painful. Handle them carefully.

Procedure

a. Copy Table 28-1 into your notebook.
b. Support one small can inside the larger on three small corks about 2 cm high. This leaves an air space under and around the can as shown in Figure 28-3.

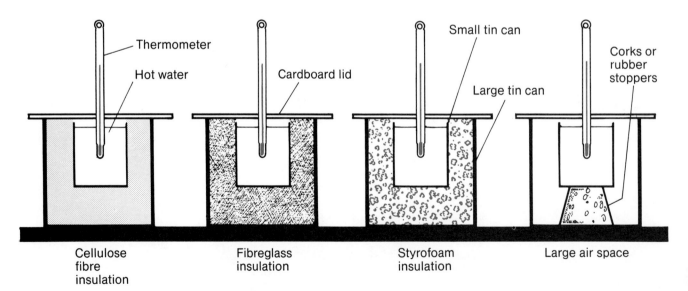

Fig. 28-3 Comparing insulating materials

c. Pack 2 cm of cellulose fibre in the bottom of one can. Put fibreglass in another and styrofoam in the third. Support three small cans on this insulation.

d. Pack cellulose fibre around the can supported on cellulose fibre. Pack fibreglass around the second and styrofoam around the third.
e. Pour hot water into each small can. Fill each can to the same depth.
f. Place a cardboard cover over each set of cans. Make a hole in the centre of each cover.
g. Insert a thermometer and record the initial temperature of the water.
h. Read the temperatures after 10 min and again after 20 min.

Table 28-1 Insulating Materials

Insulating material	R-Value per/cm	Temperature (°C)		
		At the start	After 10 min	After 20 min
Large air space	0.7			
Cellulose fibre	1.5			
Fibreglass	1.2			
Styrofoam	1.8			

Discussion

1. Why were metal cans used rather than glass beakers?
2. Which insulating material limits heat transfer the best?
3. Arrange the materials in order of their insulating abilities from best to worst.
4. How does this order compare to the order of the R values?

28.3   ACTIVITY   The Transfer of Radiant Energy

Radiation is the third method of heat transfer. **Radiation** [rad-ee-AY-shun] *is the process by which radiant energy travels.* Sunlight is one form of radiant energy. A radio wave is another. Radiation is the only way energy can travel through empty space. However, radiant energy can also travel through air.

Problem

To study the travel and absorption of radiant energy.

Materials

2 clothes pins   timer
incandescent light bulb (100 W)   matches   candle
small meat pie tin with both surfaces shiny, light
small meat pie tin with inside surface dull black and outside surface shiny, light

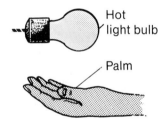

A.

Hot light bulb

Palm

Procedure

a. Set up the bulb so that it points sideways as shown in Figure 28-4,A. Turn on the light bulb.
b. Point your palm toward the bulb and about 2 cm from it. Hold your hand below, beside, and above the bulb. Determine if heat reaches your hand in each location.
c. Set the bulb so that it points down.
d. Use a candle to place a blob of wax on the two meat pie tins. Place the wax in the centre of the outside shiny surface.
e. Grasp each tin with a clothes pin. Bring the inner surface of each tin close to the light bulb (Fig. 28-4,B). Do not let the tins touch the bulb.
f. Measure the time it takes for the wax to start to melt on each tin.

Discussion

1. Describe the direction of heat travel by radiation.
2. **a)** How were the meat pie tins the same?
   b) How were the meat pie tins different?
3. In which tin did the wax melt first?
4. Which surface has more ability to absorb radiant energy?

28.4 Thermal Expansion and Measuring Temperature

Thermal Expansion

You studied the effect of heat on the volume of water in Unit 3. Most solids and liquids and all gases expand when heated. They contract when cooled. Different solids expand different amounts for the same temperature change. For example, pyrex expands less than ordinary glass. Therefore it doesn't break when heated quickly. Different liquids also expand different amounts (Fig. 28-5). Water above 20°C expands about three times as much as mercury does. Ethyl alcohol expands about six times as much as mercury.

Based on the behaviour of solids and liquids, one might predict that different gases also expand different amounts. But, in fact, most gases expand just about the same amount (Fig. 28-6). This example points out the importance of gathering evidence to test predictions. Never jump to conclusions! In science, experiments are done to collect data. The data are carefully analyzed before conclusions are reached.

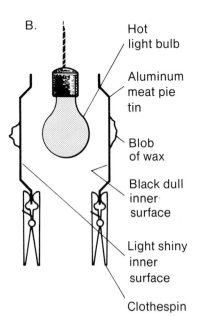

B.

Hot light bulb

Aluminum meat pie tin

Blob of wax

Black dull inner surface

Light shiny inner surface

Clothespin

Fig. 28-4 The transfer of radiant energy

Measuring Temperature

Temperature is measured using a thermometer. The expansion of

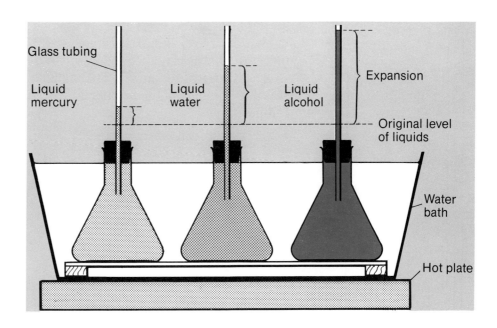

Fig. 28-5 Different liquids expand different amounts when heated.

solids, liquids, and gases is used in the design of thermometers. There are several types of thermometers.

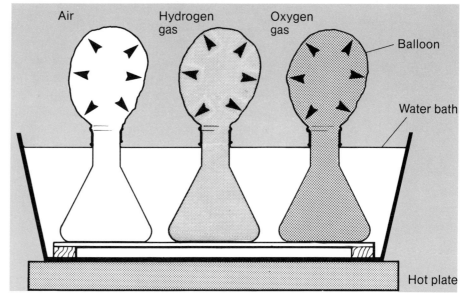

Fig. 28-6 Different gases expand the same amount when heated.

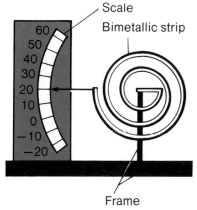

Fig. 28-7 A dial thermometer uses unequal expansion of different metals.

The **dial type thermometer** is based on the unequal expansion of different metals (Fig. 28-7). It consists of two different metals with their sides attached. This is called a bimetallic strip. The strip is wound in the form of a coil. One end is firmly attached to a frame. The other end is free to move. It takes a specific position for each temperature. Metal thermometers can withstand high temperatures. Some oven thermometers are of this type.

A **liquid thermometer** also depends on expansion. It has a thin glass bulb which is filled with a liquid (Fig. 28-8). The liquid expands

Section 28.4

up the bore of the glass tube. A scale is fixed to the tube or mounted behind it. Most liquid thermometers contain coloured alcohol or mercury. Water is not a good liquid to use. To see why, examine Table 28-2. Water freezes at 0°C. It changes to a gas at 100°C. Mercury stays a liquid to high temperatures. Alcohol stays a liquid to low temperatures.

Table 28-2 Freezing and Boiling Points of Some Liquids

Liquid	Freezing point (°C)	Boiling point (°C)
Water	0	100
Alcohol (ethyl)	−118	78
Mercury	−39	320

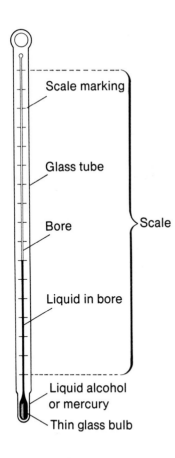

Fig. 28-8 The parts of a liquid thermometer

The Celsius Temperature Scale

To have a common temperature scale, thermometers need two reference points. These are called **fixed points**. The freezing point of water is one fixed point. The boiling point of water is the other. Once fixed points are found, the interval between them is marked off into equal sized divisions.

The Celsius scale was determined by setting the freezing point of water at 0°C. The normal boiling point was set at 100°C. The interval between was divided into 100 equal intervals. These are known as **degrees Celsius**. The Celsius scale extends below 0°C using negative numbers.

The symbol °C is used to denote a temperature interval as well as a specific temperature. Thus body temperature is quoted as 37°C. A temperature rise of thirty-seven degrees Celsius is also written as 37°C. Figure 28-9 shows the Celsius scale and indicates some common temperatures.

Section Review

1. How does heat affect the volume of solids, liquids and gases?
2. Compare the expansion of **a)** different solids, **b)** different liquids, **c)** different gases.
3. **a)** Draw and label a diagram of a dial type thermometer.
   b) Describe how it operates to measure temperature.
4. Draw and label a diagram of a liquid thermometer.
5. Which liquid, mercury or alcohol, is more suitable for measuring a large temperature range? Why?
6. What are the two fixed points on the Celsius temperature scale?
7. What two things does °C stand for?

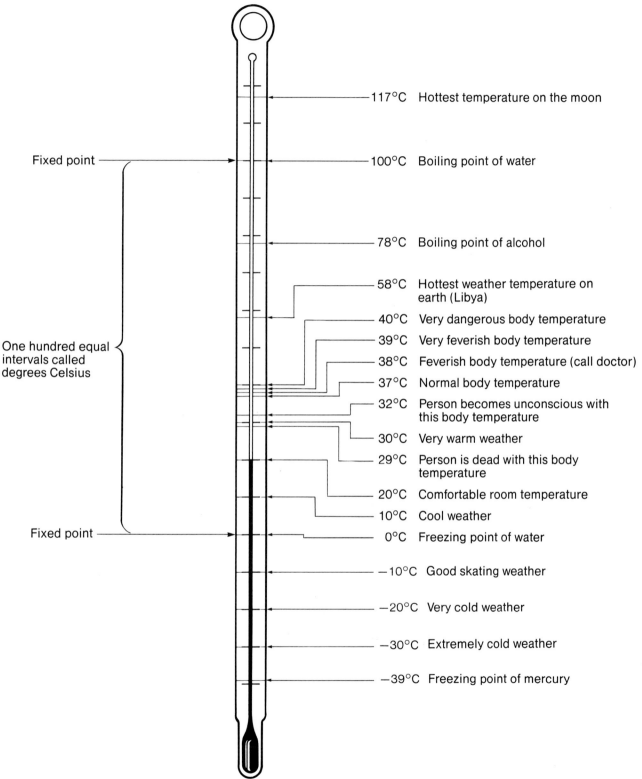

Fig. 28-9 Some common temperatures on the Celsius scale

28.5 The Kinetic Molecular Theory of Heat

What is a Theory?

Scientists make up theories to explain observations. A theory explains both everyday observations and the results of experiments. If a theory explains all the observations it is considered a good theory. If it cannot explain all the observations, it must be changed or discarded. A theory is tested by predicting the results of new experiments.

You studied the particle theory of matter in Chapter 11. You learned that all matter is made up of particles called molecules. The distance between molecules is large compared to their size. The molecules are held together by electric forces. The molecules attract one another at certain distances. If they are closer than this, they repel one another. Figure 28-10 shows an array of molecules making up a solid. The electric forces between molecules are drawn as elastic springs. Such springs do not exist. But they provide a simple and useful model. The coils of springs attract when stretched. They repel when pushed together. Molecules behave the same way.

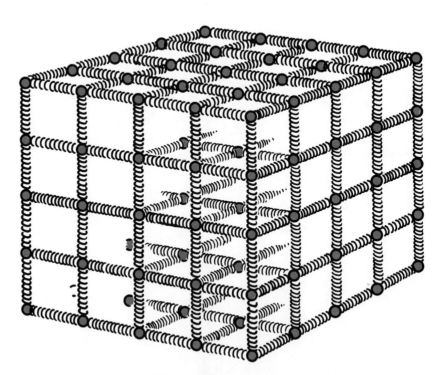

Fig. 28-10 A model of a cubic array of molecules making up a solid. The springs represent electric forces.

Kinetic Molecular Theory

The presently accepted theory of heat is called the **Kinetic Molecular Theory**. It is based on the particle theory. It says that heat is a form of energy. This energy is caused by the motion of the molecules. The molecules of matter are always moving. In Chapter 26 you learned that the energy of motion is called kinetic energy.

The Kinetic Molecular Theory can explain the source of heat. When two surfaces are rubbed together, the molecules move faster. This is sensed as heat. When a hammer pounds a nail, the molecules of the nail move faster. Suppose a Bunsen burner heats an iron rod. The molecules leaving the heat source move very fast. They collide with the molecules of iron. The iron molecules move faster. The rod heats up. The Kinetic Molecular Theory explains expansion and contraction. When matter is heated, the molecules move faster. They jostle one another and move farther apart causing the matter to expand. During cooling, the molecules move slower. The slower molecules are attracted closer together and the matter contracts.

The theory explains conduction and convection. In solids the molecules are fixed (Fig. 28-11). They cannot move from place to place. Therefore, convection does not occur. But they can vibrate back and forth. Hot molecules vibrate faster than cold ones. They cause adjacent molecules to vibrate faster. This is how heat is transferred from molecule to molecule in a solid. This is how conduction takes place.

In liquids the vibrating molecules are further apart. There is very little conduction. Liquid molecules are not fixed, so they can move from place to place. This is shown in Figure 28-12. The molecules of a hot liquid are further apart than those of a cold liquid. As a result the hot liquid is less dense and rises. This is how convection takes place.

In gases the molecules are very far apart (Fig. 28-13). Conduction does not occur. The molecules are moving very fast. Convection takes place quickly. However, if the path for the travel of the molecules is limited, convection decreases. This explains why materials such as styrofoam, which contain small air pockets, are good insulators.

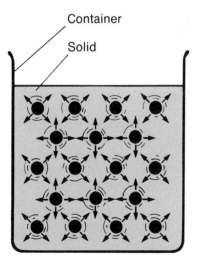

Fig. 28-11 In solids the molecules do not move from place to place. They vibrate back and forth.

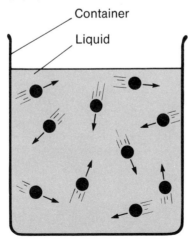

Fig. 28-12 In liquids the molecules are further apart and move from place to place.

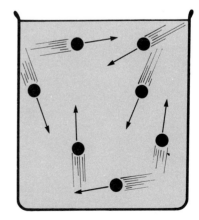

Fig. 28-13 In gases the molecules are very far apart and move very fast. They also fill the entire container.

Section Review

1. Summarize the Kinetic Molecular Theory of Heat.
2. Use the theory to explain expansion in liquids.
3. Use the theory to explain why conduction occurs in solids but not in gases.
4. Use the theory to explain why convection occurs in liquids but not in solids.

Main Ideas

1. Heat is a form of energy.
2. Heat travels in solids by conduction.
3. Heat travels in liquids and gases (fluids) by convection.
4. Heat travels in all directions in space by radiation.
5. Black, dull surfaces are the best absorbers of radiant energy.
6. All solids, liquids, and gases expand when heated.
7. Thermometers make use of expansion to measure temperature.
8. A theory is a scientist's explanation for a set of observations.
9. The particle theory says that matter is made up of particles.
10. The Kinetic Molecular Theory says that heat is caused by the motion of the particles.

Glossary

conduction	kon-DUK-shun	movement of heat energy through matter by molecules bumping into each other
conductor	kon-DUK-tor	a substance that transfers heat readily by conduction
convection	kon-VEK-shun	movement of heat energy through liquids and gases by the movement of the heated molecules
insulator	IN-suhl-ator	a substance that restricts the flow of heat by conduction
radiation	rad-ee-AY-shun	the movement of heat energy through empty space as waves

Study Questions

A. True or False

Decide whether each of the following sentences is true or false. If the sentence is false, rewrite it to make it true. (Do not write in this book.)

1. Heat is a fluid.
2. Heat travels through gases by conduction.
3. Radiant energy travels in all directions.
4. Solids, liquids, and gases contract when heated.
5. A theory is a scientist's observations of natural phenomena.

B. Completion

Complete each of the following sentences with a word or phrase which will make the sentence correct. (Do not write in this book.)
1. Heat is one form of ▓▓▓▓ .
2. Heat travels through metals by ▓▓▓▓ .
3. Heat travels through space by ▓▓▓▓ .
4. Energy from the sun heats up ▓▓▓▓ surfaces better than ▓▓▓▓ surfaces.
5. The presently accepted theory of heat is the ▓▓▓▓ Theory.

C. Multiple Choice

Each of the following statements or questions is followed by four responses. Choose the correct response in each case. (Do not write in this book.)
1. Which of the following is the end form of all forms of energy?
   - **a)** chemical
   - **b)** electrical
   - **c)** heat
   - **d)** nuclear
2. Heat is distributed through a liquid mainly by
   - **a)** conduction
   - **b)** convection
   - **c)** expansion
   - **d)** radiation
3. Where should ice be placed in a picnic cooler for the best transfer of heat?
   - **a)** at the top
   - **b)** at the bottom
   - **c)** at the sides
   - **d)** it doesn't matter
4. Which of the following is the best surface for absorbing radiant energy?
   - **a)** black, dull
   - **b)** black, shiny
   - **c)** light, dull
   - **d)** light, shiny
5. The Kinetic Molecular Theory says that the molecules of a solid
   - **a)** cannot move
   - **b)** move about slower than a gas
   - **c)** move about slower than a liquid
   - **d)** vibrate back and forth about one spot

D. Using Your Knowledge

1. Exposed hot water pipes are sometimes wrapped with fibreglass insulation. But fibreglass is a better conductor than air. Why wrap the pipes?
2. Why does a metal rod inserted into a roast make it cook faster?
3. **a)** Explain the role convection plays in the operation of a glider plane.

    b) Should the pilot fly in the morning or afternoon, or does it matter? Explain.
4. A thick glass is more likely to crack than a thin one when both are filled with hot water. Why?
5. Use the Kinetic Molecular Theory to explain why different liquids expand different amounts for the same temperature change.

E. Investigations

1. What effect might compacting fibreglass have on its insulating ability? Why? Design and do an experiment to test your prediction. Make sure you control the variables.
2. Kitchen aluminum foil has one side shiny and one side dull. Make an hypothesis about the cooking time and the way of wrapping baked potatoes. Design and do an experiment to test your prediction.
3. Predict whether a light, shiny or a dark, dull surface will give off heat better. Design and do an experiment to test your prediction.
4. Design an experiment to test the hypothesis "A rubber band expands when heated and contracts when cooled." Check the experimental design with your teacher.

29 Quantity of Heat

29.1 Temperature and Heat
29.2 Activity: Heat and Different Substances
29.3 Measuring Quantities of Heat
29.4 Activity: Heat Exchange in Mixtures
29.5 Activity: Measuring the Specific Heat Capacity of a Metal

Adding heat raises the temperature of an object. Removing heat lowers the temperature. How can we tell how much the temperature will rise? Does it make any difference what mass of material is heated? Does the material being heated make any difference? If two liquids are mixed, how is the heat transferred? You find answers to these and other questions in this chapter.

29.1 Temperature and Heat

Temperature

Temperature is a measure of the hotness or coldness of an object. Temperature determines the direction of heat flow. Heat flows from an object at a higher temperature to an object at a lower temperature. The molecules of the hotter object move faster than the molecules of the cooler object. The molecules collide. The slower molecules speed up. The faster molecules slow down (Fig. 29-1). Heat ceases to flow when the temperatures are the same. At this temperature the molecules have the same average kinetic energy.

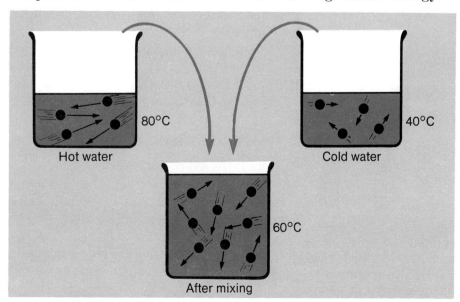

Fig. 29-1 In a hot liquid the molecules move faster than in a cold liquid. When mixed, the molecules all move at about the same speed.

Temperature is a measure of the level of heat. But temperature does not measure the quantity of heat. The quantity of heat is the amount of energy that flows from a hot object to a cooler object. A cup and a kettle of boiling water have the same temperature. Thus the water molecules in both have the same average kinetic energy. But there is more heat in the kettle than in the cup (Fig. 29-2). This is because there are more molecules in the kettle.

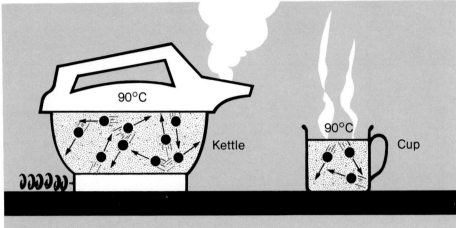

Fig. 29-2 There is more heat in a kettle of hot water than in a cup when both are at the same temperature.

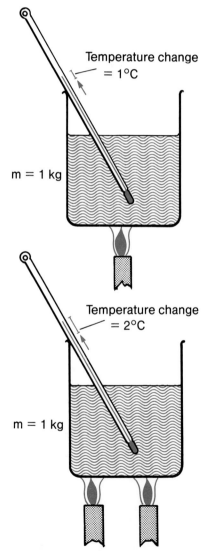

Fig. 29-3 It takes twice as much heat to produce twice as large a temperature change with the same mass.

How Heat Depends on Mass and Temperature Change

Heat is a form of energy. The unit of heat is the joule. There is no instrument for measuring the quantity of heat directly. Instead we measure the factors which affect the quantity of heat transferred. For example, we can compare quantities of heat by the temperature changes they cause. Suppose we heat 1 kg of water until its temperature rises 1°C. Then we heat the same mass of water until its temperature rises 2°C (Fig. 29-3). The quantity of heat to produce twice the temperature change is twice as large.

We can also compare quantities of heat by warming different masses through the same temperature change. Suppose we heat 1 kg of water until its temperature rises 1°C. Then we heat 2 kg of water until the temperature rises 1°C (Fig. 29-4). The heat to produce the same temperature with twice the mass is twice as large.

Two factors which must be considered to determine the quantity of heat transferred are mass and temperature change.

Section Review

1. Describe temperature.
2. Distinguish between heat and temperature.
3. Use an example to show that temperature does not measure the quantity of heat in an object.
4. In what direction does heat flow?
5. Compare the motion of the molecules of hot and cold objects.
6. How is heat transferred between different objects at different temperatures?

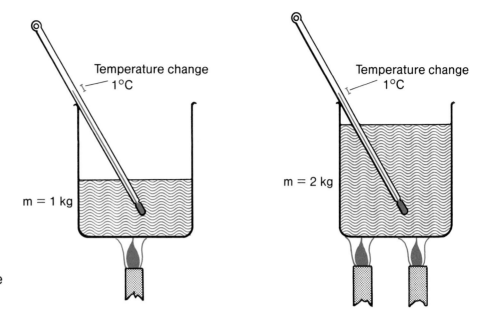

Fig. 29-4 It takes twice as much heat to heat double the mass through the same temperature change.

7. How does the mass of an object affect the quantity of heat needed to raise its temperature?
8. How does a temperature change affect the quantity of heat?

29.2   ACTIVITY   Heat and Different Substances

In this activity you will heat equal masses of two different substances. You will produce the same temperature change in water and ethylene glycol. Will the heat required be the same or different?

Problem

Does it take the same amount of heat to raise the temperature of water and ethylene glycol the same amount?

Materials

2 insulated cups   thermometer   100 mL graduated cylinder
immersion heater   stirring rod   timer

NOTE: An electric immersion heater is used to add a known quantity of heat. The heater has a known power — say 30 W. This heater produces 30 J of heat every second. In 2 s it produces 60 J.
The quantity of heat added (J) = power (W) x time (s).

Procedure

a. Copy Table 29-1 into your notebook.
b. Enter the power of your heater in the table.

c. Add 91 mL of cold ethylene glycol to an insulated cup. 91 mL of ethylene glycol have a mass of 0.1 kg.
d. Place the heater into the ethylene glycol (Fig. 29-5). Be sure the heating element is covered.

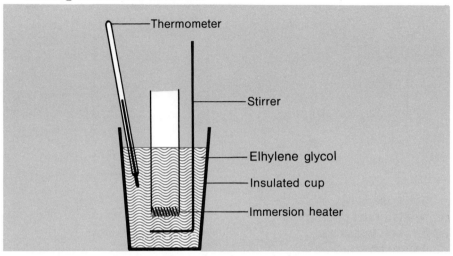

Fig. 29-5 Apparatus for studying the effect of heat on different substances

e. Record the starting temperature of the liquid.
f. Turn on the heater. Start the timer. Record the time it takes in seconds to produce a temperature change of 40°C.
g. Add 100 mL of cold water to the other insulated cup. 100 mL of water have a mass of 0.1 kg.
h. Repeat steps (d), (e), and (f) for water.
i. Calculate the heat energy added to each liquid.

Table 29-1 Heat and Different Substances

Substance	Mass (kg)	Temperature change (°C)	Heater power (W)	Time (s)	Heat energy (J)
Ethylene glycol	0.1	40			
Water	0.1	40			

Discussion

1. Compare the heat added to warm the same mass of water and ethylene glycol through the same temperature change.
2. Calculate the heat in joules needed to warm 1 kg of each substance 1°C. Your teacher will help you. This is called **specific heat capacity** [spe-SIH-fik HEET kap-A-sit-ee].
3. Obtain a class value for the specific heat capacity of each substance. Your teacher will help you.
4. The accepted value for the specific heat capacity of water is 4200 J/(kg•°C). It is 2200 J/(kg•°C) for ethylene glycol. Compare the class values with the accepted values.
5. How could the experiment be changed to get better results? (Hint: Consider the initial temperature of the liquids compared to room temperature.)

29.3 Measuring Quantities of Heat

There is no instrument for measuring the quantity of heat. Instead we measure three factors. One factor is the mass heated. Another is the temperature change. The third is the specific heat capacity.

Specific Heat Capacity

Experiments show that it takes 4200 J of heat to raise the temperature of 1 kg of water 1°C. This is true for all samples of water. It takes only 2200 J to raise the temperature of 1 kg of ethylene glycol by 1°C.

Specific heat capacity is the quantity of heat in joules needed to raise the temperature of 1 kg of a substance by 1°C. The symbol for specific heat capacity is c. The unit is the J/(kg•°C). The specific heat capacities of some common substances are listed in Table 29-2. These, together with densities and melting points, are used to identify unknown substances. All are characteristic physical properties.

Table 29-2 Specific Heat Capacities

Substance	Specific heat capacity J/(kg•°C)	Substance	Specific heat capacity J/(kg•°C)
Aluminum	900	Methanol	2500
Copper	390	Oxygen	920
Ethylene glycol	2200	Paraffin oil	2100
Gold	130	Sand	800
Ice	2100	Silver	240
Iron	450	Water	4200
Lead	130	Water vapour	2000
Magnesium	980	Zinc	390

Effects of Different Specific Heat Capacities

Examine Table 29-2. Water has the largest specific heat capacity listed. It is over 5 times that of sand. It takes more energy to heat up water than sand on a hot summer day. Water has to lose more energy than sand when it cools on a cold winter night. As a result a large body of water has a moderating effect on climate. The temperature changes more slowly near water than inland. That is one reason why people like a beach on a hot summer day. That is why some people prefer Vancouver temperatures over Calgary temperatures.

Water is the best liquid for storing heat. It stores about twice as much heat as the same mass of ethylene glycol. Thus it is the best for cooling engines in cars and factories. Can you imagine what would happen with just ethylene glycol in a radiator in a traffic jam?

Calculating the Quantity of Heat

Once we know the mass, the temperature change, and the specific heat capacity of a substance, we can calculate the quantity of heat added or removed.

Quantity of heat = mass x temperature change x specific heat capacity

$Q = m \bullet \Delta t \bullet c$ where Q is the quantity of heat in (J), m is the mass in (kg), Δt is the change in temperature in (°C), and c is the specific heat capacity in (J/(kg•°C)).

Sample Problem

Find the quantity of heat needed to raise the temperature of 500 g of methanol from 20°C to 60°C.

Given
$m = 500$ g $= 0.5$ kg
$t_1 = 20$°C
$t_2 = 60$°C
$c = 2500$ J/(kg•°C) (Table 29-2)
Required
Q
Analysis
Find Δt
Substitute in equation

Solution
$\Delta t = 60$°C - 20°C $= 40$°C
$Q = m \times \Delta t \times c$
$ = 0.5$ kg x 40°C x 2500 J/(kg•°C)
$ = 0.5 \times 40 \times 2500$ J
$ = 50\ 000$ J
$ = 50$ kJ
Statement
The quantity of heat added to the methanol is 50 kJ.

Using the same formula, we can work out the values for specific heat capacities, knowing Q, m, and Δt.

Section Review

1. **a)** Name three factors that affect the quantity of heat.
   b) How does each factor affect the quantity of heat?
2. Define specific heat capacity.
3. What is the unit for specific heat capacity?
4. How is the specific heat capacity of a substance determined?
5. The heat energy collected in a solar energy system is sometimes stored in liquids in underground tanks. Which liquid will hold more heat, ethylene glycol or water? Why?
6. Explain why the climate near a large body of water is more moderate than the climate inland.
7. Calculate the amount of heat transferred when 4.8 kg of ethylene glycol cools from 35°C to 20°C. Is heat gained or lost?

29.4 ACTIVITY Heat Exchange in Mixtures

Chilled cream cools hot coffee. The cream warms up. The coffee cools down slightly. The cream gains heat in warming. The coffee loses heat in cooling. A transfer of heat occurs when substances at different temperatures are mixed.

Problem

Suppose hot and cold water are mixed. How does the heat lost by the hot water compare to the heat gained by the cold water?

Materials

Teacher
large electric kettle
100 mL graduated cylinder
class set of styrofoam cups

Student Groups
2 thermometers
100 mL graduated cylinder
large styrofoam cup
stirrer

CAUTION: Wear safety goggles during this activity. Also, be careful with the hot water.

Procedure

a. Your teacher will boil water at the front using an electric kettle.
b. Copy Table 32-3 into your notebook.
c. Measure out 150 mL of cold water into the large styrofoam cup. Assume 150 mL have a mass of 150 g.
d. Obtain 100 mL of hot water in a small styrofoam cup from your teacher. Assume 100 mL have a mass of 100 g.
e. Take the initial temperature of the hot water and the cold water as shown in Figure 29-6. Record this in the table.

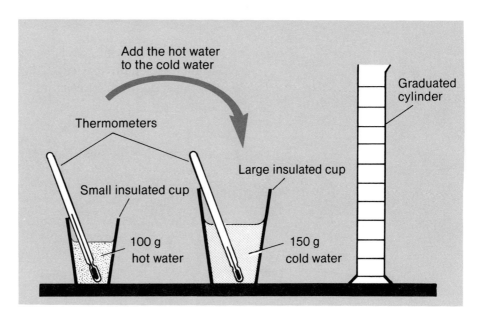

Fig. 29-6 Apparatus for studying heat exchange in mixtures

f. Add the hot water to the cold water. Stir the mixture.
g. Record the highest temperature reached by the mixture.
h. Calculate the heat lost by the hot water. Use the formula $Q = m \bullet \Delta t \bullet c$.
i. Calculate the heat gained by the cold water in the same way.

Section 29.4

Table 29-3 Heat Exchange in Mixtures

Mixture	Water	Mass (kg)	Temperature (°C) Initial	Final	Change	Specific heat capacity J/(kg•°C)	Heat exchanged (J)
Large cup	Hot					4200	
	Cold					4200	

Discussion

1. Compare the heat lost by the hot water with the heat gained by the cold water.
2. If the heat lost is not equal to the heat gained, which should be greater? Why?
3. How could you modify the activity to get better results?
4. Complete the following statement: In any mixture the heat gained by a ▬▬▬ substance in warming ▬▬▬ the heat lost by a ▬▬▬ substance in cooling. This is the **Principle of Heat Exchange** for mixtures.

29.5 ACTIVITY Measuring the Specific Heat Capacity of a Metal

Every substance has a definite specific heat capacity. In this activity you will find the specific heat capacity of some metals. To do this you will use the method of mixtures and the Principle of Heat Exchange.

Problem

Can you find the specific heat capacity of some metals?

Materials

samples of several metals (aluminum, copper, iron, lead, zinc)
ring stand, ring clamp, wire gauze
Bunsen burner   100 mL graduated cylinder   thermometer
200 mL beaker   insulated cup   balance

Procedure

a. Copy Table 29-4 into your notebook.
b. Choose one metal. Find the mass of the sample. Enter this in the table.
c. Attach a thread to the metal. Suspend it in water in a beaker (Fig. 29-7). Do not let it touch the bottom.

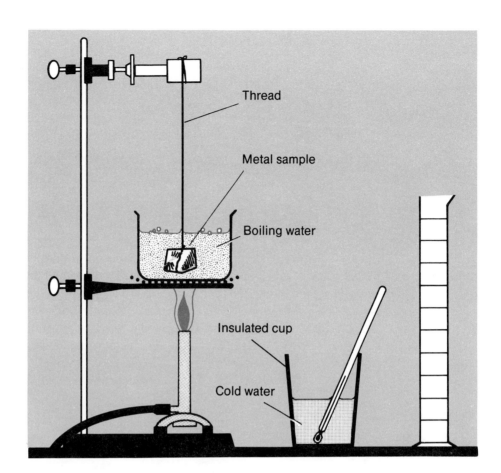

Fig. 29-7 Apparatus for specific heat capacity of a metal

d. Heat the water to the boiling point. Allow it to boil for about five minutes. Let the metal reach the temperature of the water.
e. Take the temperature of the hot water. Record this as the temperature of the metal.
f. Add 100 g of cold water to a cup. Measure and record its temperature.
g. Quickly remove the metal sample from the hot water. Put the sample in the cold water.
h. Stir the mixture gently. Record the final temperature of the mixture.
i. Use the Principle of Heat Exchange. Calculate the specific heat capacity of the metal. Use the equation $Q = m \cdot \Delta t \cdot c$. Equate the heat lost by the metal to the heat gained by the water. Then solve for c.

Table 29-4 Determining the Specific Heat Capacity of a Metal

Component of mixture	Mass (kg)	Temperature (°C)			Specific heat capacity J/(kg·°C)	Heat exchanged (J)
		Initial	Final	Change		
Metal					c	
Water	0.10				4200	

Discussion

1. Compare the value you obtained for specific heat capacity with other values obtained for the same metal. Your teacher will help you.
2. Compare the class value with the accepted value. (See Table 29-2).
3. List possible sources of error in the experiment. Indicate for each error whether it raises or lowers the value of specific heat capacity. Suggest changes to overcome the errors.

Main Ideas

1. Temperature determines the direction of heat flow.
2. Heat depends on mass, temperature change and specific heat capacity.
3. Specific heat capacity is constant for a given substance.
4. Specific heat capacity has the symbol c and units of J/(kg•°C).
5. Water has one of the largest specific heat capacities — 4200 J/(kg•°C).
6. Quantity of heat is calculated using the equation $Q = m \cdot \Delta t \cdot c$.

Glossary

specific heat capacity   spe-SIH-fik HEET kap-A-sit-ee   the quantity of heat needed to raise the temperature of a unit mass by 1°C

Study Questions

A. True or False

Decide whether each of the following sentences is true or false. If the sentence is false, rewrite it to make it true. (Do not write in this book.)

1. Temperature measures the quantity of heat in an object.
2. The larger the mass of an object heated through 1°C, the larger the quantity of heat.
3. The larger the temperature change of a 1 kg mass, the larger the quantity of heat.
4. The symbol for specific heat capacity is J/(kg•°C).
5. The equation for quantity of heat is $Q = m \cdot t \cdot c$.

B. Completion

Complete each of the following sentences with a word or phrase which will make the sentence correct. (Do not write in this book.)

1. The molecules of two objects at the same temperature have _____ kinetic energy.
2. 1800 J of heat raises the temperature of 2 kg of iron by 2°C. The same heat raises the temperature of 1 kg of iron by _____ .
3. When 1.5 kg of silver cools 2°C, it releases 720 J of heat. When 1.5 kg of silver cools 1°C, it releases _____ J of heat.
4. If the specific heat capacity of one substance is twice another, it will store _____ as much heat.
5. The Principle of Heat Exchange states that in any _____ the heat lost by the warmer body in _____ is _____ the heat gained by the cooler body in _____ .

C. Multiple Choice

Each of the following statements or questions is followed by four responses. Choose the correct response in each case. (Do not write in this book.)

1. Heat is exchanged between two objects when they have a different
   a) specific heat capacity
   b) mass
   c) temperature
   d) ability to conduct heat
2. The same mass of liquids x and y were heated using identical immersion heaters. Liquid x increased in temperature by 20°C in 4 min. Liquid y increased in temperature by 10°C in 2 min. The specific heat capacity of liquid y
   a) is larger than liquid x
   b) is the same as liquid x
   c) is smaller than liquid x
   d) cannot be compared to liquid x
3. The amount of heat needed to raise the temperature of 1 kg of a substance by 1°C is called
   a) the joule
   b) the newton
   c) the specific heat capacity
   d) the watt
4. Specific heat capacity depends on
   a) mass
   b) temperature change
   c) heat energy
   d) the substance
5. Breezes at the lake come toward the shore during the day because
   a) sand warms up faster than water
   b) water warms up faster than sand
   c) water cools down faster than sand
   d) sand cools down faster than water

D. Using Your Knowledge

1. Predict the relationship between the density and specific heat capacity of a metal. Then look up values for copper and aluminum. Try to explain the results.
2. What is the temperature change when 200 g of water gains 8400 J of heat?
3. What is the specific heat capacity if 20 kJ changes the temperature of 10 kg of a substance by 4°C?

4. Use the Kinetic Molecular Theory of Heat to explain the temperature changes which occur when a hot liquid is added to a cold liquid.
5. A mixture is made by adding 75 g of an unknown liquid at a temperature of 25°C to 60 g of water at a temperature of 90°C. The final temperature of the mixture is 65°C. Calculate the specific heat capacity of the liquid. What is the liquid? Can you be sure your identification is correct? Discuss.

E. Investigations

1. Design and carry out an experiment to compare the specific heat capacity of dry and wet sand. Would crops in dry or wet lands be hurt more by frost? Why?
2. Design and do an experiment to compare the specific heat capacity of water and a mixture of water and ethylene glycol.

30 Heat and Changes of State

30.1 Latent Heat
30.2 Activity: Specific Latent Heat of Fusion of Ice
30.3 Demonstration: Specific Latent Heat of Vapourization of Water
30.4 Applications of Latent Heat
30.5 Activity: The Energy Content of Foods
30.6 Heat Energy, Food, and Exercise

You found out in Unit 3 that heat is needed to melt ice. Why doesn't ice warm up as it melts? Why does a glacier stay around in summer? How does a sprinkler prevent frost damage?

Boiling also requires heat. What does the heat do? Why is water an excellent radiator coolant? Why is a burn from steam more severe than one from boiling water? You will find answers to these and other questions in this chapter.

30.1 Latent Heat

The temperature of a substance stays constant as it changes state. Consider water. The temperature stays at 0°C during melting. During vapourization it stays at 100°C. The heat added changes ice to water and water to water vapour. It does not change the temperature. *Heat which causes a change in state is called* **latent heat** [LAY-tent]. The word *latent* means *hidden*. The heat is hidden because no change in temperature occurs.

Specific Latent Heat of Fusion

You may live in a region where there is snow in winter. A strange thing happens in the spring. Temperatures well above freezing occur for several days. The air temperature may be above 20°C every day. But at the end of this hot spell there is still snow on the ground. Apparently it takes much more heat to melt the snow than to warm the air.

A definite amount of heat is required to change the state of a substance. *The quantity of heat needed to change 1 kg of a substance from the solid state to the liquid state is called the* **specific latent heat of fusion** [FU-shun]. Remember that the temperature of ice stays constant while it melts (Fig. 30-1,A).

Fig. 30-1 Latent heat causes a change in state, not a change in temperature.

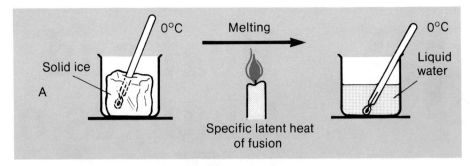

Specific Latent Heat of Vapourization

Have you ever been burned by the steam from a boiling kettle? A burn is much more severe from steam than from boiling water. Yet you found in Unit 3 that the temperature of the water and the steam are the same. There must be more heat energy in the steam than in the boiling water. A definite amount of heat is required to change a substance from a liquid to a gas. *The quantity of heat required to change 1 kg of a substance from the liquid state to the gaseous state is called the* **specific latent heat of vapourization** [VA-por-iz-A-shun]. Remember that the temperature of boiling water stays constant while it vapourizes (Fig. 30-1,B).

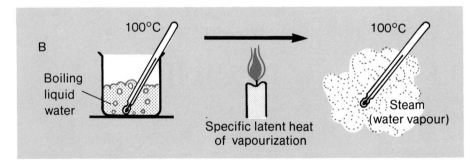

Section Review

1. Describe the temperature of a substance as it changes state.
2. Define latent heat.
3. Define specific latent heat of fusion.
4. Define specific latent heat of vapourization.

30.2 ACTIVITY Specific Latent Heat of Fusion of Ice

It takes more heat to melt 1 kg of ice than to warm 1 kg of water by 1°C. But during melting, no change in the temperature occurs.

Problem

How much energy is needed to melt 1 kg of ice?

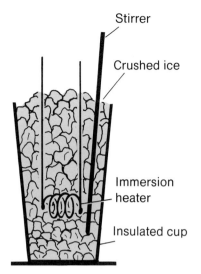

Fig. 30-2 Apparatus for melting crushed ice

Materials

immersion heater balance crushed ice
insulated cup timer

Procedure

a. Copy Table 30-1 into your notebook.
b. Measure and record the mass of the cup.
c. Place the immersion heater in the cup as shown in Figure 30-2.
d. Fill the cup heaping full with crushed ice.
e. Turn on the heater and the timer at the same time.
f. Keep the immersion heater covered with ice and water. To do this push down on the ice. You may also have to add more ice.
g. Stir the mixture continuously until the ice is all gone.
h. Record the time it takes to completely melt the ice.
i. Measure and record the mass of the cup and water.
j. Calculate the mass of the ice.
k. Calculate the heat added.
l. Use the data to find the heat needed to melt 1 kg of ice.

Table 30-1 Finding the Specific Latent Heat of Fusion of Ice

Heater power = ▨ W	Mass of cup and ice = ▨ g = ▨ kg
Melting time = ▨ s	Mass of cup = ▨ g = ▨ kg
Heat added = ▨ kJ	Mass of ice = ▨ kg

Discussion

1. How much heat would you need to melt 1 kg of ice?
2. Compare your value with other values in the class. Your teacher will help you get a class value.
3. The accepted value for the specific latent heat of fusion of ice is 336 kJ/kg. Compare the class value with the accepted value.
4. Why is the class value different from the accepted value?
5. List sources of error in the experiment. Suggest ways to reduce these errors.

30.3 DEMONSTRATION Specific Latent Heat of Vapourization of Water

It takes more heat to vapourize 1 kg of water than to warm 1 kg of water by 1°C. But during vapourization no change in temperature occurs.

Problem

How much energy is needed to vapourize 1 kg of water?

Materials

electric kettle   timer   top loading balance (2 kg)
CAUTION: Your teacher will demonstrate this experiment.

Procedure

a. Copy Table 30-2 into your notebook.
b. Record the power of the kettle in watts in the table.
c. Fill the kettle about half full of water as shown in Figure 30-3.

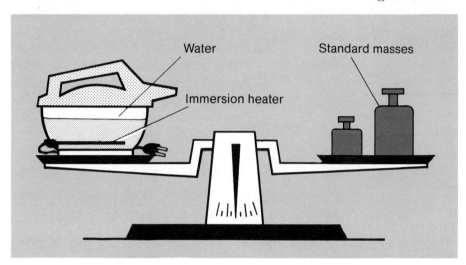

Fig. 30-3 Apparatus for specific latent heat of vapourization of water

d. Measure and record the mass of the kettle and water in kilograms.
e. Plug in the kettle.
f. When the water begins to boil, start the timer.
g. Boil the kettle for about 5 min. CAUTION: Make sure water always covers the heating coil.
h. Unplug the kettle and stop the timer.
i. Record the time the water boiled in seconds.
j. Measure and record the mass of the kettle and hot water.
k. Calculate the mass of water that has vapourized.
l. Calculate the heat energy used to vapourize the water.
m. Use the data to calculate the heat needed to vapourize 1 kg of water.

Table 30-2 Finding the Specific Latent Heat of Vapourization of Water

Kettle power = ▓▓▓ W	Mass of kettle and cold water = ▓▓▓ g = ▓▓▓ kg
Boiling time = ▓▓▓ s	Mass of kettle and hot water = ▓▓▓ g = ▓▓▓ kg
Heat added = ▓▓▓ kJ	Mass of water vapourized = ▓▓▓ kg

Discussion

1. How much heat would you need to vapourize 1 kg of water?

2. Compare your value with other values in the class. Your teacher will help you get a class value.
3. The accepted value for the specific latent heat of vapourization of water is 2268 kJ/kg. Compare the class value with the accepted value.
4. Why is the class value different than the accepted value?
5. List sources of error in the experiment. Suggest ways to reduce the errors.

30.4 Applications of Latent Heat

Latent Heat of Fusion of Ice

Water has one of the largest specific heat capacities. It also has one of the largest specific latent heats of fusion. For this reason ice is an excellent refrigerant to use in picnic coolers. When the ice melts it takes heat from its surroundings. It takes 336 000 J to melt 1 kg of ice. This heat comes from the contents of the cooler.

When water freezes it gives heat to its surroundings. This explains why large containers of water are placed in root cellars during cold weather. As the cellar cools, the water in the containers cool. In order to freeze 1 kg of water, 336 kJ of heat must leave the water. This heat is available to keep the root cellar above freezing. This also explains why sprinkler systems are used to prevent frost damage to crops such as strawberries (Fig. 30-4).

Fig. 30-4 A water sprinkler system can prevent frost damage.

Latent Heat of Vapourization of Water

When water vapourizes it takes heat from its surroundings. Wet swimmers feel cool when they step into the open air. Every kilogram of water needs 2 268 000 J of energy to evaporate. If only 10 mL of water evaporates from the skin, that's 22 680 J. Much of this heat comes from the body. As a result the skin cools down.

When steam condenses it gives heat to its surroundings. This explains why a burn from steam is much worse than a burn from boiling water. If only 1 mL of water condenses, 2268 J of energy are transferred to the skin. That alone is enough to cause a severe burn. The condensed steam also cools from 100°C to body temperature of 37°C. This transfers another 265 J to the skin. So don't pass your arm across the spout of a boiling kettle.

Section Review

1. What is the direction of heat transfer when ice melts? when water freezes?
2. Why is ice an excellent refrigerant?
3. Why is a sprinkler system used on strawberry crops when there is a frost warning?
4. What is the direction of heat transfer when steam condenses? when water vapourizes?
5. Why do wet swimmers out in the air feel cool?
6. Why is a steam burn so severe?

30.5 ACTIVITY The Energy Content of Foods

Foods contain a form of potential energy. The body "burns" food to do work. Some of the food's potential energy is stored in the muscles. Some becomes kinetic energy as the arms, legs, and blood move. Much of the energy is changed to heat energy. This energy is used to replace the heat lost from the body. Without it our body temperature would not stay at 37°C.

In this activity you will burn some foods. You will use the heat to increase the temperature of water.

Problem

How many joules of potential energy are in 1 kg of some foods?

Materials

50 mL beaker
2 paper clips
beaker tongs
balance
peanuts, walnuts, shredded wheat
10 mL graduated cylinder
sheet of aluminum foil 10 cm x 15 cm

CAUTION: Wear safety goggles during this activity.

Procedure

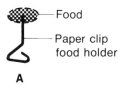

a. Copy Table 30-3 into your notebook.
b. Make a food holder out of a paper clip by bending it as shown in Figure 30-5,A.
c. Make a heat funnel by folding the sheet of aluminum foil to form a cone with an opening at both ends. Fasten the ends together with a paper clip as shown in Figure 30-5,B.
d. Find the mass of a piece of walnut in grams. Insert the end of the paper clip holder into the sample. Be careful not to puncture your hand. Do not lose any of the food!
e. Place 30 mL of cold water in the 50 mL beaker. Record the temperature of the water.
f. Ignite the food with a match. Place the heat funnel over the food as shown in Figure 30-5,C. Hold the bottom of the beaker in the tip of the flame until the food is burned to an ash. Support the beaker using the tongs.
g. Record the highest temperature reached by the water.
h. Repeat steps (c) to (f) for a peanut and for a spoon-sized shredded wheat.
i. Calculate the energy content in joules per kilogram of each food. Determine the class values.

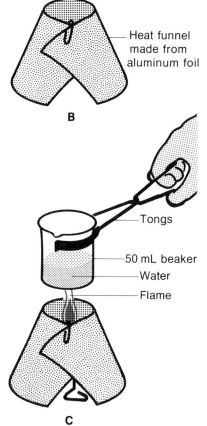

Fig. 30-5 Apparatus for studying the energy content of foods

Table 30-3 Energy Content of Some Foods

Food Burned			Water Heated				
			Temperature (°C)				
Food	Mass (kg)	Mass (kg)	Initial	Final	Change	Specific heat capacity (J/kg·°C)	Heat from food (J)
Walnut						4200	
Peanut						4200	
Spoon-sized shredded wheat						4200	

Discussion

1. What is the class value for the energy content of walnuts, peanuts, and spoon-sized shredded wheat in joules per kilogram (J/kg)?
2. the accepted values are: walnuts 27 300 kJ/kg; peanuts 24 600 kJ/kg; spoon-sized shredded wheat 14 900 kJ/kg. Compare the class values with the accepted values.

30.6 Heat Energy, Food, and Exercise

Specific Heat of Combustion

Foods are compounds which contain carbon and hydrogen. They store the sun's energy as chemical potential energy. When completely burned in oxygen, foods produce carbon dioxide, water, and heat energy. *The heat energy produced when 1 kg of a substance is burned is called the* **specific heat of combustion** [kom-BUST-shun]. The specific heat of combustion of substances is measured in kilojoules per kilogram. The complete combustion of 1 kg of spoon-sized shredded wheat produces 14 900 kJ of heat. 1 kg of walnuts produces 27 300 kJ. Walnuts have about twice as much chemical potential energy as shredded wheat. Now you know why people on diets should not eat nuts. Some heats of combustion for foods are given in Table 30-4.

Table 30-4 Specific Heats of Combustion of Some Foods

Food	Specific heat of combustion (kJ/kg)
Bacon, raw	27 900
Beef, t-bone steak	16 700
Bread, white	11 300
whole wheat	10 200
Carrot, raw	1 800
Cereal, oatmeal	16 400
shredded wheat	14 900
Eggs, fried	9 100
hard boiled	6 800
Peanuts	24 600
Potatoes, boiled	3 200
Potato chips	23 900

Food and Exercise

A 50 kg boy at the age of 13-15 needs about 13 000 kJ of energy daily for good nutrition. A 50 kg girl at the same age needs 11 000 kJ. This energy is used for growth and activity. Different exercises make your body burn joules at different rates. By exercising you get the same result as cutting down your food intake. For example, you could burn as many joules as are found in a piece of bread by walking for 15 min. About 10 min of bicycling, 7 min of swimming or 4 min of running do the same thing. The rate of energy use for different forms of activity are shown in Table 30-5.

Table 30-5 Activity and Energy Use

Activity (70 kg person)	Rate of energy use (kJ/min)
Reclining	6
Walking fast	22
Bicycling	35
Swimming	47
Running	82

Section Review

1. Define specific heat of combustion.
2. What are the units of specific heat of combustion?
3. A person reclines for 15 min. The person then bicycles for 15 min to a swimming pool. The swim lasts 30 min. How much energy is used in the hour?

Main Ideas

1. Latent heat causes a change in state rather than a change in temperature.
2. The amount of heat needed to change the state of 1 kg of a substance is called the specific latent heat.
3. It takes about 80 times as much heat to melt ice as to warm water 1°C.
4. It takes about 540 times as much heat to vapourize water as to warm water 1°C.
5. The specific heat of combustion of a substance is the amount of heat released when 1 kg completely burns.

Glossary

latent	LAY-tent	hidden or concealed
fusion	FU-shun	the change in state in which a solid changes to a liquid
vapourization	VA-por-iz-A-shun	the change in state in which a liquid changes to a vapour or gas
combustion	kom-BUST-shun	the process of burning

Study Questions

A. True or False

Decide whether each of the following sentences is true or false. If the sentence is false, rewrite it to make it true. (Do not write in this book.)

1. The heat which causes a change in state is called specific heat.
2. It takes more heat to melt 1 kg of ice than to change the temperature of water by 1°C.
3. Ice is useful in a cooler because it has a very large specific heat capacity.
4. The heat needed to change the temperature of water from 0°C to 100°C is larger than the heat required to vapourize it.
5. The specific heat of combustion of peanuts is less than that of walnuts.

B. Completion

Complete each of the following sentences with a word or phrase which will make the sentence correct. (Do not write in this book.)

1. Specific latent heat of vapourization is the amount of heat required to change 1 kg of a substance from the ▒▒▒ to the ▒▒▒ state.
2. It takes ▒▒▒ heat to raise the temperature of water by 1°C than to vapourize it.
3. A boiling water burn is ▒▒▒ severe than a steam burn.
4. When ice melts, heat travels ▒▒▒ the ice ▒▒▒ the surroundings.
5. Specific heat of combustion is the amount of heat ▒▒▒ when ▒▒▒ kg of a substance completely burns.

C. Multiple Choice

Each of the following statements or questions is followed by four responses. Choose the correct response in each case. (Do not write in this book.)

1. The term latent heat means that which of the following will *not* change?
   a) state   b) temperature   c) mass   d) potential energy
2. How much heat does 1 kg of ice give up when it freezes?
   a) 2268 kJ   b) 336 kJ   c) 4.2 kJ   d) 2.1 kJ
3. What effect does evaporation have on the temperature of a substance?
   a) increase the temperature
   b) decrease the temperature
   c) does not affect the temperature
   d) the effect depends on the substance

4. During a heavy snowfall it becomes warmer outside because of
   a) evaporation   **b)** condensation   **c)** freezing   **d)** melting
5. Which of the following activities makes your body burn joules the fastest?
   a) bicycling   **b)** running   **c)** swimming   **d)** walking fast

D. Using Your Knowledge

1. Why do glaciers stay around in summer when it's warm enough for grass and flowers to grow nearby?
2. The high specific latent heat of vapourization of water contributes to its usefulness as a cooling agent in the radiators of cars. Explain why.
3. Some hot water radiators have a single pipe connected to them. The steam comes to the radiator and the water leaves in the same pipe. The steam and the water are both at a temperature of 100°C. Where does the heat come from to warm the room?
4. One kilogram of fat stores 30 000 kJ of energy. Suppose you take in 500 kJ more energy each day than you burn off through physical activity. (This is the energy in a small bag of potato chips.)
   a) How many kilograms will you gain in 36 d?
   b) Assume that you are carrying 5 kg of excess fat. How many days of running 20 min per day will it take you to get rid of the excess fat?

E. Investigations

1. Keep track of the food you eat during a typical day. Also record your activity for the day. Consult a cook book to find the energy content of the food. Calculate how long it will take you to burn off each meal.
2. Design an experiment to compare the energy content of different kinds of cereals. For example, you could compare corn flakes, rice crispies, and bran flakes.
3. Design an experiment using the method of mixtures to find the specific latent heat of fusion of ice.

Unit 10

CHAPTER 31
Electrostatics

CHAPTER 32
Sources of Current Electricity

CHAPTER 33
Series and Parallel Circuits

Electricity

Electricity is important in our lives. It heats homes; it lights neon signs and streets; it helps start cars; it runs microcomputers; it does work lifting heavy objects.

You may have seen a large crane suspend a disc above a heap of scrap metal. Suddenly the metal leaps to the disc. Pieces appear to defy the force of gravity. Why? The disc is an electromagnet. The crane operator has directed electricity to the disc. The disc becomes magnetic and attracts the metal. Turn off the electricity and the electromagnet ceases to operate.

In this unit you will study electricity. First you will learn about electric charge at rest. Then you will study electric charge in motion. Finally you will learn the properties of series and parallel circuits. Learn as much as you can about electricity because it is affecting our lives more and more every day.

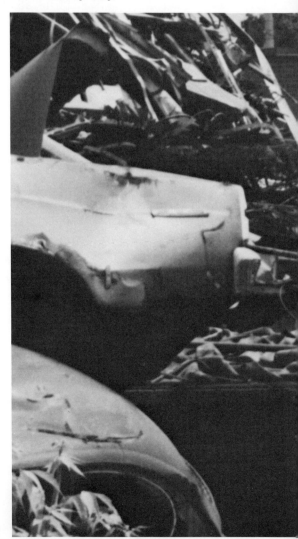

Fig. 31-0 A powerful electromagnet at work

31 Electrostatics

31.1 Static Electricity
31.2 Activity: The Kinds of Electric Charge
31.3 The Law of Electric Charges
31.4 The Electron Theory of Charge
31.5 Activity: Charging a Metal Leaf Electroscope by Contact
31.6 Activity: Charging a Metal Leaf Electroscope by Induction

Why does a comb attract hair sometimes but not other times? Why does a rubber balloon sometimes stick to the wall? What happens when objects are charged? How can we tell the kind of charge on an object? What is the theory to explain charging? You will find answers to these and other questions in this chapter.

31.1 Static Electricity

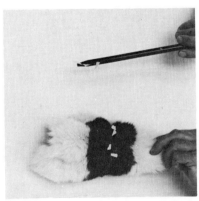

Fig. 31-1 A charged object attracts a neutral object. Both the ebonite rod and fur are charged.

Electrostatics

Electricity is a general term which refers to electric charge. *Electric charge at rest is called* **static electricity**. The attraction between a comb and combed hair results from static electricity. Clothes cling together in a clothes dryer because of static electricity. *The study of electric charge at rest is called* **electrostatics** [ee-LEK-tro-STA-tiks].

How Static Electricity is Produced

Static electricity is produced by rubbing different substances together. The first person to study static electricity was a Greek. The Greek philosopher Thales lived about 600 B.C. He noticed that when a brownish solid called amber was rubbed with fur it behaved strangely. Rubbed amber attracted small pieces of paper, straw, and wood shavings. Without rubbing, there was no attraction.

When two different materials are rubbed together, both become charged. Both an ebonite rod and the fur used to rub it become charged. A glass rod and a silk cloth also become charged when rubbed together.

How a Charged Object Affects a Neutral Object

An object without a charge is said to be **neutral**. Small pieces of paper and sawdust are neutral. A comb rubbed through hair is **charged**. A charged comb attracts neutral paper. An ebonite rod rubbed with fur attracts paper. The paper is also attracted to the fur (Fig. 31-1). Indeed, any charged object attracts any neutral object. Two neutral objects do not attract.

Section Review

1. Define static electricity.
2. Define electrostatics.
3. How is static electricity produced?
4. How does a neutral object affect another neutral object?
5. How does a charged object affect a neutral object?

31.2 ACTIVITY The Kinds of Electric Charge

In this activity you will study the kinds of charge. How many different kinds of charge are there? Do two rubbed materials get the same kind of charge?

Problem

What different kinds of charge are there?

Materials

two ebonite rods	stirrup and support	silk cloth
insulating thread	two glass rods	folded cardboard
fur	metre stick	C clamp

Procedure

a. Copy Table 31-1 into your notebook.
b. Charge an ebonite rod by rubbing it with fur. Suspend the ebonite rod level in the stirrup as shown in Figure 31-2.

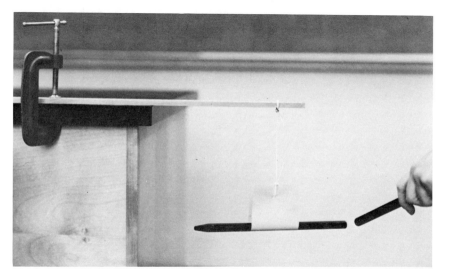

Fig. 31-2 Testing for kinds of electric charge

c. Charge a second ebonite rod with fur.
d. Bring the charged fur near one end of the suspended ebonite rod. Record whether the rod is attracted or repelled.
e. Bring the second charged ebonite rod near one end of the suspended ebonite rod. Record what happens.
f. Charge a glass rod by rubbing it with silk.
g. Bring the charged silk near one end of the suspended ebonite rod. Record what happens.
h. Bring the charged glass rod near one end of the suspended ebonite rod. Record what happens.
i. Place a charged glass rod in the stirrup. See what happens if another charged glass rod is brought near one end.

Table 31-1 How Charged Objects Affect Each Other

Suspended object	Hand-held charged object	Observations
Ebonite rod	Fur	
Ebonite rod	Ebonite rod	
Ebonite rod	Silk	
Ebonite rod	Glass rod	
Glass rod	Glass rod	

Discussion

1. Use the observations to show that there are two kinds of electric charge.
2. How do you know that two substances get different charges when they are rubbed together?
3. How do like charges affect each other?
4. How do unlike charges affect each other?

31.3 The Law of Electric Charges

The Law of Charges

A charged object attracts a neutral object. Charged objects also affect each other. Two charged ebonite rods repel. Two charged glass rods repel. But not all charged objects repel. Ebonite and the fur used to rub it attract. Both are charged. There must be a different charge on the ebonite than on the fur. The same is true of glass rubbed with silk. Indeed any two materials rubbed together get different kinds of charge.

Scientists have stated a law to summarize these observations. **The Law of Electric Charges** states *like charges repel and unlike charges attract.*

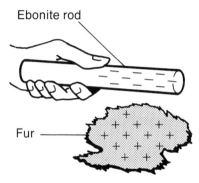

A. An ebonite rod charged with fur is negative. The fur is positive.

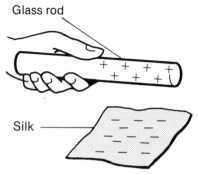

B. A glass rod charged with silk is positive. The silk is negative.

Fig. 31-3 The two kinds of changes

Kinds of Charge

There are only two kinds of charges. We now call them positive and negative. The charge on an ebonite rod rubbed with fur is negative (Fig. 31-3,A). The charge on a glass rod rubbed with silk is positive (Fig. 31-3,B).

Testing for Charge

The charge on an object is found by seeing how it affects an object with a known charge. If the two repel, the charges are alike. If they attract, the charges are opposite. If you think the object is neutral, test it with another neutral object. A neutral object does not attract another neutral object. But a neutral object is attracted by an object with either a positive or negative charge.

Section Review

1. State the Law of Electric Charges.
2. Name the two kinds of electric charges.
3. Give an example of each kind of charge.
4. How can you find the kind of charge on an object?
5. Pieces of straw are attracted to both a negative and a positive rod. What charge is on the straw? How do you know?

31.4 The Electron Theory of Charge

Difference Between a Law and a Theory

A scientific law summarizes many observations. For example, electric charges always affect each other in the same way. The Law of Electric Charges summarizes how the electric charges affect each other. A scientific theory explains observations. It also helps us make predictions. The modern electron theory of charge explains the behaviour of charged objects. It explains the charging of objects. It also explains the two kinds of charge.

The Electron Theory of Charge

The modern electron theory says that matter consists of atoms. Each atom is made up of many smaller particles. Three of these are the **proton** [PRO-ton], the **neutron** [NEW-tron] and the **electron** [ee-LEK-tron]. Protons are heavier than electrons. Each proton has one unit of positive charge. The proton is found in the nucleus of the atom. The neutron is also found in the nucleus. It is neutral. An electron is much

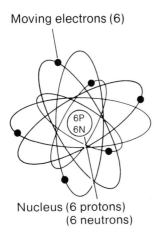

Fig. 31-4 A simple model of a carbon atom

lighter than the proton. Each electron has one unit of negative charge. Electrons move at high speeds. They form a cloud-like region around the nucleus.

All atoms are neutral. Thus each atom has the same number of protons and electrons. A simple model of a carbon atom is shown in Figure 31-4. A carbon atom has six protons and six neutrons in the nucleus. It has six electrons moving around the nucleus.

How the Electron Theory Explains Charging by Rubbing

The action of rubbing removes loosely held electrons from atoms. But protons cannot be removed by rubbing. When two different materials are rubbed together, electrons move from one to the other. One material gains electrons. The other loses electrons. The material which gains electrons gets a negative charge. (Remember, electrons have a negative charge.) The material which loses electrons gets a positive charge. Thus a material that has too few electrons has a positive charge. A material that has a surplus of electrons has a negative charge. The charging process is shown in Figure 31-5.

The electron theory explains why rubbed materials get opposite charges. One material loses electrons. The other gains electrons. Thus one becomes positive and the other negative. The theory even predicts that rubbed materials get the same quantity of charge. Scientists have found this to be true.

Section Review

1. **a)** What is the difference between an observation and a law?
   b) What is one difference between a law and a theory?
2. **a)** What particles in the atom are responsible for electric charge?
   b) Which particle moves? Which particle stays relatively fixed?
3. Describe the number and kind of charged particles in each of the following:
   a) a neutral object **b)** a positive object **c)** a negative object

31.5 ACTIVITY Charging a Metal Leaf Electroscope by Contact

An **electroscope** [ee-LEK-tro-skope] is used to detect and identify charge. The parts of a metal leaf electroscope are shown in Figure 31-6. It consists of an insulating case, a metal rod and sphere, and a small metal leaf. The leaf is attached at its upper end to the metal rod. The leaf is enclosed in a case. The case prevents moving air from damaging the leaf. In this activity you will study the parts of the electroscope. Then you will give the electroscope a known charge by contact. Finally, you will use the electroscope to detect and identify charge.

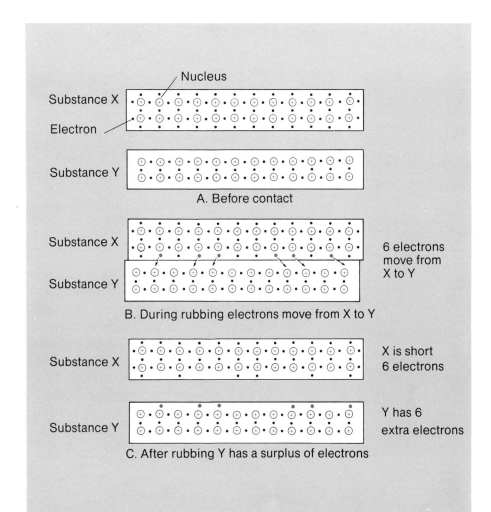

Fig. 31-5 Charging by rubbing. The rubbing transfers electrons from one substance to another.

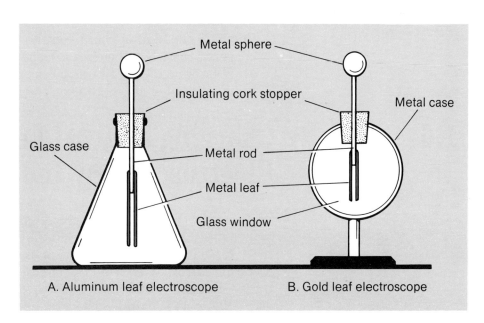

Fig. 31-6 Metal leaf electroscopes

Section 31.5 385

Problem

How can an electroscope be charged by contact?

Materials

| metal leaf electroscope | glass rod silk | ebonite rod fur | comb wool |

Procedure

a. Examine the metal leaf electroscope. Find the parts shown in Figure 31-7.
b. Touch the metal sphere with your fingers. This will neutralize the electroscope.
c. Rub an ebonite rod with fur to make the rod negative. Approach but do not touch the sphere of the electroscope with the negative rod. Watch the metal leaf. Remove the rod. Record your observations.
d. Touch the sphere with the negative rod. Watch the metal leaf. Remove the rod. Record your observations.
e. Approach but do not touch the charged electroscope with a negative rod. Record your observations.
f. Repeat step (e) with a positive rod.
g. Charge a comb by rubbing it with wool.
h. Use the charged electroscope to identify the charge on the comb and on the wool.

Discussion

1. a) State what was observed during each step of charging the metal leaf electroscope by contact.
   b) Use diagrams to illustrate your answer. Your teacher will help you.
2. Compare the kind of charge on the electroscope with that on the charging rod when the electroscope is charged by contact.
3. a) What kind of charge was on the comb? Why?
   b) What kind of charge was on the wool?

31.6 ACTIVITY Charging a Metal Leaf Electroscope by Induction

We can give the leaf and sphere of a neutral electroscope opposite charges without touching them (Fig. 31-7,A). Bring a negative rod close to, but not touching, the sphere. The leaf spreads apart. Electrons are repelled from the sphere to the leaf (Fig. 31-7,B). The sphere becomes positive. The leaf becomes negative. *The process of producing a charge on an object using a nearby charged object is*

Fig. 31-7 Giving an electroscope a temporary charge by induction

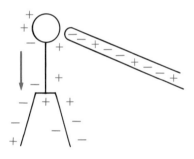

A. A neutral metal leaf electroscope showing some electrons and protons.

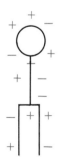

B. Electrons repelled from sphere to leaves. Sphere has an induced charge of positive two. Leaf has an induced charge of negative two.

C. The electrons have returned to their original location. All parts of the electroscope are neutral.

called **induction** [in-DUK-shun]. The charge produced is called an **induced charge**. The induced charge in this example is temporary. When the charged rod is removed, the shifted electrons return to the sphere (Fig. 31-7,C). The leaf collapses.

Problem

How can you produce a permanent charge on the electroscope by induction?

Materials

metal leaf electroscope   ebonite rod   fur   glass rod   silk

Procedure

a. Touch the sphere of the electroscope with your fingers. This will neutralize it.
b. Approach but do not touch the sphere of the electroscope with a negative rod. Watch the metal leaf.
c. Without removing the rod, touch the sphere with your finger as shown in Figure 31-8. Watch the metal leaf.

Fig. 31-8 Charging an electroscope permanently by induction

d. With the rod still near, remove your finger from the sphere. Then remove the negative rod. Record your observations.
e. Approach but do not touch the sphere with a positive rod (glass). Remove it. Record your observations.
f. Repeat step (e) using a negative rod.

Discussion

1. Define induction.
2. Define induced charge.
3. State what was observed during each step of charging the metal leaf electroscope by induction. Use diagrams to illustrate your answer. Your teacher will help you.

Section 31.6

4. Compare the charge on the electroscope with the charge on the rod when an electroscope is being given a permanent charge by induction.
5. Describe how to give the electroscope a negative charge using a glass rod and silk.
6. You touched the sphere with your finger in steps (a) and (c). Why?

Main Ideas

1. Electrostatics is the study of electric charge at rest.
2. Static electricity is produced by rubbing different materials together.
3. A charged object attracts a neutral object.
4. There are two kinds of charge: positive and negative.
5. The Law of Electric Charges states that like charges repel and unlike charges attract.
6. Two particles in the atom, electrons and protons, are responsible for electric charge.
7. An electroscope can be charged by contact or by induction.

Glossary

electrostatics	ee-LEK-tro-STA-tiks	the study of electric charge at rest
electron	ee-LEK-tron	a particle in the atom having a negative charge
proton	PRO-ton	a particle in the atom having a positive charge
electroscope	ee-LEK-tro-skope	a device for sensing and identifying electric charge
induction	in-DUK-shun	the process of charging a conductor using a charged object which is nearby but not touching the conductor

Study Questions

A. True or False

Decide whether each of the following sentences is true or false. If the sentence is false, rewrite it to make it true. (Do not write in this book.)
1. Static charge is electric charge which is moving.
2. A charged object attracts a neutral object.

3. Like charges attract and unlike charges repel.
4. The electron and the neutron are responsible for electric charge.
5. An electroscope charged by induction gets the same kind of charge as the charging rod.

B. Completion

Complete each of the following sentences with a word or phrase which will make the sentence correct. (Do not write in this book.)
1. Electrostatics is the study of _____.
2. Static electricity is produced by _____.
3. The two kinds of charge are _____ and _____.
4. When a positive rod approaches a negative electroscope the leaf _____.
5. During charging by friction, contact, or induction, the particles that move are the _____. The _____ do not move.

C. Multiple Choice

Each of the following statements or questions is followed by four responses. Choose the correct response in each case. (Do not write in this book.)
1. A neutral ebonite rod is rubbed with a neutral fur. Which of the following receive a charge?
   a) ebonite rod only
   b) fur only
   c) both the ebonite rod and the fur
   d) neither the ebonite rod nor the fur
2. A glass rod is rubbed with silk. An ebonite rod is rubbed with fur. Which of the following describes the charge on the materials after rubbing?

	Ebonite	Fur	Glass	Silk
a)	−	−	+	+
b)	+	+	−	−
c)	+	−	+	−
d)	−	+	+	−

3. Which of the following statements correctly describes a scientific law? A scientific law
   a) summarizes observations only
   b) explains observations only
   c) summarizes and explains observations
   d) is the same as a scientific theory
4. A metal leaf electroscope is being charged positively by contact. Which particles move? In what direction?
   a) Electrons move from the sphere to the metal leaves.
   b) Electrons move from the sphere to the charged rod.
   c) Protons move from the sphere to the metal leaves.
   d) Protons move from the charged rod to the sphere.
5. To charge an electroscope positively by induction, the sphere is touched with a finger. The finger
   a) neutralizes the electroscope
   b) repels electrons to the leaf
   c) conducts electrons to ground
   d) conducts protons from ground

D. Using Your Knowledge

1. A rubber balloon is rubbed with wool. The balloon repels an ebonite rod rubbed with fur.
   a) What charge is on the balloon? the wool?
   b) Explain why a charged balloon will stick to a wall on a dry day.
2. Lycopodium powder consists of the spores of a club moss. These spores are neutral and are very small. If a negative ebonite rod is placed in the powder, spores cling to the rod. However, a short time later they fly off in all directions. Explain why this happens.
3. Three ping pong balls are suspended from the corners of a triangle using insulating threads. They attract each other. What charges are on the ping pong balls?
4. Four small styrofoam spheres A, B, C, and D are hung from the corners of a square. They are suspended by insulating threads. Spheres D and C repel. Spheres A, B, and C attract each other. If A is positive, what is the electric charge on B, C, and D?
5. Use the Electron Theory of Charge to explain why there is no charge on the inside of a charged hollow metal sphere.

E. Investigations

1. Do a library search to find out what is meant by grounding a conductor and what is meant by shielding.
2. Scientists have made an **electrostatic series** to show the electric charge different pairs of materials receive when rubbed together. Do a library search to find an electrostatic series. Use it to predict the charge when a steel rod is rubbed with rubber. Do an experiment to test your prediction.
3. The Electron Theory of Charge predicts that two objects rubbed together get the same amount of charge. However the charges are opposite in kind. Design and do an experiment using a metal leaf electroscope to test this prediction.
4. Do a library search to find the contribution to the Electron Theory of Charge made by the following people: Dr. William Gilbert (1544-1603); John Canton (1718-1772); and Benjamin Franklin (1706-1790).
5. Some materials let electrons move better than others. Conductors let electrons move easily. Insulators hinder the flow of electrons. Design and do an experiment to classify materials as conductors or insulators. Use an electroscope in your experiment.

32 Sources of Current Electricity

32.1 Electric Potential and Electric Current
32.2 Activity: Factors Affecting Current from a Voltaic Cell
32.3 Sources of Current Electricity
32.4 Activity: Charging and Discharging a Lead Storage Cell
32.5 Storage Batteries

Electric charge in motion is called **current electricity**. In this chapter you will study sources of current electricity. These include wet cells, dry cells, and batteries. You will find out how to use a voltmeter and ammeter to measure electricity.

32.1 Electric Potential and Electric Current

Static electricity is produced by rubbing two materials together. The person doing the rubbing does work. The work separates electrons from atoms. One material gains electrons and becomes negative. The other material loses electrons and becomes positive.

Chemical energy can also be used to do work to separate electrons from atoms. Some of the work done to separate the charge is stored in the separated surplus electrons as potential energy. *The energy stored in a group of electrons is called* **electric potential**. The electric potential of the surplus electrons on a negative material is similar to the potential energy stored in raised water. If the water is released, it flows from a high to a low level. If the surplus electrons are free to move, a current flows from an area with an excess of electrons to an area deficient in electrons. The greater the electric potential of the surplus electrons, the larger the electric current. Electric potential is measured in **volts**. The symbol for volt is **V**. A flashlight D dry cell has an electric potential of about 1.5 V (dc). A car battery has an electric potential of about 12 V (dc).

Measuring Electric Potential

A **voltmeter** [VOLT-meet-er] measures electric potential. When connected between the terminals of a source of electricity, the needle on the dial moves. The needle points to the number of volts. Each voltmeter has a different scale. It must be connected correctly to prevent damage. Your teacher will show you how to connect and read a voltmeter.

Electric Current

A water current consists of moving water. An electric current consists of moving electrons. Electrons flow through the conducting wires. The rate of flow of electrons through a wire is measured in **amperes** [AM-pir]. The symbol for ampere is **A**. A current of about one ampere (1 A) flows through a 100 W bulb. One ampere is a flow of about 6 000 000 000 000 000 000 electrons past a point each second.

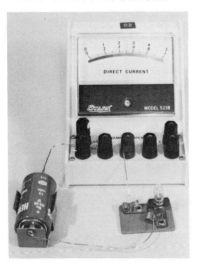

Fig. 32-1 An ammeter measures electric current.

Measuring Electric Current

An **ammeter** [AM-meet-er] measures electric current. When electric current flows through the meter, the needle points to the number of amperes. A typical ammeter is shown in Figure 32-1. It must be connected correctly to prevent damage. Your teacher will show you how to connect and read an ammeter.

Section Review

1. What is electric potential?
2. What is the unit and symbol for electric potential?
3. Name the instrument used to measure electric potential.
4. What is electric current?
5. What is the unit and symbol for electric current?
6. Name the instrument used to measure electric current.

32.2 ACTIVITY Factors Affecting Current from a Voltaic Cell

A voltaic cell is a source of electricity. It makes an electric current from a chemical reaction. In this activity you will make and study a voltaic cell. You will find out the factors needed to produce an electric current. And you will use an ammeter and a voltmeter to measure electricity.

Problem

What factors affect the electric current from a voltaic cell?

Materials

voltaic cell
steel wool
tap water
2 copper strips
1 zinc strip
2 conducting wires
potassium dichromate
dilute sulfuric acid
ammeter (0-1 A)
voltmeter (0-1.5 V)

CAUTION: Sulfuric acid is corrosive. Wear safety goggles during this activity.

Procedure

a. Copy Table 32-1 into your notebook.
b. Polish the metal strips using steel wool.
c. Half fill the voltaic cell with dilute sulfuric acid solution (Fig. 32-2).

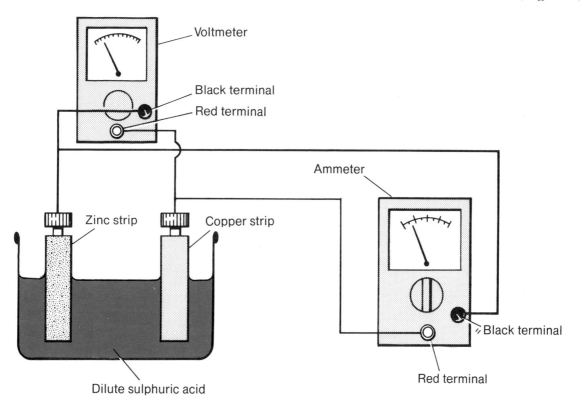

Fig. 32-2 A simple voltaic cell connected to an ammeter and a voltmeter

d. Place a zinc strip and a copper strip in the solution.
e. Connect the zinc strip to the black terminal of the ammeter. Connect the copper strip to the red terminal. Also connect the voltmeter as shown.
f. Record the reading for electric current on the ammeter. Also record the reading for electric potential on the voltmeter. Your teacher will help you read the meters.
g. Watch the surface of the metal strips for bubbles. Record any action.
h. Allow the cell to operate for about 5 min. Read the meters again. Also, note whether bubbles are present.

Section 32.2 393

i. Wipe away any bubbles from the metal strips. Repeat steps (f) and (g).
j. Again let the cell operate for another 5 min. Then add a spoonful of potassium dichromate to the acid solution. Repeat steps (f) and (g).
k. Replace the zinc strip with another copper strip. Place the two copper strips in the solution used in (j). Repeat steps (f) and (g).
l. Rinse the metal strips with tap water. Replace the solution in the voltaic cell with tap water. Repeat steps (f) and (g) using a zinc and copper strip.

Table 32-1 Current from a Voltaic Cell

Factor changed	Metal strips	Solution	Electric current (A)	Electric potential (V)	Action at strips
—	Copper and zinc	Sulfuric acid			
Leave 5 min	"	"			
Wipe bubbles off	"	"			
Add potassium dichromate	"	Sulfuric acid + potassium dichromate			
Use identical strips	Copper and copper	"			
Change solution	Copper and zinc	Tap water			

Discussion

1. What did you observe on the surface of the copper strip?
2. What happened to the electric current and electric potential as time passed?
3. What effect does adding potassium dichromate have on the cell?
4. Which combination of factors produced the best voltaic cell?

32.3 Sources of Current Electricity

There are several sources of current electricity. The first continuous source was discovered by an Italian physicist Alessandro Volta. This source of current electricity was named the voltaic cell in honour of Volta.

The Voltaic Cell

A **voltaic cell** consists of two different metal plates in a solution. The plates are called **electrodes** [ee-LEK-trodes]. The solution is called the electrolyte [ee-LEK-tro-lite]. The solution may be an acid, a base, or a salt dissolved in water. This is called a **wet cell** because of this solution.

One kind of voltaic cell has copper and zinc electrodes. A chemical reaction takes place between the electrodes and sulfuric acid. Chemical energy separates the positive and negative charges. The copper gets a positive charge. The zinc gets a negative charge. If the electrodes are connected outside by a conductor as shown in Figure 32-3, electrons start to flow through the conductor. Electrons move from the negative zinc electrode to the positive copper electrode. The electric current continues as long as the chemical action continues.

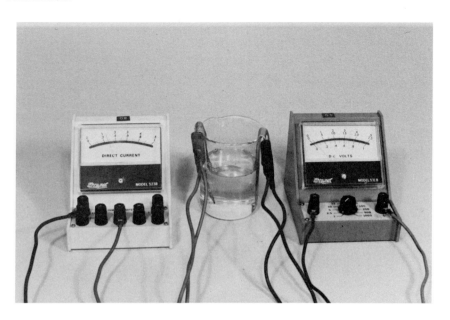

Fig. 32-3 Electrons flow from the zinc electrode through the ammeter to the copper electrode in a voltaic cell.

The voltaic cell has a number of faults which limit its use. The zinc electrode is used up. This is because zinc reacts with the acid and goes into solution. Also, bubbles of hydrogen gas form around the copper electrode. This allows less of the copper to be in contact with the acid. Therefore the action of the cell slows down. As a result, less electric current flows. This can be overcome by removing the hydrogen. One way is to wipe the hydrogen off the electrode; another is to use a chemical. It reacts with the hydrogen to produce water. One chemical used is potassium dichromate. The main fault is the danger of spilling the acid solution. Acid can eat through clothes and burn skin. This fault led to the invention of the dry cell.

The Dry Cell

The structure of the dry cell is shown in Figure 32-4. The cell has a carbon electrode and a zinc electrode. The zinc electrode is the cylindrical container around the outside. The carbon electrode is the central rod. Between the carbon and zinc electrodes is a moist paste of ammonium chloride. This salt is the electrolyte. Carbon powder is mixed with the electrolyte to improve conduction. The chemical action within the cell removes electrons from the carbon and gives them to the zinc. Zinc becomes the negative electrode. The carbon becomes the positive electrode. A chemical called manganese dioxide prevents the collection of hydrogen gas on the surface of the carbon electrode. A metal cap in contact with the carbon electrode is the positive terminal. The negative terminal is another metal cap connected to the zinc container.

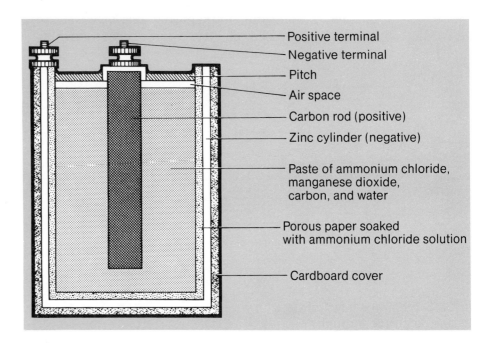

Fig. 32-4 Cross section of a dry cell

Section Review

1. Make a labelled diagram of a simple voltaic cell. Mark the direction of electron flow in the conducting wire and in the solution.
2. What causes the decrease in the current from a voltaic cell? How is this overcome?
3. Make a labelled diagram of a dry cell. Describe the function of each part.
4. What is one advantage of a dry cell over a voltaic cell?

32.4 ACTIVITY Charging and Discharging a Lead Storage Cell

Both the simple voltaic cell and the dry cell have a short life span. They stop producing electricity for two reasons. The electrolyte dries out. The zinc electrode is used up. These cells are called **primary cells** because they cannot be charged again.

The lead storage cell overcomes these faults. Electrical energy is stored as chemical energy while the cell is being charged. As the chemical energy is used to produce electrical energy, the cell discharges. But the cell can be charged again. Therefore it is called a **secondary cell**.

In this activity you will study how a lead storage cell works. You will observe the changes that take place during charging and discharging.

Problem

How does a lead storage cell work?

Materials

2 lead strip electrodes dilute sulfuric acid
steel wool flashlight bulb and socket, 1.5 V (dc)
battery jar 1-6 V (dc) power supply
2 conducting wires marker

CAUTION: Sulfuric acid is corrosive. Wear safety goggles during this activity.

Procedure

a. Copy Table 32-2 into your notebook.
b. Clean the lead strips with steel wool.
c. Mark one lead strip + and the other − with the marker.
d. Half fill a battery jar with dilute sulfuric acid.
e. Place the lead strips in the electrolyte.
f. Attach a red conducting wire to the lead strip marked +.
g. Attach a black conducting wire to the lead strip marked −.
h. Connect the wires to the flashlight bulb. Record your observations. Disconnect the bulb.
i. Connect the wires to the power supply as shown in Figure 32-5.

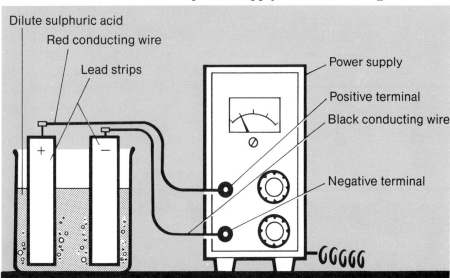

Fig. 32-5 A lead acid storage cell being charged

Section 32.4 397

j. Plug in the power supply for about 3 min. Record any action at the electrodes.
k. Replace the power supply with the bulb. Record your observations of the bulb. Also record any action you see at the electrodes.
l. Repeat steps (i), (j), and (k) if you have time.

Table 32-2 The Operation of the Lead Storage Cell

Step	Observations		
	Positive electrode	Negative electrode	Lamp
Cell uncharged			
Cell charging			
Cell discharging			

Discussion

1. Describe the surface of the lead strips before the cell was charged.
2. What changes took place in the lead strips during the charging?
3. How do you know the strips are different from one another after charging?
4. What changes took place in the lead strips and bulb during discharging?
5. What is one difference between a wet or dry cell and a storage cell?

32.5 Storage Batteries

The Difference Between a Cell and a Battery

A **cell** has two electrodes made of different materials. For example, the electrodes may be made of zinc and copper. The electrodes do not touch. They are covered with an electrolyte. Each electrode is connected to the outside by a terminal. One terminal is positive. The other is negative. This is shown in Figure 32-6,A.

A **battery** is a number of cells connected together with two sets of electrodes. Each set consists of several parallel plates connected together (Fig. 32-6,B). They look like the leaves of a book. The electrodes in each set are made of the same material. But the electrodes in different sets consists of different materials. One set is connected to the positive terminal of the battery. The other set is connected to the negative terminal. The two sets of electrodes do not touch. But they are covered by the same electrolyte. Thus a battery consists of a number of cells connected together. A battery is stronger than the cell.

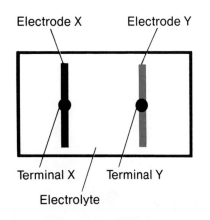

A. Top view of a cell

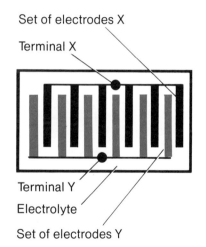

B. Top view of a battery showing the plates making up a set of electrodes

Fig. 32-6 The difference between a cell and a battery

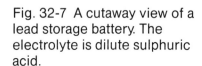

Fig. 32-7 A cutaway view of a lead storage battery. The electrolyte is dilute sulphuric acid.

The Lead Storage Battery

The lead storage battery is used in cars. When charged, one set of electrodes is made of black lead dioxide. It is connected to the positive terminal. The other set of electrodes is made of grey spongy lead. It is connected to the negative terminal. The two sets of electrodes are covered with dilute sulfuric acid. A cutaway view of a battery is shown in Figure 32-7. When the terminals are connected to a load like the headlamps, a current flows. The battery acts like a voltaic cell. Both sets of plates undergo a chemical change. Both begin to change to lead sulfate. When both sets of plates change completely to lead sulfate, the current stops.

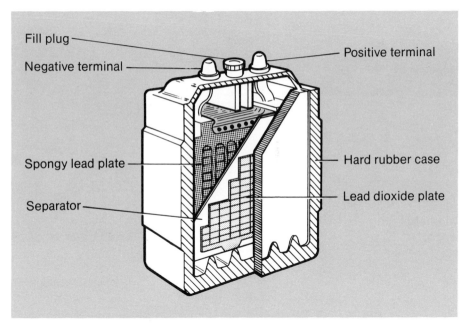

Checking the Charge of a Battery

The charge of a battery is determined by measuring the density of the electrolyte. The electrolyte consists of sulfuric acid and water. Water has a density of 1000 g/L. Pure sulfuric acid has a density of 1800 g/L. When charged, the electrolyte has a density of about 1300 g/L. When discharged it has a density of about 1200 g/L. As the battery discharges more water is produced. The density becomes closer to the density of water. During charging more sulfuric acid is produced. The density approaches that of pure sulfuric acid. Thus the density of the electrolyte is a measure of the charge of the battery. The density of the electrolyte is measured using an instrument called a **battery hydrometer** [hy-DRAM-e-ter] (Fig. 32-8).

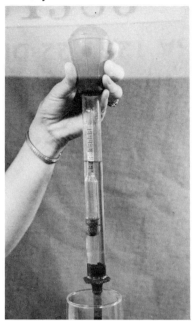

Fig. 32-8 A battery hydrometer is used to check the charge of a lead acid storage battery.

The Need for New Storage Batteries

The lead storage battery has two problems. Lead is one of the densest metals. Even a small amount of lead is very heavy. It is also costly for the amount of energy it can store. A car battery contains about 9 kg of lead. It stores about the same energy as 150 mL of gasoline. This is enough energy to move a car about 1.5 km. It would take about 500 kg of storage batteries to move the lightest car 80 km (Fig. 32-9). Thus battery-run cars are not very practical. But engineers and scientists are searching for lighter and more efficient storage batteries. One of these is the mercury battery. Another is the lithium-sulfur battery. A third is the sodium-sulfur battery. Better storage batteries may make the electric car economical in the 1980s.

Fig. 32-9 A typical electric automobile, the *Lectric Leopard*

Section Review

1. What is the main difference between a storage cell and a storage battery?
2. Compare a simple voltaic cell and a lead storage battery. Use the following headings: electrolyte; kind of electrodes; number of electrodes; size of current; ability to be recharged.
3. Refer to the densities of water and sulfuric acid. Describe how density is used to measure the charge of a battery.
4. Why is the lead storage battery not suitable for powering electric cars?

Main Ideas

1. A voltmeter measures electric potential. An ammeter measures electric current.
2. A simple voltaic cell consists of two electrodes and an electrolyte.
3. A voltaic cell produces electricity using chemical energy.

4. A dry cell is similar to the voltaic cell except the electrolyte is a paste.
5. Primary cells, such as the dry cell and voltaic cell, cannot be recharged.
6. Secondary cells, such as the lead storage cell, can be recharged.
7. A battery consists of a number of cells connected together.

Glossary

ammeter	AM-meet-er	an instrument to measure electric current
ampere	AM-pir	the rate of flow of electrons through a wire
cell		a source of electricity with two electrodes
battery		a number of cells connected together
electrodes	ee-LEK-trodes	the terminals of a source of electricity
electrolyte	ee-LEK-tro-lite	a solution that conducts electricity
voltmeter	VOLT-meet-er	a instrument for measuring electric potential

Study Questions

A. True or False

Decide whether each of the following sentences is true or false. If the sentence is false, rewrite it to make it true. (Do not write in this book.)
1. The electrodes in a voltaic cell are different.
2. Once used, a voltaic cell can be recharged.
3. A dry cell must contain moisture before it will work.
4. When charged, the electrodes in a lead storage cell are the same.
5. A lead storage battery is made up of a number of primary cells connected together.

B. Completion

Complete each of the following sentences with a word or phrase which will make the sentence correct. (Do not write in this book.)
1. A voltaic cell is an example of a ▬▬▬ cell.
2. Hydrogen gas on the ▬▬▬ electrode causes the current from a voltaic cell to ▬▬▬ .
3. A dry cell no longer works once the ▬▬▬ is used up.
4. The negative electrode in a dry cell is made of ▬▬▬ .
5. The electrolyte in a lead storage battery is ▬▬▬ .

C. Multiple Choice

Each of the following statements or questions is followed by four responses. Choose the correct response in each case. (Do not write in this book.)

1. What parts are needed to keep a voltaic cell producing current?
   a) two copper electrodes and dilute sulfuric acid
   b) a zinc electrode, a copper electrode, and tap water
   c) a zinc electrode, a copper electrode, and dilute sulfuric acid
   d) a zinc electrode, a copper electrode, dilute sulfuric acid and potassium dichromate
2. The addition of potassium dichromate to a voltaic cell
   a) causes the current to increase
   b) keeps the current the same
   c) causes the current to decrease
   d) has no effect on the cell
3. A chemical reaction takes place between the electrodes and the electrolyte in a dry cell. The chemical reaction
   a) creates more electrons at the zinc electrode than the carbon electrode
   b) creates more protons at the zinc electrode than the carbon electrode
   c) transfers electrons from the zinc electrode to the carbon electrode
   d) transfers electrons from the carbon electrode to the zinc electrode
4. During charging of a lead storage cell
   a) both electrodes turn the same colour
   b) there is a chemical reaction at both electrodes
   c) both electrodes change to lead
   d) the amount of water in the cell increases
5. Which of the following statements explains the change in density of a lead storage battery during discharging? The density
   a) increases because the amount of sulfuric acid increases
   b) decreases because the amount of sulfuric acid increases
   c) increases because the amount of sulfuric acid decreases
   d) decreases because the amount of sulfuric acid decreases

D. Using Your Knowledge

1. Describe the energy conversion that takes place in a wet or dry cell.
2. Describe the energy conversions that take place in a storage battery.
3. A surge of current is observed when a voltaic cell is first connected to an ammeter. Why?
4. What is the reason for the sharp pain if a piece of aluminum foil is held between teeth in which there are fillings?
5. How many lead storage batteries are needed to move the lightest electric car 80 km?

E. Investigations

1. Predict the effect on the current from a voltaic cell of using aluminum, lead, or carbon in place of the copper electrode. Design and do an experiment to test your prediction.
2. Make a voltaic cell in your mouth. Clean a penny and a nickel with soap and water. Connect them with a conducting wire. Put one on top of your tongue and the other under it. What do you feel?
3. Use a potato to detect the electric current from a good dry cell. Slice a potato in half. Attach conducting wires to the dry cell. Push the bare ends closer together into the potato. Do not let them touch. Describe what you see.
4. Do a library search to find the work being done to produce better storage batteries.
5. Obtain a used dry cell. With a hacksaw blade cut the cell from top to bottom down the middle. CAUTION: Do not touch the inside. The materials can cause burns. Study the parts of the cell. Soak the cell in a solution of ammonium chloride for about half an hour. See if it will light a bulb.

33 Series and Parallel Circuits

33.1 Activity: Properties of a Circuit
33.2 A Simple Electric Circuit
33.3 Activity: Light Bulbs in Series
33.4 Activity: Light Bulbs in Parallel
33.5 Connecting Loads
33.6 Activity: Dry Cells in Series
33.7 Activity: Dry Cells in Parallel
33.8 Connecting Sources

An electrical circuit has a source of electric current, conducting wires, and a switch. It also has a load. A load changes electric energy into other forms of energy such as heat and light. It also limits the current. These parts are connected together. They may be connected in different ways. In this chapter you will study two ways, series and parallel. You will find the difference between a series and a parallel circuit. You will also study a short circuit. Finally, you will find out how a fuse prevents overheating of electrical wiring.

33.1   ACTIVITY   Properties of a Simple Circuit

An electric **circuit** [SIR-cut] is the path that an electric current follows. In this activity you will connect a simple circuit. You will learn where to place a switch and how to connect an ammeter. You will compare the current at various places. Finally, you will compare the current when different light bulbs are used.

Problem

What are the properties of a simple electric circuit?

Materials

1 D dry cell
1 D dry cell holder
1 switch
1 bulb holder
1 #41 WB bulb ($V = 2.5$ V)
1 #48 PB bulb ($V = 2.0$ V)
4 conducting wires
1 ammeter

Procedure

a. Copy Table 33-1 into your notebook.

b. Connect the circuit shown in Figure 33-1,A. Use a #41 WB bulb.

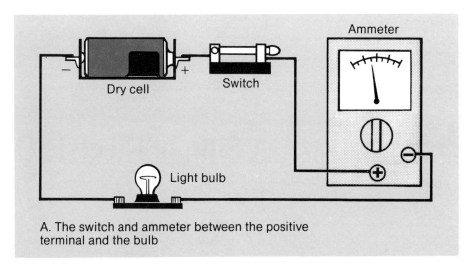

Fig. 33-1 Properties of a simple electric circuit

A. The switch and ammeter between the positive terminal and the bulb

c. Close the switch. Record the brightness of the bulb and the current.
d. Connect the circuit shown in Figure 33-1,B. Repeat step (c).

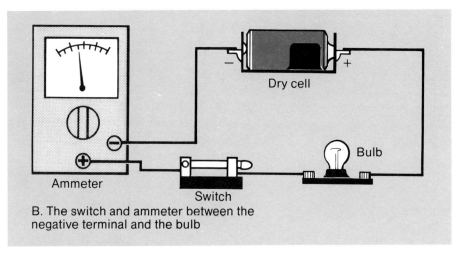

B. The switch and ammeter between the negative terminal and the bulb

e. With the ammeter reading a current, unscrew the #41 WB bulb. Repeat step (c).
f. Replace the #41 WB bulb with the #48 PB bulb. Repeat step (c).

Table 33-1 Properties of a Simple Circuit

Location of switch and ammeter	Bulb type	Bulb brightness	Electric current (A)
Between positive terminal and bulb	#41 WB		
Between negative terminal and bulb	#41 WB		
Between negative terminal and bulb	#48 PB		

Section 33.1

Discussion

1. Where can a switch be placed in a simple circuit?
2. Compare the current at various places in the circuit.
3. Explain the reading on the ammeter when the bulb was unscrewed.
4. Which light bulb decreases the current the most?
5. Which light bulb is the largest load?

33.2 A Simple Electric Circuit

A. Photograph showing a short circuit

The Switch

A **switch** controls the movement of electrons in a circuit. With the switch closed, the circuit is complete and the electrons have a path to follow. With the switch open, the circuit is incomplete. Electrons cannot flow through an air gap. In a simple circuit it doesn't matter where the switch is placed.

The Load

A **load** changes electric energy to other forms of energy. A light bulb changes electricity to light and heat. A motor changes electricity to kinetic energy and heat. The load limits the size of the electric current. The larger the load, the smaller the current. Without a load, the current is very large. This causes overheating of the conducting wires.

The Short Circuit

Sometimes the insulation on conducting wires breaks down. The bare wires touch, completing the circuit. The current flows but it does not pass through the load. This is called a **short circuit**. Figure 33-2 shows a short circuit. Without a load, the current becomes very large. A large current causes the wires to become very hot. They may either melt or cause a fire in nearby materials. The dangers caused by a short circuit are prevented by a fuse.

B. Symbols showing a short circuit. Bare wires are touching. The current becomes very large because it bypasses the lamp.

Fig. 33-2 A photograph and symbol diagram showing a short circuit.

The Fuse

The **fuse** is a safety device. It acts as a switch. Fuses are connected into the circuit close to the source. All the current flowing in the circuit passes through the fuse. When too much current flows, the fuse burns out. This creates an air gap. No more current flows. The construction of a plug fuse is shown in Figure 33-3. The essential part is the fuse wire. It has a low melting point.

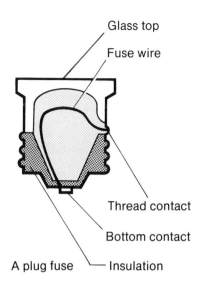

Fig. 33-3 A fuse. The current flows through the fuse wire connected between the bottom contact and the thread contact.

Symbols

Symbols are used to represent the parts and wiring details of an electric circuit. It is simpler to show them using symbols than to draw diagrams. Some of the most common symbols are shown in Figure 33-4.

Section Review

1. Draw a diagram using symbols for a simple circuit. Show a source, a switch, a lamp, a fuse, an ammeter, and conducting wires.
2. Describe the function of each part of the electric circuit you drew.
3. a) What causes a short circuit?
   b) What danger results from a short circuit?
4. a) What is a fuse?
   b) How does a fuse protect a circuit?

Fig. 33-4 Some electrical symbols

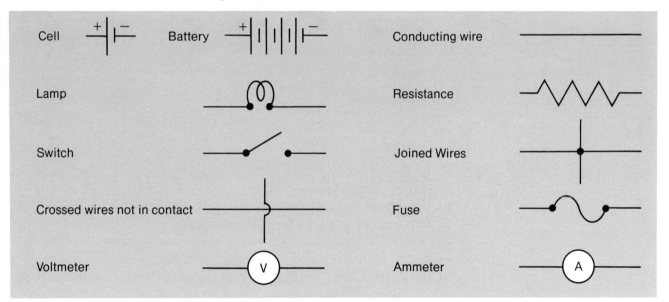

33.3 ACTIVITY Light Bulbs in Series

A **series circuit** [SEE-rees] has only one path through which current can flow. A series circuit is similar to a race track. All cars in a race follow the same path. So do all electrons in a series circuit. In this activity you will study light bulbs in series.

Section 33.3 407

Problem

How do light bulbs in series behave?

Materials

1 D dry cell
1 D dry cell holder
1 switch
1 ammeter

3 bulbs
3 bulb holders
6 connecting leads

Procedure

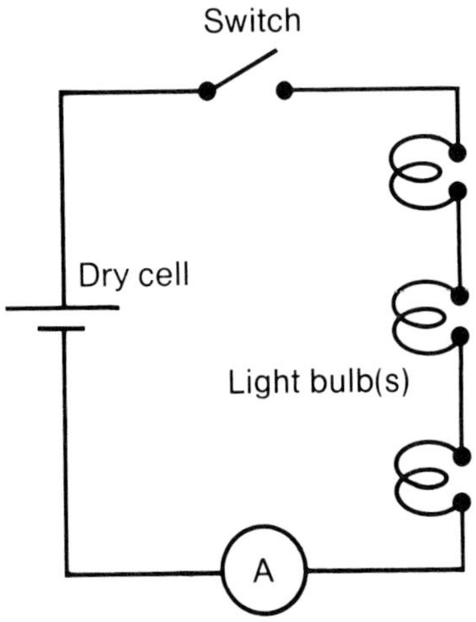

a. Copy Table 33-2 into your notebook.
b. Draw the circuit shown in Figure 33-5 in your notebook.
c. Connect the circuit shown in Figure 33-5. Use only one bulb. Close the switch.
d. Note the brightness of the bulb. Record the current.
e. Connect a second bulb in series. Repeat step (c).
f. Connect a third bulb in series. Repeat step (c).
g. Unscrew one bulb. Record the effect on the current and the bulb.

Table 33-2 Light Bulbs Connected in Series

Bulbs connected	Bulb brightness	Electric current (A)
1		
2		
3		
3 (1 unscrewed)		

Fig. 33-5 Light bulbs in series

Discussion

1. As more bulbs are connected in series, what happens to
   a) the brightness of each bulb?
   b) the size of the current?
   c) the number of paths for the current?
   d) the size of the load?
2. In a series circuit, if one light bulb burns out, what happens to
   a) the current?   b) the brightness of the remaining bulbs?

33.4 ACTIVITY Light Bulbs in Parallel

A **parallel circuit** [PA-ra-lel] has two or more paths through which electrons can flow. These are called **branches**. A parallel circuit is like streets around several city blocks. Cars leaving a house can take different paths before returning home. In like manner, electrons leaving the source take different paths before returning to the source. Look at Figure 33-6. Three light bulbs are connected in parallel. Do you think the current from the dry cell will increase, decrease, or remain the same? What will happen to the brightness of the bulbs? Write predictions in your notebook. Then do the activity to find out.

Problem

How do light bulbs in parallel behave?

Materials

1 D dry cell
1 D dry cell holder
1 switch

1 ammeter
3 bulbs

3 bulb holders
8 connecting leads

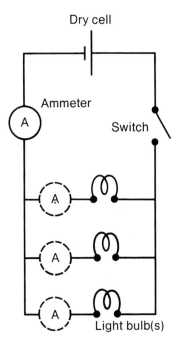

Fig. 33-6 Light bulbs in parallel

Procedure

a. Copy Table 33-3 into your notebook.
b. Draw the circuit shown in Figure 33-6 in your notebook.
c. Connect the circuit shown in Figure 33-6. Use only one bulb. Close the switch.
d. Note the brightness of the bulb. Record the reading in the main circuit using the ammeter.
e. Connect a second bulb in parallel. Compare and record the brightness.
f. Record the current in the main circuit. Then record the current in each branch circuit as shown.
g. Connect a third bulb in parallel.
h. Record the effect on the current in the main circuit and in each branch circuit.
i. Unscrew one bulb. Record the effect on the current in the main circuit.

Table 33-3 Light Bulbs Connected in Parallel

Bulbs connected	Bulb brightness	Reading on ammeter (A)		
		Main circuit	Branch circuits	
1				
2				
3				
3 (1 unscrewed)				

Discussion

1. As more bulbs are connected in parallel, what happens to the
   a) brightness of each bulb?
   b) current in the main circuit?
   c) current in the branches?
   d) number of paths for current?
   e) size of the total load?

2. In a parallel circuit, if one light bulb burns out, what happens to
   a) the current in the main circuit?
   b) the brightness of the remaining bulbs?

33.5 Connecting Loads

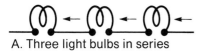
A. Three light bulbs in series

B. One light bulb burned out in a series circuit stops the current

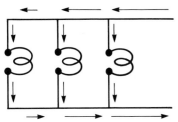
C. Three light bulbs in parallel

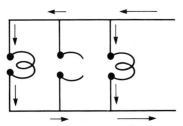
D. One light bulb burned out in a parallel circuit decreases the current

Fig. 33-7 Connections of light bulbs

Series Connection

A series circuit has only one path for the current. Thus the current in every part of the circuit is the same. A series connection of three light bulbs is shown in Figure 33-7,A. Three bulbs connected in series are three times the load of one. As the load increases, the current decreases. Thus as more bulbs are connected, each bulb glows less brightly. If one bulb burns out, all the bulbs go out. The broken bulb creates an incomplete circuit. A burned out bulb is shown in Figure 33-7,B. Some older versions of Christmas tree lights were connected in series. What problem would this cause?

Parallel Connection

A parallel circuit has two or more branches for the current. Consider two light bulbs connected in parallel. The current leaving the source splits into parts. Part of the current goes through each bulb. Thus it is easier for the current to flow. It is just like a single lane highway dividing into two lanes. It is easier for cars to travel in two lanes than in one. The current from the source increases.

Three bulbs connected in parallel are shown in Figure 33-7,C. The current flowing through one bulb is not affected by the other bulbs. Thus the addition of more bulbs in parallel does not affect bulb brightness. Also if one bulb goes out, there is still a complete circuit for the current. The remaining bulbs stay lit. A burned out bulb is shown in Figure 33-7,D. Most lights today, including household lights, are connected in parallel.

Section Review

1. a) Define a series circuit.
   b) Define a parallel circuit.
2. How do you know that house lights are connected in parallel?
3. a) Draw a diagram showing two bulbs connected in series to a battery.
   b) Draw a diagram to show two bulbs connected in parallel to a battery.
   c) Write a paragraph that describes how the circuits drawn in (a) and (b) are different.
   d) Which connection would produce more light? Why?

33.6 ACTIVITY Dry Cells in Series

In this activity you will connect dry cells in series. You will then study how the number of dry cells affects bulb brightness. You will use a voltmeter to measure the electric potential of the source. You will also use an ammeter to measure the current leaving the source. Finally, you will see what happens if a dry cell is connected backward to another.

Problem
How do dry cells in series work?

Materials

3 D dry cells
3 D dry cell holders
1 switch
1 ammeter

1 bulb
1 bulb holder
1 voltmeter
6 connecting leads

Procedure

a. Copy Table 33-4 into your notebook.
b. Connect the circuit shown in Figure 33-8. Start with one dry cell.

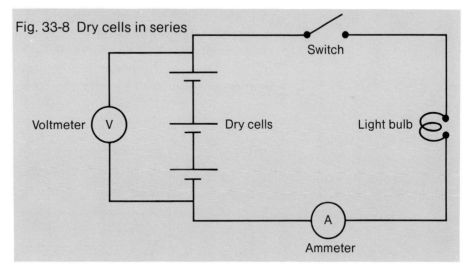

Fig. 33-8 Dry cells in series

c. Close the switch. Note the brightness of the bulb. Record the reading on the voltmeter and ammeter.
d. Connect a second dry cell in series (positive to negative). Connect the positive terminal of one dry cell to the negative terminal of the next. Repeat step (c).
e. Connect three dry cells in series. Repeat step (c).
f. Reverse one of the dry cells. Connect two positive terminals together.
g. Repeat step (f) with the two negative terminals joined together.

Table 33-4 Dry Cells Connected in Series

Number of dry cells	Bulb brightness	Electric current (A)	Electric potential (V)
1			
2 — positive to negative			
3 — positive to negative			
3—2 positive to positive			
3—2 negative to negative			

Discussion

1. As more dry cells are connected in series, what happens to the
   a) bulb brightness?   **b)** current?   **c)** electric potential?
2. In a series circuit, if one dry cell is backwards to the others, what happens to
   a) the current?   **b)** the electric potential?

33.7 ACTIVITY Dry Cells in Parallel

When dry cells are connected in parallel, like terminals are joined as in Figure 33-9. What effect do you think increasing the number of dry cells in parallel has on the brightness of the light bulb? on the electric potential? on the total current? What will happen if one dry cell is connected backwards to another? Write your predictions in your notebook. Do an activity similar to 33.6 to test your predictions.

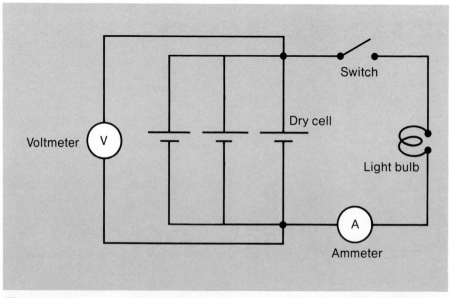

Fig. 33-9 Dry cells in parallel

Discussion

1. As more dry cells are connected in parallel, what happens to
   a) the brightness of the bulb?
   b) the current?
   c) the electric potential?
2. If one dry cell is backwards to the others in a parallel circuit, what happens to
   a) the current?
   b) the electric potential?

33.8 Connecting Sources

Series Connection

Three dry cells connected in series are shown in Figure 33-10,A. Each positive terminal is connected to a negative terminal. Connecting dry cells in series increases the electric potential of the source. The volts add. Thus more current flows through the load. If the load is a light bulb, it gets brighter. Suppose one dry cell is placed backwards as shown in Figure 33-10,B. The potential decreases to zero. The current that flows also decreases to zero. Any time a dry cell is connected backwards in a series circuit, the electric potential decreases and the electric current decreases.

Parallel Connection

Three dry cells connected in parallel are shown in Figure 33-10,C. All the positive terminals are connected together. So are all the negative terminals. Connecting dry cells in parallel does not change the electric potential. Nor does it change the size of the current. A light bulb glows with the same brightness. But each dry cell gives only part of the current. With three dry cells, each dry cell gives one third of the total current. Thus each dry cell lasts three times as long. If one dry cell is placed backwards as shown in Figure 33-10,D, the electric potential decreases to zero. The current that flows also decreases to zero. Two good dry cells in parallel with opposite terminals connected produce no electric potential and no current. If a dry cell is connected backwards in a parallel circuit, the electric potential and current decrease.

Section Review

1. Describe how dry cells are connected in series. Use a diagram in your answer.
2. Describe how dry cells are connected in parallel. Use a diagram in your answer.
3. Suppose that three 1.5 V (dc) dry cells power a light bulb. Compare the results if the dry cells are connected in series and in parallel. Explain.

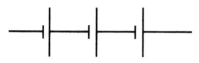

A. Three dry cells in series. The electric potentials add.

B. Two dry cells connected backwards in series. The electric potentials cancel.

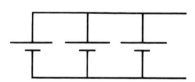

C. Three dry cells in parallel. The electric potential is the same as with one.

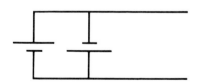

D. Two dry cells connected backwards in parallel. The electric potentials cancel.

Fig. 33-10 Connections of dry cells

Main Ideas

1. A circuit consists of a source, conducting wires, a switch, and a load.
2. A load converts electric energy to other forms of energy.
3. A fuse prevents overheating caused by a short circuit.
4. Light bulbs connected in series increase the load and decrease the current.
5. Light bulbs connected in parallel decrease the load and increase the current.
6. The electric potentials add if dry cells are connected in series.
7. Dry cells connected in series increase the current through the load.
8. The electric potential stays the same if dry cells are connected in parallel.
9. Dry cells connected in parallel do not change the current through the load.
10. Two dry cells reversed produce no electric potential and no current.

Glossary

circuit	SIR-cut	the path an electric current follows
series circuit	SEE-rees	a circuit with only one current path
parallel circuit	PA-ra-lel	a branching circuit with more than one current path

Study Questions

A. True or False

Decide whether each of the following sentences is true or false. If the sentence is false, rewrite it to make it true. (Do not write in this book.)

1. A circuit must have a load.
2. A short circuit results if a load is present in a circuit.
3. Light bulbs connected in parallel decrease the current from the source.
4. The current from two dry cells connected in series is the same as from one.
5. If one of two dry cells in parallel is reversed, the current decreases.

B. Completion

Complete each of the following sentences with a word or phrase

which will make the sentence correct. (Do not write in this book.)
1. A short circuit causes the current to ▓▓▓ .
2. A load converts ▓▓▓ to other forms of energy.
3. The more light bulbs connected in series, the ▓▓▓ the current.
4. Connecting dry cells in parallel ▓▓▓ the electric potential.
5. If one of two dry cells in series is reversed, the electric potential ▓▓▓ .

C. Multiple Choice

Each of the following statements or questions is followed by four responses. Choose the correct response in each case. (Do not write in this book.)

1. A symbolic diagram of a simple electric circuit is shown. Which symbol represents the load?
2. Where must the switch be placed in a simple circuit?
    a) next to the positive terminal   c) next to the load
    b) next to the negative terminal   d) anywhere in the circuit
3. Which of the following statements describes the current in a series circuit? The current is
    a) the same at every point
    b) largest through the load
    c) largest next to the positive terminal
    d) largest next to the negative terminal
4. Two bulbs x and y are connected in parallel to a dry cell. The switch is closed. If bulb x is then unscrewed, what will happen to the brightness of bulb y?
    a) increase   b) decrease   c) remain the same   d) become zero
5. Three dry cells x, y, and z are connected in series to one light bulb. The switch is closed. If one dry cell is then reversed, what will happen to the current?
    a) increase   b) decrease   c) remain the same   d) become zero

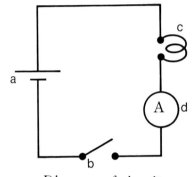

Diagram of circuit

D. Using Your Knowledge

1. Suppose that you want to build a circuit with a source, a motor, and a lamp. The lamp must tell you if the motor stops working. Draw the circuit.
2. Why should a plug type fuse never be replaced with a copper penny? CAUTION: Do not try it!
3. Compare the size of the current a load receives from four dry cells in series with that received from four dry cells in parallel.
4. Compare the time a bulb will stay lit connected to two dry cells in series with the time a bulb will stay lit connected to four dry cells in series.
5. Draw a circuit with two dry cells, a switch, and three light bulbs. If a cell or a bulb becomes defective the circuit must still work.

E. Investigations

1. Do a library search to find out what a circuit breaker is and how it operates. CAUTION: Do not touch the electrical circuits in your home.
2. Get some fuse wire from your teacher. Design a series circuit using dry cells, a switch, conducting wires, fuse wire, and a light bulb to show the action of a fuse.
3. Figure 33-11 shows two electrical circuits using symbols. Figure 33-11,A shows two different light bulbs A and B connected in series. Figure 33-11,B shows the same two different light bulbs A and B connected in parallel. Bulb A is a greater load than bulb B.
    a) Which bulb will glow the brightest when the bulbs are connected in series? Why?
    b) Which bulb will glow the brightest when the bulbs are connected in parallel? Why?
    c) Design and do an experiment to test your predictions.
4. Design and do an experiment to test the ability of different materials to conduct electricity. Use dry cells, an ammeter, a light bulb, and a switch to test the materials.

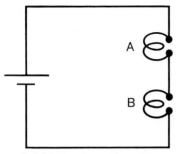

A. Two different light bulbs connected in series

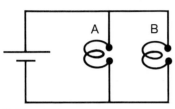

B. Two different light bulbs connected in parallel

Fig. 33-11 Comparing the brightness of different light bulbs connected in series and in parallel

Unit 11

Energy: A Critical Resource

CHAPTER 34
The Nature of Energy

CHAPTER 35
Non-Renewable Energy Sources

CHAPTER 36
Renewable Energy Sources

CHAPTER 37
Conserving Energy for Heating

CHAPTER 38
Conserving Electricity and Gasoline

A steady stream of cars rolls along the expressway. Each car gobbles up some of our remaining fossil fuel. Part of the fuel is used to move the car and passengers. Part produces electrical energy which runs radios, headlights, horns and air conditioners. But about nine-tenths is wasted by friction. In the end, all the energy stored in the fossil fuel becomes waste heat.

In North America we are using energy at an alarming rate. Each man, woman and child uses about four hundred thousand million joules per year. This is equal to about 200 tankfuls of gasoline. Today we use 70 times as much energy per person as the early pioneers. Our population has also increased. As a result, we are facing a severe energy crisis.

Energy has the ability to do work. Without energy, work cannot be done. Fossil fuels such as oil and natural gas are running out. Thus we must learn to harness new sources of energy. Also we must learn to use them wisely. In this unit you will study the forms and nature of energy. You will also study the kinds and sources of non-renewable and renewable energy. Then you will study ways to use energy wisely.

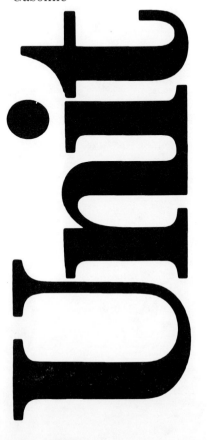

Fig. 34-0 Cars on a busy expressway

34 The Nature of Energy

34.1 Demonstration: Some Forms of Energy
34.2 The Kinds and Forms of Energy
34.3 Energy Conversions and the Conservation Law
34.4 Activity: Energy Conversions in Machines
34.5 The Characteristics of Energy

Imagine a motorcycle moving downhill at 50 km/h. The lights are on and the horn is blaring. How many different forms of energy does it demonstrate? What are some of the characteristics of energy? You will study the kinds, forms, and characteristics of energy in this chapter.

34.1 DEMONSTRATION Some Forms of Energy

Energy can do work. It can be used to exert a force. It can make things move. It can also produce heat. In this activity you will study several forms of energy.

Part 1 Radiant Energy

Materials

light source   Crooke's radiometer

Procedure

Shine light on Crooke's radiometer (Fig. 34-1). Record what happens.

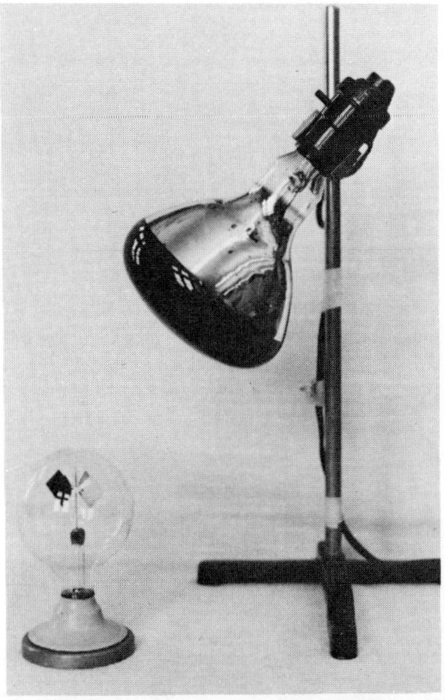

Part 2 Chemical Energy

Materials

metal can with metal lid   small funnel   candle and match
lycopodium powder   rubber bulb
CAUTION: Wear safety goggles during this activity.

Fig. 34-1 Radiant energy can do work

420 Chapter 34

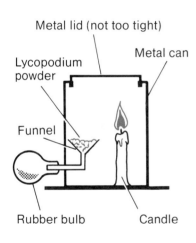

Fig. 34-2 Chemical energy can do work when released.

Procedure

a. Assemble the metal can, funnel, and rubber bulb as shown in Figure 34-2.
b. Place 10 mL of lycopodium powder in the funnel.
c. Light the candle. Then place the lid on the can.
d. Squeeze the rubber ball. Record what happens.

Part 3 Heat Energy

Materials

Hero's engine   Bunsen burner   water   ring stand and ring clamp
CAUTION: Wear safety goggles during this activity.

Procedure

a. Set up the apparatus as shown in Figure 34-3.
b. Fill Hero's engine about 1/3 full of water.
c. Bring the water to a gentle boil.
d. Continue boiling and record what happens.

Part 4 Electric Energy

Materials

Wimhurst machine   metalized pith ball on an insulated thread

Fig. 34-4 Electric energy can do work.

Fig. 34-3 Heat energy can do work.

Procedure

a. Charge the Wimhurst machine shown in Figure 34-4.

Section 34.1   421

b. Suspend the pith ball between the two plates connected to the terminals of the Wimhurst machine.
c. Record what happens.

Part 5 Sound Energy

Materials

two identical tuning forks on resonance boxes   rubber hammer

Procedure

a. Place two identical tuning forks mounted on resonance boxes about 0.5 m apart. Make sure the open ends of the resonance boxes face each other as shown in Figure 34-5.

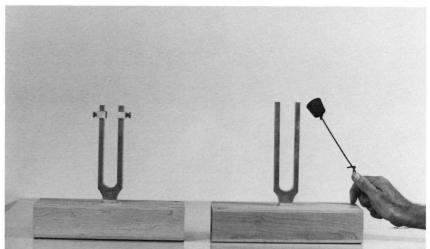

Fig. 34-5 Sound energy can do work.

b. Strike one tuning fork hard with the rubber hammer.
c. After about 5 s stop the tuning fork from vibrating.
d. Listen to the other tuning fork. Record your observations.

Discussion

1. What can energy do?
2. How do you know that light, heat, electricity, and sound are forms of energy?
3. How do you know that lycopodium powder contains energy?
4. How is the energy in lycopodium powder different from the energy in a tuning fork?

34.2 The Kinds and Forms of Energy

Energy is defined as the ability to do work. There are two basic kinds of energy: potential energy and kinetic energy.

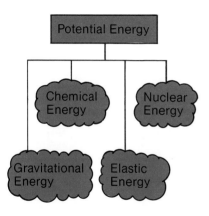

Fig. 34-6 Forms of potential energy

Fig. 34-7 The raised ram of a pile driver has gravitational energy.

Potential Energy

Potential energy [po-TEN-shall] *is the energy that can be stored in an object for a long period of time.* Some objects store energy because they are raised above the earth. Some store energy because they are forced out of shape. Others store energy because of the makeup of their molecules. Still others store energy in their atoms. These four main forms of potential energy are shown in Figure 34-6. No work is done by potential energy until it is changed to kinetic energy.

1. *Gravitational Energy*

   Gravitational energy *is the potential energy an object has as a result of its distance above the earth.* A raised object can do work. For example, the raised ram of a pile driver stores energy (Fig. 34-7). When released, the ram falls. It exerts a force on the pile. This force drives the pile into the ground. The larger the raised mass, the more gravitational energy it has. The higher the mass, the more gravitational energy it has.

2. *Elastic Energy*

   Elastic energy *is the energy stored in an object when it is forced out of its normal shape.* Objects can be compressed, stretched or twisted. Some return to their normal shape when released. One example is a stretched bow. An arrow when released from a stretched bow travels far and fast.

3. *Chemical Energy*

   Chemical energy *is stored in molecules.* Foods and fuels are made of molecules. A hamburger stores about 2 000 000 J. A tankful of gasoline contains about 2 000 000 000 J of chemical energy. Each molecule has many atoms joined together. The atoms are held together by chemical bonds (Fig. 34-8). Chemical energy is the energy stored in these chemical bonds. When these bonds are broken, energy is released.

4. *Nuclear Energy*

   Nuclear energy *is the energy stored in the nucleus of an atom.* It is released when a large nucleus breaks apart. This process is called **nuclear fission**. Atoms of Uranium 235 undergo fission in the CANDU reactor. The atoms in one fuel bundle produce as much heat energy as six railway cars of coal. Nuclear energy is also released when two or more small nuclei join together. This is called **nuclear fusion**. Nuclear fusion occurs in the sun. Hydrogen nuclei in the sun fuse to form helium nuclei. Fusion is the source of the sun's radiant energy.

Kinetic Energy

Kinetic energy [ki-NE-tic] *is energy of motion.* Moving objects, particles, and waves carry kinetic energy. They can all do work. A falling hammer can drive a nail into wood. Moving electrons can turn motors. Moving molecules in hot steam can turn turbines. Moving molecules in a sound wave can vibrate the ear drum. Radiant energy can eject electrons from metals. Five main forms of kinetic energy

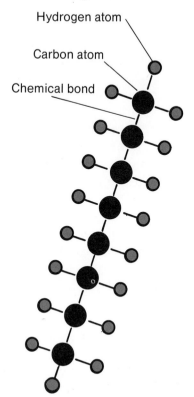

Fig. 34-8 A molecule of gasoline (octane). It has chemical energy stored in the chemical bonds which hold the atoms together.

are shown in Figure 34-9. All forms of potential energy must change to one or more of these forms of kinetic energy before work can be done.

1. *Bulk Kinetic Energy*
   Bulk kinetic energy *is the energy due to the motion of a whole object.* The larger the mass of a moving object, the more bulk kinetic energy it has. The faster it moves, the more bulk kinetic energy it has. A moving car has bulk kinetic energy. A small car moving quickly can have as much bulk kinetic energy as a larger car moving slowly.

2. *Electric Energy*
   Electric energy *is the energy carried by moving electric charges.* Lightning is the result of moving electric charges. The energy in a bolt of lightning can split the trunk of a tree.

3. *Heat Energy*
   Heat energy *is the energy an object has as a result of the random motion of its molecules.* The molecules in a hot object move faster than the molecules in a cold object. Different molecules move in different directions. Because no two molecules have to move in exactly the same way, we say the motion is random. Figure 34-10 shows the random motion of molecules.

4. *Sound Energy*
   Sound energy *is energy that is carried from molecule to molecule by longitudinal waves.* A vibrating drum produces sound energy. The drum causes molecules of air to vibrate. These molecules cause others to vibrate. The sound travels from the drum to a listener without the air molecules moving along with it. Sound energy travels the length of a football field in about a third of a second.

5. *Radiant Energy*
   Radiant energy [RAY-di-ant] *is energy which travels as electromagnetic waves.* Radiant energy can travel through a vacuum. The sun is our largest source of radiant energy. Part of this radiant energy is visible to the human eye. This is called light. Other parts are not visible to the eye such as ultraviolet, infrared, and radio waves. Radiant energy travels very fast. It takes about 8 min for the radiant energy from the sun to travel to the earth.

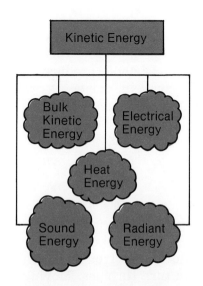

Fig. 34-9 Forms of kinetic energy

Section Review

1. Define the term energy.
2. Name the two basic kinds of energy.
3. a) What is potential energy?
   b) Name four forms of potential energy. Give an example of each.
4. a) What is kinetic energy?
   b) Name five forms of kinetic energy. Give an example of each.

5. Nine examples of energy are listed below. Decide which form of energy each has.
   a) a moving hockey puck
   b) a hacksaw blade bent to one side
   c) electrons flowing along a conductor
   d) a bag of peanuts
   e) a piece of lead hit several times with a hammer
   f) a uranium bundle
   g) a flashlight bulb which is turned on
   h) a guitar string which is plucked
   i) a stone sitting on a post

34.3 Energy Conversions and the Conservation Law

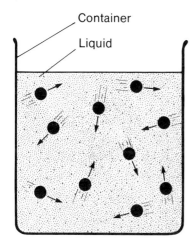

Fig. 34-10 Heat energy is the energy an object has due to the random motion of its molecules.

Energy Conversions in a Pendulum

The pendulum in Figure 34-11 shows energy being converted from one form to another. Energy is converted from gravitational energy to bulk kinetic energy. Then it is converted back to gravitational energy. When the mass of the pendulum is pulled from its rest position to one side (point A) it rises. A force raises the mass through a height h. This force does work. The work done is stored in the mass as gravitational energy.

When the mass is released from point A, some of this gravitational energy changes to bulk kinetic energy. The further the mass falls, the more bulk kinetic energy it gains and the more gravitational energy it loses. At the bottom of the swing (point B), all the gravitational energy stored in the raised mass is gone. The mass now has its greatest speed and greatest bulk kinetic energy.

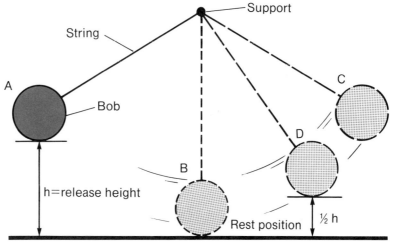

Fig. 34-11 Energy conversions in a pendulum

During the rest of the swing, the bulk kinetic energy changes back into gravitational energy. At the far side of its swing (point C) all the bulk kinetic energy has changed back into gravitational energy. This conversion repeats itself many times.

If the pendulum is allowed to swing for some time, it stops. Its bulk kinetic energy becomes zero. The mass is also back at the height it started. Thus it has no gravitational energy. Some of the stored energy has been given to the air molecules hit by the pendulum. The molecules of the support have gained the rest. Bulk kinetic and gravitational energy have become heat energy.

The Law of Conservation of Energy

Scientists have studied energy conversions in simple machines like levers and in complex machines like cars. The results show that we cannot get more useful energy out of a system than has been put into it. In fact, we cannot break even. This is because of friction; it limits the efficiency of energy conversions to less than 100%.

Even when friction is present, the energy which disappears is not lost. It becomes heat energy in the nearby molecules of matter. The heat energy gained by these molecules is exactly the amount of bulk kinetic and gravitational energy which disappears.

The study of the forms of energy and energy conversions has given us one of the great laws of science. This law is called the **Law of Conservation of Energy**. It states: *Energy cannot be created or destroyed. It can be changed from one form to another. But the total amount of energy in the universe stays constant.*

Fig. 34-12 A demonstration of energy conservation. Does the person need to move back when the mass returns?

Section Review

1. Figure 34-11 shows the pendulum at various points in its swing.
   a) Where does the mass have its greatest height? its greatest gravitational energy?
   b) Where does the mass have its greatest speed? its greatest bulk kinetic energy?
   c) Describe the energy conversion taking place as the mass moves from A to B; from B to C.
   d) Compare the gravitational energy and bulk kinetic energy at point D.
   e) Write a word equation for the total energy the mass has at any point in its swing.
2. a) State the Law of Conservation of Energy.
   b) Figure 34-12 shows a person demonstrating energy conservation. He is releasing a heavy pendulum mass. The mass has been pulled so it touches the nose. When the mass returns, does the person need to move back to be safe? Explain your answer.

34.4 ACTIVITY Energy Conversions in Machines

In this activity you will study the forms of energy demonstrated by two machines. You will also study the energy changes that take place in the machines.

Problem

What energy changes take place in candle chimes and a drinking bird?

Part 1 Candle Chimes

Materials

candle chimes   matches   candles

Procedure

a. Set up the candle chimes as shown in Figure 34-13.
b. Light the candles. Observe what happens.
c. Note the forms of energy demonstrated by the machine.
d. Note the energy conversions that are taking place.

Part 2 Drinking Bird

Materials

drinking bird   glass   light source   water

Procedure

a. Set up the drinking bird as shown in Figure 34-14.

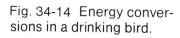

Fig. 34-13 Energy conversions in candle chimes.

Fig. 34-14 Energy conversions in a drinking bird.

Section 34.4   427

b. Wet the head of a bird.
c. Turn on the light source. Place it near the bird. Observe what happens.
d. Note the forms of energy demonstrated by the set-up.
e. Note the energy conversions that are taking place.

Discussion

1. List the observations you made for each machine.
2. List the kinds and forms of energy demonstrated by each machine.
3. What energy conversions took place in each machine?
4. Do you think that energy is conserved? Discuss.

34.5 The Characteristics of Energy

There are five characteristics of energy:

1. **Energy is transferred from one object to another.** This happens whenever work is done. When a person lifts an object from the floor to a table, the person does work on the object. The object gains gravitational energy. The person loses chemical energy. The chemical energy was originally stored in food the person ate. It is transferred to the raised object. It is stored as gravitational energy.

2. **Energy comes in many forms.** Energy takes the form of bulk kinetic energy in moving objects. It is gravitational energy in raised objects. It is elastic energy in deformed objects. And it is chemical energy in molecules of food and fuel. It is stored as nuclear energy in the nuclei of atoms.

3. **Energy can be stored for long periods of time.** This stored energy can be used to do work at any future date. An object can be raised to a height and left there for years. But it still has the ability to do work when released. Chemical energy was stored in fossil fuels millions of years ago. It is being released to do work for us today. Scientists are experimenting to find better ways of storing energy in batteries. Their findings may help us store energy from the sun and wind in the future.

4. **Energy can be changed from one form to another.** A car's engine uses the chemical energy in gasoline. It is changed to heat energy. Some of the heat energy is changed to bulk kinetic energy in the engine. Some is changed to gravitational energy as the car climbs a hill. Some becomes electric energy from the generator. Some becomes radiant energy from the headlights. And some becomes sound energy from the horn.

5. **Energy is always conserved.** Energy can be transferred, stored, or changed from one form to another. But the total energy stays the same. But useful energy is not conserved. Unfortunately all the chemical energy in the gasoline a car uses eventually becomes waste heat. It is so widely dispersed that it is no longer useful to us.

Section Review

1. List five characteristics of energy.
2. Give an example for each characteristic. Give different examples from those in the text.

Main Ideas

1. Energy is the ability to do work.
2. There are two basic kinds of energy: kinetic and potential.
3. Potential energy is stored energy.
4. The forms of potential energy are: gravitational, elastic, chemical, and nuclear.
5. Kinetic energy is energy of motion.
6. The forms of kinetic energy are: bulk kinetic, electric, heat, sound, and radiant.
7. Energy can be changed from one form to another.
8. Energy is always conserved.
9. Energy can be transferred from one object to another.

Glossary

kinetic energy	ki-NE-tic	energy of motion
potential energy	po-TEN-shall	stored energy due to position, shape, or composition

Study Questions

A. True or False

Decide whether each of the following sentences is true or false. If the sentence is false, rewrite it to make it true. (Do not write in this book.)

1. Energy and work mean the same thing.
2. Chemical energy is energy stored in the bonds of molecules.
3. Nuclear energy is energy stored in the nucleus of an atom.
4. A radio wave is an example of sound energy.
5. Some energy is lost when energy changes from one form to another.

B. Completion

Complete each of the following sentences with a word or phrase which will make the sentence correct. (Do not write in this book.)
1. The final form of all forms of energy is _____ .
2. The two basic kinds of energy are _____ and _____ .

3. Potential energy is any form of energy which is _____ .
4. When energy is passed from one object to another we say it is _____ .
5. The Law of Conservation of Energy states that the total amount of energy in the universe _____ .

C. Multiple Choice

Each of the following statements or questions is followed by four responses. Choose the correct response in each case. (Do not write in this book.)

1. The two basic kinds of energy are
   a) elastic and gravitational
   b) radiant and nuclear
   c) chemical and electric
   d) kinetic and potential
2. Which of the following is a form of **potential** energy?
   a) electric   b) sound   c) elastic   d) heat
3. Chemical energy is
   a) stored in the nucleus of an atom
   b) stored in the bonds holding atoms together
   c) a form of nuclear energy
   d) a form of kinetic energy
4. If an object is moving faster than another of the same mass, it has more
   a) heat energy
   b) bulk kinetic energy
   c) gravitational energy
   d) sound energy
5. A vibrating pendulum slows down and stops. The fact that it stops shows that
   a) energy is stored
   b) energy is transferred
   c) energy is converted
   d) friction is present

D. Using Your Knowledge

1. How are energy and work the same? How are they different?
2. Distinguish between the bulk kinetic energy and the heat energy of an object.
3. Describe the energy changes that would happen to you when riding a roller coaster. Use sketches in your answer.
4. The electric energy needed for some bicycle headlights is made by a generator touching the tire. A bicycle salesman claims that such a bicycle is harder to pedal when the light is on than if it is burned out. Do you agree or disagree? Why?

E. Investigations

1. Use the toaster at home to make a piece of toast. Eat the toast. Then list all the forms of energy and energy conversions involved in the activity.

2. Get an elastic spring and a 500 g mass from your teacher. Design and do an experiment to determine the forms of energy and energy conversions taking place as it vibrates. Study how well it shows the Law of Conservation of Energy.
3. Describe the energy conversions in playing baseball. Consider pitching, hitting, and catching the ball.

35 Non-Renewable Energy Sources

35.1 Conventional Fossil Fuels
35.2 Non-Conventional Fossil Fuels
35.3 Activity: Separating Bitumen and Sand
35.4 Activity: The Composition of Tar Sand
35.5 Nuclear Fission
35.6 Nuclear Fusion
35.7 Generating Electricity Using Heat Energy

North Americans enjoy a high standard of living. At present this is achieved by using huge amounts of energy. Although we consume about 40% of the world energy, we make up only 7% of world population.

Over 95% of the energy we use today is non-renewable. **Non-renewable energy** *is energy that is used up faster than it is replaced.* It's like a decreasing back account; money is withdrawn and spent faster than deposited. There are two kinds of non-renewable energy: fossil fuels and nuclear fuels.

35.1 Conventional Fossil Fuels

Coal, natural gas, and oil are fossil fuels. **Fossil fuels** [FOS-il] were formed from the remains of plants and animals.

Coal

Origin. Coal formed from vast forests. Wood, decayed by bacteria, formed peat bogs. The bogs were covered by ancient seas. The shale and limestone which settled out of the water covered the bogs. This prevented oxygen from reaching the peat. The bacteria kept working. The peat was compressed and heated. Different kinds of coal were formed depending on the pressure and carbon content. First came lignite; then bituminous coal and, finally, anthracite. Figure 35-1 shows the volume change as these kinds of coal formed.

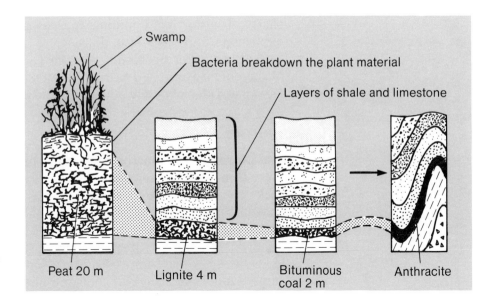

Fig. 35-1 Coal takes hundreds of millions of years to form.

Reserves and Problems. Canada has coal in the Northwest Territories, Nova Scotia, Saskatchewan, Alberta, and British Columbia. Much of our coal lies close to the surface. It is beneath the grazing and grain growing lands of the Mid-West. The coal can be mined by removing the overburden. This process is called **strip mining** (Fig. 35-2). Strip mining is the easy way to mine the coal but may destroy valuable farm land.

Fig. 35-2 Giant loading shovels are used to strip mine coal.

Other deposits of coal lie deep beneath the surface. One of the largest reserves may be under the Cabot Strait between Cape Breton and Newfoundland. Deep coal is mined using underground

Section 35.1 433

shafts. Underground mining increases the risk of mine cave-ins, explosions, and black-lung disease.

Coal has other problems. It is costly to mine and transport. It is inconvenient to use. Also, coal pollutes. When burned it releases fly ash and oxides of sulfur, nitrogen, and carbon.

Petroleum and Natural Gas

The word **petroleum** [pe-TRO-le-um] means rock oil. Petroleum is formed in layers of **sedimentary rock** [sed-i-MEN-tar-e]. Sedimentary rock was formed from sand and clay. Sand and clay present in ancient seas settled to the bottom as did plant and animal organisms. These formed layers of sediment. The sediment eventually became hundreds of metres thick in places. It became so heavy that the lower layers were compressed into sedimentary rock. The sand became sandstone; the clay became shale and the dead organisms became petroleum and natural gas.

Petroleum and natural gas tend to collect in the sandstone. Thus sandstone is called the **reservoir rock** [RES-e-vor]. The shale rock is called the **cap rock** since it traps the oil and gas. Figure 35-3 shows how a dome of shale rock traps the oil and gas in the sandstone.

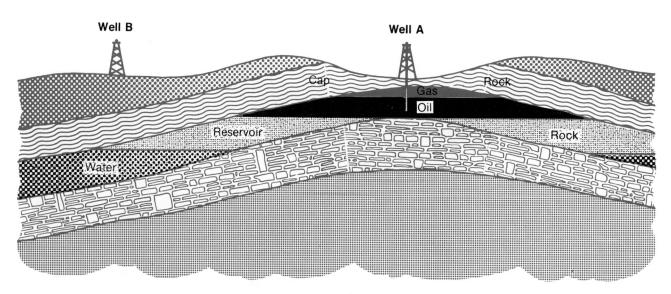

Fig. 35-3 A common method by which petroleum is trapped in the earth

Reserves and Problems Because petroleum and natural gas are running out, we are looking for new reserves. New reserves have been found in the Rockies. Others have been found in the barrens of the north. Canada's largest reserve may be on the Continental Shelf east of St. John's, Newfoundland (Fig. 35-4).

Drilling for oil in the ocean and the cold north is dangerous and expensive. The pollution risks are great. Stormy seas can cause disastrous spills. Pipelines buried in the north need special insulation to prevent heated oil from melting the permafrost. Also, pipelines may destroy plant and animal life.

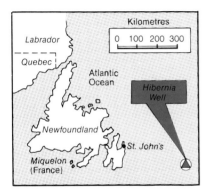

Fig. 35-4 Huge reserves of oil may lie under the ocean off Newfoundland.

Section Review

1. What is a non-renewable resource?
2. **a)** Name the fossil fuels.
   b) Why are fossil fuels non-renewable energy sources?
3. **a)** How was coal formed?
   b) Where are the deposits of coal in Canada?
4. What are the problems with: **a)** strip mining coal; **b)** mining coal underground; **c)** burning coal?
5. **a)** How was petroleum formed?
   b) What are the problems in obtaining and transporting oil from the far north and from the Continental Shelf?

35.2 Non-Conventional Fossil Fuels

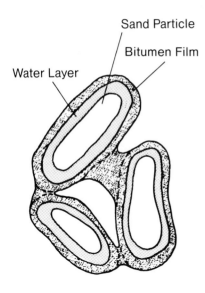

Fig. 35-5 Cross section of a tar sand sample

Tar Sand

Tar sand is a mixture of sand, water and tar. The thick, black tar is called **bitumen** [bi-TU-men]. It fills the pores between the sand particles (Fig. 35-5). The bitumen is too thick to pump to the surface without heating.

There are huge deposits of tar sand in Northern Alberta (Fig. 35-6). About one-third of the tar sand is close to the surface. Layers up to 50 m thick are located beneath overburden (Fig. 35-7). This tar sand is surface mined by removing the overburden. Today bitumen is separated from the sand using heat and chemicals. But in the future it may be possible to use bacteria instead. Bacteria are cheap and cause no ecological problems.

About two-thirds of the tar sand is too deep to surface mine. This tar sand is heated underground. One method uses **steam injection** (Fig. 35-8). The steam changes bitumen to a liquid. The liquid flows to a nearby well. Here it is pumped to the surface.

The problems in mining and extracting bitumen from tar sand are enormous. Surface mining disturbs the environment. Extraction processes release smelly gases. Sand tailings take up more space than the original tar sand. Also it is difficult and costly to reclaim the overburden.

Shale Rock

A kind of rock called **oil shale** traps a thick, heavy oil called **kerogen** [KER-e-jen]. Canada's largest deposit is located near Albert County,

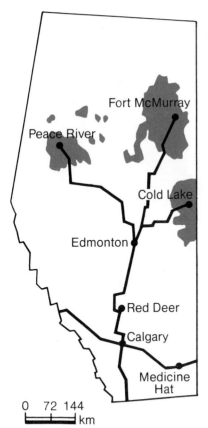

Fig. 35-6 The shaded areas on this map of Alberta show where the tar sands are located.

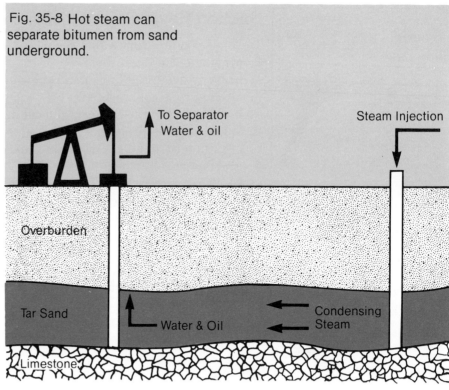

Fig. 35-8 Hot steam can separate bitumen from sand underground.

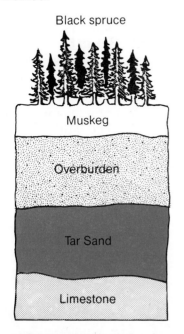

Fig. 35-7 Tar sands profile

New Brunswick. Ways have already been found to extract kerogen. The mined rock is ground up, then mixed with water and heated to 500°C. This process uses a lot of energy. The water becomes salty and pollutes rivers. Also, ground up rock takes up more space than the original rock. Fortunately, phosphate and ammonium sulfate may be by-products of the process in New Brunswick. These are both used to make fertilizer. As the price of oil rises, it may become economical to harness this vast resource.

Section Review

1. **a)** What is tar sand?
   b) Where are Canada's reserves of tar sand?
   c) How is oil extracted from tar sand?
   d) Describe two problems with mining and extracting oil from tar sand.

2. **a)** What is oil shale?
   b) Where is the largest deposit of oil shale in Canada?
   c) How is kerogen extracted from the oil shale?
   d) Describe two problems with mining and extracting kerogen from oil shale.

35.3 ACTIVITY Separating Bitumen and Sand

Tar sand is a mixture of sand, water, and tar. In this activity you will separate the tar (bitumen) from sand and water. You will use a method similar to that used in industry.

Problem
How can we get the tar out of tar sand?

Materials

tar sand
balance
400 mL metal can
newspaper

sodium hydroxide solution
Bunsen burner
stirrer
ring stand, iron ring, wire gauze

styrofoam cup
plastic spoon
safety glasses

CAUTION: Wear safety glasses. Do not overheat.

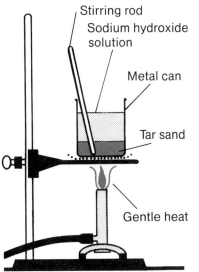

Fig. 35-9 Separating bitumen and sand

Procedure
a. Set up the apparatus as shown in Figure 35-9.
b. Measure and record the mass of the metal can.
c. Add 100 g of tar sand to the metal can.
d. Stir in 250 mL of sodium hydroxide solution.
e. Gently heat the mixture to the boiling point.
f. Continue gently heating and stirring for about 15 min.
g. Turn off the heat.
h. Measure and record the mass of a styrofoam cup.
i. Skim the bitumen off the surface into the cup.
j. Let the bitumen cool and thicken for about 5 min. Observe it carefully.
k. Pour off the excess water from the bitumen. Let it dry overnight.
l. Pour off the excess water from the sand tailings. Let it dry overnight.
m. Keep the bitumen and sand tailings for the next activity.

Discussion
1. Describe the smell and appearance of bitumen.
2. Summarize the method used to separate bitumen and sand.
3. What problems do you see with this separation method?

35.4 ACTIVITY The Composition of Tar Sand

In this activity you will calculate how much bitumen there is in 100 g of tar sand. You also compare the appearance of tar sand, bitumen, and sand tailings.

Problem

What percentage of tar sand is tar (bitumen)?

Materials

balance
tar sand
microscope slide
microscope (low power)
sand tailings and metal can
bitumen and cup
from Activity 35.3

Procedure

a. Copy Table 35-1 into your notebook.
b. Record the mass of the cup and the can from Activity 35.3 in the table.
c. Find the mass of the cup plus bitumen. Record it in the table.
d. Find the mass of the can plus sand tailings. Record it in the table.
e. Calculate the mass of bitumen obtained from 100 g of tar sand.
f. Calculate the percentage of bitumen contained in the tar sand.
g. Calculate the mass of tailings obtained from 100 g of tar sand.
h. Calculate the percentage of sand tailings in the tar sand.
i. Place 1 or 2 grains of tar sand and sand tailings side by side on a microscope slide.
j. Study the grains under low power. Compare them.

Table 35-1 Composition of Tar Sand

Bitumen		Sand tailings	
Mass of cup + bitumen =	g	Mass of can + tailings =	g
Mass of cup =	g	Mass of can =	g
Mass of bitumen =	g	Mass of tailings =	g
% of bitumen =		% of sand tailings =	

Discussion

1. What percentage of your 100 g sample of tar sand was bitumen?
2. Compare your results to others in the class. Explain the results.
3. What percentage of your sample of tar sand was sand tailings?
4. How much of your tar sand sample was water?
5. Compare the appearance of tar sand and the sand tailings as seen under the microscope. Account for the difference.

35.5 Nuclear Fission

Demand for energy is increasing. Yet we have a limited supply of oil and natural gas in Canada. Tar sand is difficult and costly to extract. As a result we are turning more and more to nuclear energy.

The nuclear reactors in use today produce heat by a process called **nuclear fission** (Fig. 35-10). The fuel used is mainly uranium. Canada has large reserves of uranium in British Columbia, Saskatchewan, and Ontario.

Fig. 35-10 A nuclear generating station uses nuclear energy to produce heat. The heat turns water to steam. The steam is used to drive the turbine that runs the electric generator.

Fig. 35-11 Fission of uranium 235. Three neutrons are emitted. If one of these neutrons is slowed down and then collides with another nucleus, the reaction continues.

A fission reactor splits the nucleus of a uranium atom into smaller parts. In Canadian reactors, a slow neutron splits the atom. An atom of barium and an atom of krypton result. Also, two or three fast neutrons are released (Fig. 35-11). The original uranium atom and the slow neutron have a certain mass. The atoms and fast neutrons produced in the reaction have less mass. The lost mass is converted to energy. The fast neutrons are slowed down using a liquid called the **moderator**. The moderator in Canadian reactors is heavy water. These slow neutrons are then used to split other atoms of uranium. A **chain reaction** results when enough slow neutrons are produced to continue the reaction. Energy in the form of heat is produced constantly.

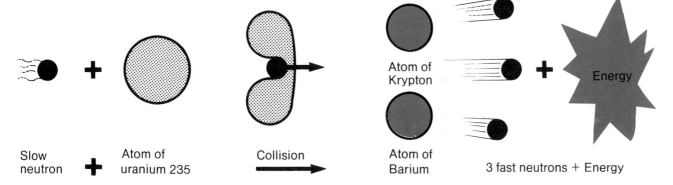

The energy released by nuclear fission is used to make electricity. Generating electricity by fission has two advantages. First, a small amount of nuclear fuel yields a lot of energy. The atoms in one fuel bundle (m = 20 kg) produce as much heat energy as six railway cars of coal. Second, fission plants do not release gases to pollute the atmosphere.

But there are disadvantages. Electric plants fueled by fission give off a lot of waste heat. There is also the risk of escape of radioactivity to the outside. And a secure place has not yet been found for the long term storage of the radioactive spent fuel. By the year 2000 we may have 130 000 t of spent fuel.

Section Review

1. Define fission.
2. Where does the energy come from during fission?
3. Where are Canada's reserves of uranium?
4. What are the advantages of nuclear fission?
5. What are the disadvantages of nuclear fission?

35.6 Nuclear Fusion

Nuclear fusion is the process of combining the nuclei of smaller atoms to form a new nucleus. Hydrogen has the smallest nucleus. Under very high temperatures such as those existing on the sun, the positive nuclei of hydrogen atoms move at very high speeds. These speeds enable them to overcome the forces of electrical repulsion between them. The nuclei collide and fuse (join) to form a heavier nucleus of helium. A neutron is released in the process (Fig. 35-12). The mass of the nucleus of helium and the released neutron is less than the mass of the two original heavy hydrogen nuclei. The lost mass is converted into energy.

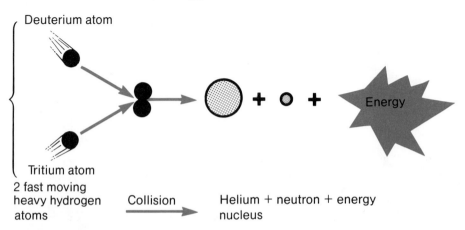

Fig. 35-12 Fusion of two forms of heavy hydrogen (deuterium and tritium) to form helium and release energy.

The harnessing of nuclear fusion using the hydrogen on earth has a number of advantages. It would yield an almost unlimited supply of energy. Also, the process of fusion is almost non-polluting.

Several difficulties must be overcome before nuclear fusion can be used as an energy source. Temperatures of about three thousand million degrees (3.0×10^9 °C) are needed. Such high temperatures are hard to produce. Also hydrogen nuclei at such a high temperature are difficult to confine.

Section Review

1. Define fusion.
2. Where does the energy come from during fusion?
3. What difficulties are scientists having in harnessing fusion?
4. What advantages does fusion have over fission as an energy source?

35.7 Generating Electricity Using Heat Energy

Fossil Fueled Power Plant

The parts of a fossil fueled power station are shown in Figure 35-13. Heat is produced by burning fossil fuels. The heat changes water into steam. Then the steam is passed through a steam turbine. Each steam turbine has many wheels. Each wheel has many paddle-like blades. The moving molecules of water vapour cause the turbine to rotate. The turbine turns a generator. The generator changes bulk kinetic energy into electric energy.

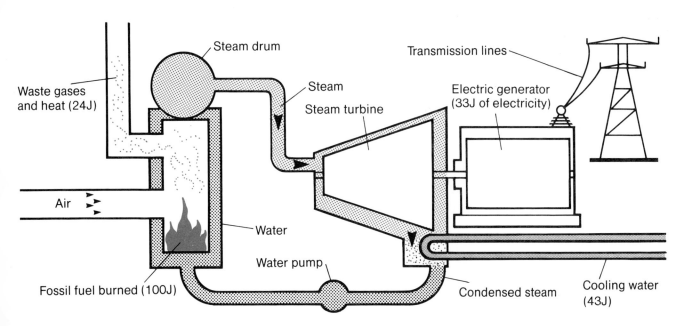

Fig. 35-13 Elements of a thermal power system which is only 33% efficient.

Generating electricity in a fossil fueled power plant has disadvantages. Two forms of energy leave the plant. One is electricity from the generator. The other is waste heat from the steam turbine and hot gases. The waste heat causes the fossil fueled generator plants to be only about 33% efficient. In addition, dangerous gases such as nitrogen dioxide, sulfur dioxide, and carbon monoxide are given off into the atmosphere. Also, if coal is used, about 9% of the coal becomes solid waste.

Nuclear Fission Power Plant

Figure 35-14 is a diagram of a CANDU nuclear fission generating station. Heat is produced when uranium atoms are split in the fuel rods. This heats the heavy water moderator. Heat is transferred to ordinary water and used to produce steam and then electricity. At present nuclear stations emit more heat to the water used for cooling than the total from fossil fueled plants. Nuclear plants are about 30% efficient.

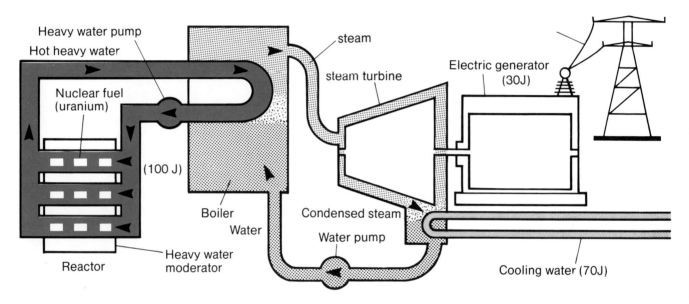

Fig. 35-14 The CANDU nuclear power generating system (simplified).

Section Review

1. What is the source of energy to generate electricity in
   a) a fossil fueled power plant   **b)** a nuclear power plant
2. Which is more efficient, a fossil fueled plant or a nuclear plant. Why?

Main Ideas

1. Non-renewable energy sources are used faster than they are replaced.
2. Fossil fuels and nuclear fuels are non-renewable energy sources.
3. Fossil fuels were formed from the remains of decayed plants and animals.

4. Bitumen, coal, natural gas, and petroleum are fossil fuels.
5. Finding, mining, extracting, and using new reserves of fossil fuels in North America are difficult and may harm the environment.
6. Nuclear reactions change part of the mass of a nucleus directly into energy.

Glossary

chain reaction		a neutron causes a uranium atom to fly apart. This frees other neutrons to continue the reaction.
fossil fuel	FOS-il	the remains of organisms embedded in the surface of the earth which are used as fuel (e.g. coal, oil, natural gas)
tar sand		a loose grained rock material held together by bitumen and water

Study Questions

A. True or False

Decide whether each of the following sentences is true or false. If the sentence is false, rewrite it to make it true. (Do not write in this book.)

1. Most of the energy humans use today is non-renewable.
2. Coal has smaller reserves than other fossil fuels.
3. The tar sand located in Alberta is a mixture of oil, gas, and sand.
4. Nuclear fusion is the process of breaking down a larger nucleus into two smaller nuclei releasing neutrons and energy.
5. Nuclear fission fueled electric plants generate less waste heat than fossil fueled electric plants.

B. Completion

Complete each of the following sentences with a word or phrase which will make the sentence correct. (Do not write in this book.)

1. Non-renewable energy is energy which is _____ .
2. The two main kinds of non-renewable energy are _____ and _____ .
3. The word petroleum means _____ .
4. Nuclear reactors in use today change mass to energy by a process called _____ .
5. Before nuclear _____ is used to generate electricity, scientists must learn how to create very high _____ .

C. Multiple Choice

Each of the following statements or questions is followed by four responses. Choose the correct response in each case. (Do not write in this book.)

1. About what percentage of the world's energy do North Americans consume?
   a) 10%   b) 20%   c) 30%   d) 40%
2. Which of the following materials would occupy the greatest space during the decay of wood to form coal?
   a) anthracite coal
   b) bituminous coal
   c) lignite coal
   d) peat
3. Tar sand consists of
   a) tar only
   b) bitumen only
   c) tar and sand only
   d) tar, sand, and water
4. One example of a nuclear fuel is
   a) coal   b) bitumen   c) kerogen   d) uranium
5. Which of the following statements is true for nuclear fission?
   a) The mass of the products is less than the mass of the reactants.
   b) The mass of the products is the same as the mass of the reactants.
   c) The mass of the products is greater than the mass of the reactants.
   d) Energy is changed to mass during nuclear fission.

D. Using Your Knowledge

1. Assume that the city nearest you is planning to build another fossil fueled plant. Which fossil fuel would you encourage them to use? Why?
2. Environmental problems occur both at the point of production and at the point of consumption of coal. Explain this statement.
3. Nuclear reactors have a better safety record than fossil fueled electric power stations. Why, then, are so many people worried about the increased use of nuclear energy?

E. Investigations

1. Research the processes used to obtain oil from the Athabasca Tar Sands and the deep, heavy oil deposits in Western Canada. Compare the cost of these processes with the cost of obtaining oil from traditional sources.
2. Do a library search to find out how the nuclear industry proposes to protect us against long-lived radioactive wastes. Prepare a report on your findings.

36 Renewable Energy Sources

36.1 Energy from the Sun
36.2 Activity: The Operation of a Flat-Plate Collector
36.3 Activity: The Operation of a Solar Cell
36.4 Energy from Water
36.5 Energy from Wind
36.6 Energy from Heat in the Earth's Crust
36.7 Energy from Hydrogen

More than 90% of the energy we use comes from fossil fuels. Another 5% of the world's energy comes from the fission of nuclear fuels, mostly uranium. Before this non-renewable energy runs out, we must learn to use renewable energy sources. **Renewable energy** *is energy which is replaced as fast or faster than it is used up.*

36.1 Energy from the Sun

Except for nuclear energy, sunlight is the primary source of all our energy. Radiant energy from the sun is called **solar energy**.

Passive Solar Systems

This is the simplest and least expensive way to use solar energy for heating buildings. The building itself is designed to collect solar energy. The building has most of its windows facing south. They permit solar energy to enter the building in winter when the sun is low in the sky. Heavy stone or brick walls inside store the heat in the daytime. They radiate heat to the inside at night. Insulating blinds pulled down at night decrease heat loss to the outside.

Active Solar Systems

Buildings with active solar systems have special features to collect, store, and circulate energy from the sun (Fig. 36-1). These include flat-plate collectors, storage tanks, pumps, and fans. A **flat-plate collector** is shown in Figure 36-2. Flat-plate collectors are placed on a roof with a southern exposure. The flat-plate collector changes solar energy directly to heat energy. It is covered by a transparent surface. It has one or more air chambers. The other parts are a coated plate, tubes, and insulation. The flat-plate collector makes use of the **greenhouse effect**. The glass surface lets in the sun's short wave length radiant energy. The coated plate absorbs the short wave length radiant energy and heats up. It gives off long wave length radiant energy. The long wave length radiant energy cannot pass through the glass. Thus the energy is trapped in an air chamber under the glass. The chamber and the coated collector plate heat up. The pipes under the coated metal plate carry a fluid such as water. This fluid conducts the heat to a storage area. Heat from the storage area is used when needed to heat the home.

Reflectors

Large curved reflectors can be used to collect and concentrate solar energy. One example is the solar furnace at Mont-Louis in the French Pyrenees. This reflector consists of thousands of small mirrors. The mirrors concentrate sunlight on a small area. Temperatures high enough to melt metals are produced.

Fig. 36-1 A solar-heated home. The roof contains solar collectors to trap heat energy from the sun's rays.

Fig. 36-2 A flat-plate solar collector

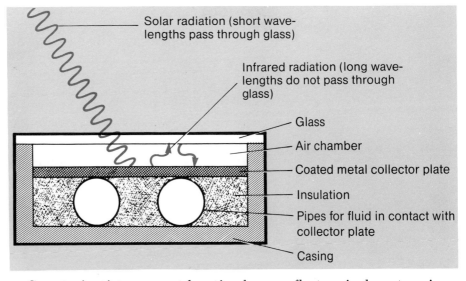

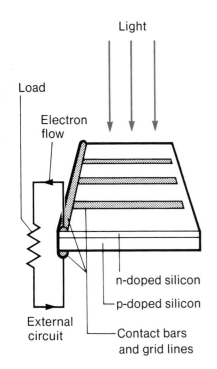

Fig. 36-3 A silicon solar cell connected to an external circuit. Light separates positive and negative charges. A surplus of electrons builds up in the n-doped silicon. A deficiency builds up in the p-doped silicon. Electrons flow through the load from the n region to the p region.

Some scientists suggest locating large reflectors in desert regions. The solar energy collected by reflectors could be used to produce steam. The steam could be used to produce electric energy. Other scientists propose a system of space mirrors. The mirrors would reflect extra solar energy to selected regions on earth. The solar energy could prevent frost damage and increase crop production.

Solar Cells

Solar energy can be changed directly to electric energy using **solar cells**. One kind of solar cell is made of silicon (Fig. 36-3). Light shines on the silicon cells forcing some electrons from the n-doped silicon to the p-doped silicon. These moving electrons produce an electric current. The greater the number of silicon cells connected together, the greater the current.

There are several problems with using solar cells today to power our appliances. The efficiency of modern solar cells is less than 15%. Also they are very costly to produce. But the price may come down as a result of new technology and mass production.

Photosynthesis

Plants grow by absorbing radiant energy. Radiant energy is converted to chemical energy by a process called **photosynthesis** [fo-to-SIN-the-sis]. The plant material produced is called **biomass** [BI-o-mas]. About 1% of the biomass produced on earth is used as food. Some of the rest could be used as fuel.

Wood could be used to power electric generating stations or to produce methanol. Methanol can run engines at a cost comparable to gasoline.

Methane can be made from plant and animal wastes. It is produced when organic material contained in manure decays in the absence of oxygen. Methane is natural gas. It can be burned to heat homes. Methane can also be changed to methanol to run engines.

Section Review

1. Why is sunlight described as a primary source of energy?
2. **a)** What is a passive solar system?
   b) What is the greenhouse effect?
   c) Describe how the greenhouse effect is used in a flat-plate collector.
   d) What colour should the coated surface have for best absorption?
3. Describe how solar reflectors are used to collect radiant energy.
4. **a)** What energy conversion takes place in a silicon solar cell?
   b) Why are silicon solar cells not used to generate electricity for homes today?
5. Photosynthesis is less than 1% efficient. Yet it is becoming more important as an energy source. Why?

36.2 ACTIVITY The Operation of a Flat-Plate Collector

In this activity you will build a simple flat-plate collector. Then you will study some factors that affects its operation. Does the material making up the transparent cover matter? What effect does the colour of the absorbing surface have? What effect does the angle the sun makes with the surface have? Write predictions in your notebook. Then do the activity to check your predictions.

Problem

How does a flat-plate collector work?

Materials

shoe box (14 cm x 29 cm x 9 cm)
styrofoam base (14 cm x 29 cm x 2 cm)
2 styrofoam sides (7 cm x 29 cm x 2 cm)
2 styrofoam ends (7 cm x 10 cm x 2 cm)
sheet of aluminum foil — shiny (10 cm x 25 cm)
sheet of aluminum foil — painted black (10 cm x 25 cm)
glass sheet (14 cm x 29 cm)
thermometer
heat lamp and holder

Procedure

a. Copy Table 36-1 into your notebook.
b. Insulate the inside of the shoe box using the styrofoam (Fig. 36-4).
c. Place the heat lamp about 40 cm directly above the box.
d. Stick a thermometer through a hole in the side of the box.
e. Place a sheet of shiny aluminum on the bottom of the box.
f. Cover the top of the box with glass.
g. Read the temperature of the air inside the box.

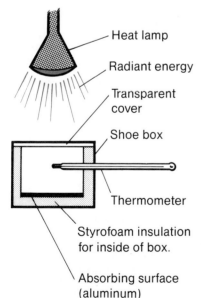

Fig. 36-4 The operation of a flat-plate collector

h. Turn on the lamp. Record the temperature after 3 min.
i. Remove the glass. Then cool the inside of the box.
j. Replace the shiny aluminum with the blackened aluminum. Repeat steps (f), (g), (h), and (i).
k. Leave the glass cover off the box. Repeat steps (g), (h), and (i).
l. Tilt the box at an angle of 45° to the light. Repeat steps (f), (g), (h), and (i).
m. Compare the temperature change of the air for each set of conditions.

Table 36-1   A Flat-Plate Collector

Conditions			Temperature °C		
Cover	Absorbing surface	Angle of light to surface	Initial	Final	Change
glass	shiny	90°			
glass	black	90°			
none	black	90°			
glass	black	45°			

Discussion

1. Which surface is best for absorbing heat energy? Why?
2. Compare the temperature change with and without the glass cover.
3. How does the angle the light makes with the surface affect the results?
4. What combination of conditions produces the best flat-plate collector?

36.3   ACTIVITY   The Operation of a Solar Cell

In this activity you will study some factors that affect the operation of a solar cell. Write in your notebook the factors that you think affect how much electricity it produces. Then do the activity to check your prediction.

Problem

How does a solar cell work?

Materials

socket   solar cell   small electric motor with fan blades
60 W bulb   connecting leads   converging lens
100 W bulb   opaque screen

Section 36.3   449

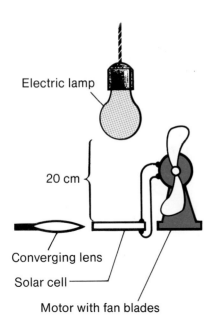

Fig. 36-5 A solar cell connected to a motor

Procedure

a. Copy Table 36-2 into your notebook.
b. Connect the solar cell to the motor.
c. Screw a 60 W bulb into the socket. Position the bulb 20 cm above the solar cell as shown in Figure 36-5.
d. Turn on the bulb. Record whether the fan blades go fast or slow.
e. Repeat (d), but use a 100 W bulb.
f. Repeat (d), but cover half the surface of the solar cell with the opaque screen.
g. Repeat (d), but concentrate the light onto the solar cell using a converging lens.
h. Repeat (d), but shine the light on the solar cell at an angle of 45°. Keep it the same average distance from the bulb.

Table 36-2 The Operation of a Solar Cell

Power of light source (W)	Conditions			Speed of electric motor (fast, medium, slow)
	Area of solar cell exposed	Angle of light to solar cell surface	Concentration of light	
60	all	90°	diffuse	
100	all	90°	diffuse	
100	half	90°	diffuse	
100	all	90°	concentrated	
100	all	45°	diffuse	

Discussion

1. What effects do the following conditions have on the operation of the solar cell?
   a) the brightness of the light
   b) the area of the solar cell exposed
   c) the angle of the solar cell to the direction of the light
   d) the concentration of the light
2. What combination of conditions produces the most electricity using the solar cell?

36.4 Energy from Water

Energy from Rivers

A hydro-electric power plant changes the gravitational energy of dammed water to electric energy. Figure 36-6 shows how this is done. The potential energy in the dammed water is changed to kinetic energy as the water flows down through large tubes. Turbines at the bottom are turned by the moving water. They gain bulk kinetic energy. The turbines turn electric generators. The bulk kinetic energy is changed into electric energy. This electric energy is distributed to customers by transmission lines.

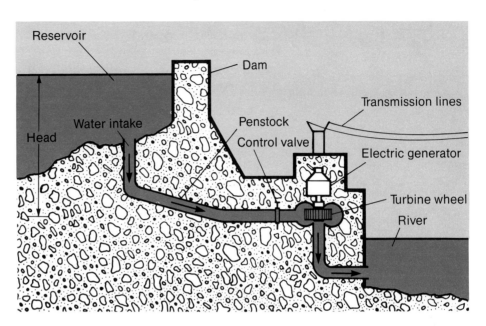

Fig. 36-6 A hydro-electric power system. A system like this is 85% efficient.

Only one-third of North America's hydro-electric potential is being used. There are problems in harnessing the rest. Small projects are not economical. And major products such as the James Bay project in Quebec cost thousands of millions of dollars. There are also environmental concerns. Large reservoirs flood river valleys destroying native settlements, plants, and animals. In addition, decaying matter from flooded vegetation interferes with the hatching of fish eggs.

Energy from Tides

One site suitable for tidal power plants exists at the Cumberland Basin between New Brunswick and Nova Scotia. The Cumberland Basin is at the northeast end of the Bay of Fundy. Differences in the water level between high and low tides on the Bay of Fundy are as great as 15 m. Water will be trapped behind dams at high tide. The water can be released at low tide to flow through a hydro-electric power plant. The gravitational energy stored in the water will be changed to electric energy.

There are several problems in harnessing tides. It would cost millions of dollars to harness the Bay of Fundy tides. Tidal dams cause sediment to build up. This causes navigational hazards. Also, unless locks are built, ships will be unable to pass through the constantly changing waters. Tides occur at periodic intervals. Therefore electricity must be stored and/or used from other sources when tidal activity is low.

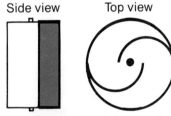

A. Savonius rotor water pumping windmill

B. Multivane water pumping windmill

Energy from Temperature Change with Depth in Oceans

The oceans receive and store much of the solar energy reaching the earth. Solar energy heats up the upper surface of the tropical oceans, leaving the deep water cold. Thus a temperature difference exists between the top and bottom water. Scientists are looking for an economical way to extract the heat from the warmer water. But removing the heat is difficult and expensive. The heat must be removed from huge volumes of water. Also, transporting the energy to distant markets is costly.

Section Review

1. What energy conversions take place in a hydro-electric power plant?
2. a) Describe how the tides can be harnessed.
   b) What problems exist in harnessing the tides?
3. What problems exist in harnessing the heat stored in the surface water of tropical oceans?

36.5 Energy from Wind

Research is now taking place to harness wind more efficiently. Windmills and wind turbines are used to harness winds. Figure 36-7 shows some vertical and horizontal axis windmills and turbines. The Savonius Rotor (Fig. 36-7,A) and the Multivane (Fig. 36-7,B) turn at slow speeds. Therefore they are useful for pumping water. The Darrieus Rotor (Fig. 36-7,C) and the three bladed wind turbine (Fig. 36-7,D) turn at high speeds. They are useful for generating electricity. Any region with an average annual wind speed above 11 km/h is a good region for converting wind energy to electricity. Figure 36-8 shows promising regions in Canada.

Because wind is irregular, some means of storing the electric energy generated is needed. Storage batteries can be charged. Water can be pumped from one level to another. The electricity can also be used to split water into hydrogen and oxygen gas. Another problem is the noise and sight pollution. Imagine thousands of windmills dotting the landscape!

C. Darrieus rotor electricity generating windmill

D. Modern three-blade electricity generating windmill

Fig. 36-7 Some vertical and horizontal axis windmills

Section Review

1. What two uses can be made of wind machines?
2. What are three ways to store electric energy generated by wind turbines?
3. What is the pollution problem with capturing the energy in wind?

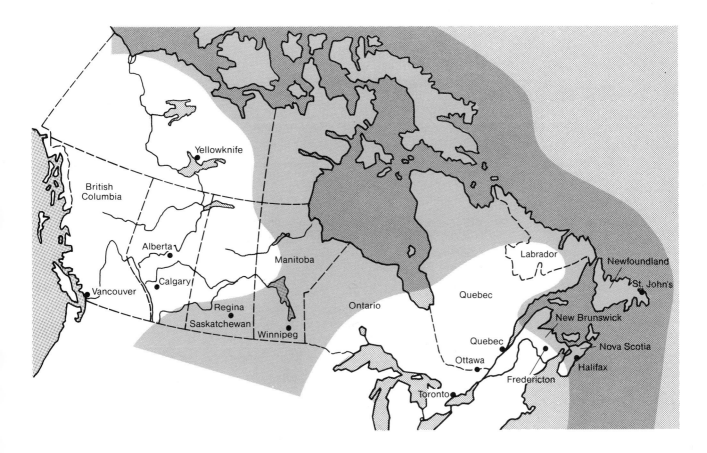

Fig. 36-8 Promising areas in Canada for wind generation.

36.6 Energy from Heat in the Earth's Crust

The core of the earth is molten rock. The temperature of this molten rock is over 1000°C. The temperature of the earth increases between 10°C and 30°C for each one kilometre depth from the surface. The heat inside the earth is called **geothermal energy** [GEE-O-THUR-mal]. Some of this heat is in the form of pockets of steam. Steam can be used to heat buildings. This is already being done in Iceland. Steam can also turn steam turbines to generate electricity. This is being done in San Francisco, U.S.A. Scientists also hope to harness the heat in the dry rock beneath the surface. To do this, the earth's crust must be drilled to great depths as shown in Figure 36-9. Water is circulated through the hot rock and returns as steam. The steam is used to generate electricity. But there are problems. The hot water brings salts to the surface. Also, conventional drilling bits overheat and break.

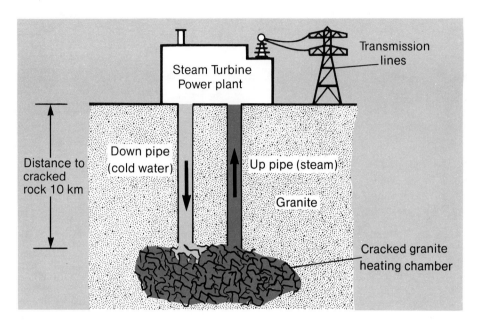

Fig. 36-9 A geothermal power system

Section Review

1. What is geothermal energy?
2. What is the approximate temperature of the earth at a depth of 10 km?
3. Describe one pollution problem with this source of energy.

36.7 Energy from Hydrogen

Hydrogen is one fuel that may replace fossil fuels. Although hydrogen can be obtained from coal, water is the cheapest source. When an electric current is passed through water, the water breaks apart. Two gases, hydrogen and oxygen, are formed.

Hydrogen has several advantages as a fuel. Hydrogen does not pollute when it burns. The end product is water. Thus hydrogen is a renewable source of energy. Hydrogen releases large amounts of energy during burning. More than three times the energy is released by a kilogram of hydrogen as by a kilogram of gasoline. Hydrogen can be transported easily and stored for long periods of time. Hydrogen, however, has several disadvantages. It is costly to produce hydrogen from water using electricity. Stored hydrogen can explode. Also, the material for storing hydrogen in transportation vehicles is costly.

Section Review

1. Name two sources of hydrogen.

2. State five advantages of using hydrogen as a fuel.
3. State three problems with using hydrogen as a fuel.

Main Ideas

1. Renewable energy sources are replaced as fast or faster than they are used.
2. The following are renewable energy: radiant energy from the sun, bulk kinetic and gravitational energy in water, heat energy in oceans, bulk kinetic energy in winds, heat energy in the earth's crust, and chemical energy stored in plants by photosynthesis.
3. Radiant energy can be changed to heat using solar reflectors and flat-plate collectors.
4. Radiant energy can be changed to electric energy using solar cells.

Glossary

greenhouse effect		the way a greenhouse uses transparent glass or plastic to trap radiant energy from the sun
hydro-electric power	HI-dro-e-LEK-trik	electricity generated by harnessing the energy in falling water
photosynthesis	fo-to-SIN-the-sis	the process by which chlorophyll in green plants changes radiant energy to chemical energy
tide		the rise and fall in the level of coastal waters which occurs twice a day
wind turbine	TUR-bin	a device used to change the kinetic energy of the wind to bulk kinetic energy and then to electric energy

Study Questions

A. True or False

Decide whether each of the following sentences is true or false. If the sentence is false, rewrite it to make it true. (Do not write in this book.)
1. Renewable energy is used as fast or faster than it is replaced.

2. Curved reflectors can be used to boil water and melt metals.
3. Radiant energy from the sun cannot be changed directly into electrical energy.
4. Harnessing the energy in moving water is inexpensive.
5. Geothermal energy is radiant energy which is stored in the ocean's surface as heat.

B. Completion

Complete each of the following sentences with a word or phrase which will make the sentence correct. (Do not write in this book.)
1. Most of the energy we use today comes from ▮▮▮▮ sources.
2. Reflectors and flat-plate collectors are used to capture ▮▮▮▮ energy.
3. Solar cells change radiant energy directly into ▮▮▮▮ energy.
4. Windmills which rotate at slow speeds are used for ▮▮▮▮ .
5. Hydrogen produced from water is ▮▮▮▮ expensive than electric energy.

C. Multiple Choice

Each of the following statements or questions is followed by four responses. Choose the correct response in each case. (Do not write in this book.)
1. Which of the following energy sources is not powered by the sun?
   a) rivers   b) winds   c) radiant   d) geothermal
2. Solar cells convert radiant energy to
   a) chemical energy   c) electric energy
   b) heat energy   d) nuclear energy
3. The glass surface of a flat-plate collector traps radiant energy because it
   a) absorbs short wave length radiation
   b) reflects short wave length radiation
   c) absorbs long wave length radiation
   d) reflects long wave length radiation
4. Photosynthesis changes radiant energy to
   a) heat energy   c) electric energy
   b) chemical energy   d) nuclear energy
5. Which of the following is not a way to store the energy from wind?
   a) as chemical energy in storage batteries
   b) as gravitational energy in raised water
   c) as chemical energy in the form of hydrogen
   d) as sound energy in turning windmills

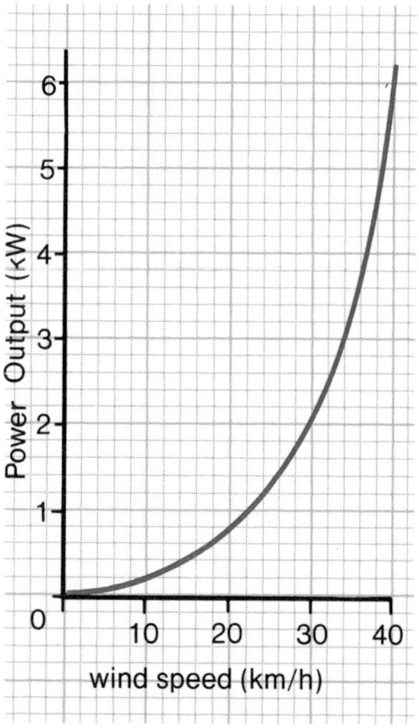

Fig. 36-10 Power output against windspeed for a Darrieus rotor electricity generating wind turbine

D. Using Your Knowledge

1. How does renewable energy give us greater self-reliance, both as a country and as individuals?

2. The power output of a Darrieus Rotor windmill changes with speed as shown in Figure 36-10.
   a) Find the power output in kW at a wind speed of 20 km/h; 40 km/h.
   b) As the wind speed doubles, what happens to the power output?
3. Assume that the average rate that solar energy falls on the roof of a house is 200 W/m². The roof has a useful surface area of 150 m². The maximum efficiency of solar cells is 15%.
   a) What is the rate in watts that solar energy strikes this roof?
   b) How much electrical power can be generated by solar cells on this roof?
   c) If solar cells cost $500/m², what will it cost to cover this roof with solar cells?

E. Investigations

1. Visit a house that uses flat-plate collectors to meet part or all of its heating needs. Find out the advantages and disadvantages of using this method to heat a home.
2. Do a library search to find out what causes the tides. Also find out why the Bay of Fundy has such high tides.
3. Do a library search to find out what materials are used for safely storing hydrogen in transportation vehicles. Find out the problems with using hydrogen in these vehicles.

37 Conserving Energy for Heating

37.1 Decreasing Heat Transfer
37.2 Insulation and R Values
37.3 Activity: The Cost and Characteristics of Insulation
37.4 Retrofitting and Heating a House
37.5 Activity: Retrofitting an Old House

We must conserve energy. We cannot afford to let energy escape as waste heat. Insulating buildings is one of the best ways to conserve energy. You will find ways to decrease heat losses in this chapter.

37.1 Decreasing Heat Transfer

About 15% of the energy used in Canada goes into heating homes. A third of this can be saved by decreasing heat transfer. Heat escapes by conduction, convection, and radiation. Let us see how they can be decreased.

Conduction

Heat is transferred through solids by conduction. During conduction, heat is passed from molecule to molecule. The heat moves from a warm to a cooler region. But the molecules do not. Conduction is reduced by using proper insulation such as air. Therefore a layer of air is often left between the inside (warm) and outside (cool) walls of a house. But the air molecules in the layer must not be allowed to move from place to place. If they move, they transfer heat by convection.

Convection

During convection, molecules of air move from place to place carrying heat with them. Convection accounts for about 35% of the heat that escapes from a home.

In an uninsulated wall space, air molecules move freely (Fig. 37-1). They move back and forth between the inside and outside surface.

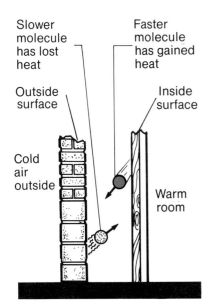

Fig. 37-1 Molecules free to move in partitions transfer heat by convection.

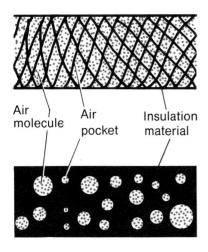

Fig. 37-2 Fibres of insulation trap air in small pockets. They also separate one air pocket from the next. This prevents air molecules from moving freely from place to place.

At the inside surface they pick up heat. They speed up. At the outside surface they give up heat. They slow down. The inside surface of an uninsulated wall can be 8°C cooler than an insulated wall. Insulation placed in the wall space lessens convection. It stops the molecules from moving back and forth (Fig. 37-2).

Some of the warm air in a home escapes to the outside through cracks. Cold air enters to take its place. A one centimetre crack under a door lets as much heat escape as a hole in the wall the size of your two fists! Caulking and weather stripping seal cracks (Fig. 37-3). Storm doors and windows also help.

Fig. 37-3 Weather stripping is used on a door to seal cracks.

Air Exchange in a Home. A house can be sealed too tightly. Poisonous gases can build up. Radon is one of these gases. Much of Canada has a high uranium content in rocks. Rock materials are used to build homes. Uranium emits radon gas. About 10% of lung cancers are now caused by indoor radon. Tightly sealing homes increases the amount of radon indoors.

Formaldehyde is another poisonous substance. It is used in making rugs, curtains, furniture and urea formaldehyde insulation. It, too, can build up. Even home cooking can be dangerous. Gas stoves not vented to the outside can produce dangerous levels of carbon monoxide. A build up of this gas causes headaches, burning eyes, and exhaustion.

Moisture, too, can be trapped in a home. Cooking, showers, baths, and breathing add moisture to the air. Warm air holds more moisture than cold air. When warm air comes in contact with a cold surface it loses moisture. You have likely seen moisture on the inside surface of windows. Some moisture is healthy for you. But the humidity is too high if, on a cold day, you see a lot of water on the inside. The humidity is correct if only a slight mist is seen.

As you can see, there needs to be some air leakage. There should be enough leakge to completely change the air in a house every 5 to 6 h. But this is no excuse for leaving large cracks. Most older homes lose warm air far too quickly.

Radiation

Any warm object radiates heat. If you sit near a campfire on a cold night you feel radiant heat. Radiant heat does not travel by conduction or convection. It travels as waves in the same way as light. If someone comes between you and the fire, the waves are blocked off. You feel cold.

Insulation placed in walls helps stop heat waves from travelling between the inside and outside. A thin film of aluminum foil inside the wall also helps. It reflects radiant energy back into the home in winter and to the outside in summer. A reflective film on the windows serves the same purpose.

Section Review

1. What percentage of the energy Canadians consume is used for heating homes?
2. Describe three methods of heat transfer.
3. How can conduction losses from a home be decreased?
4. **a)** Describe how insulation is constructed to decrease convection.
   b) How can convection losses from a home be decreased?
5. **a)** Outline three reasons why a house should not be completely sealed.
   b) How often should the air inside a house be exchanged with the outside air?
6. Describe how radiation losses from a home can be decreased.

37.2 Insulation and R Values

R Values of Insulating Materials

Materials differ in their ability to prevent heat transfer. For example, 1 cm of styrofoam transfers about half as much heat as 1 cm of wood shavings. We say that styrofoam has a greater resistance value than wood shavings. **Resistance value** is a measurement of the resistance of a material to heat transfer. It is given the symbol **R**.

Look for the symbol on bags of insulation (Fig. 37-4). The higher the R value, the less heat escapes. The R value of styrofoam is about twice the R value of wood shavings.

Fig. 37-4 The R value is marked on the outside of bags of insulation.

The R values for 1 cm of some common insulation materials are shown in Table 37-1.

Table 37-1   R Values for One Centimetre of Some Insulation Materials

Material	R value of 1 cm (R/cm)	Material	R value of 1 cm (R/cm)
wood shavings	0.9	batt fibreglass	1.3
vermiculite	0.9	cellulose fibre	1.5
loose rock wood	1.1	expanded polystyrene	1.5
loose fibreglass	1.2	styrofoam	1.8
batt rock wool	1.3	polyurethane slabs	2.2

Calculating the R Values of Different Thicknesses of Insulation

One kind of insulation may be thicker than another. But the total R value of a material is easily calculated. You just multiply its thickness in centimetres by the R value per centimetre.

Example Problem

Find the total R value of a batt of fibreglass insulation having a thickness d of 10 cm.

Section 37.2

Given
$d = 10$ cm
R = 1.3/cm
Required
R_{total}

Analysis
Units are consistent

Solution
$R_{total} = d \times R$
$= 10$ cm $\times 1.3$/cm
$= 13$

Statement
The batt of fibreglass has a total R value of 13.

R Values of Building Materials

The R values of some common building materials are shown in Table 37-2.

Table 37-2 R Values of Building Materials

Material	Thickness (cm)	R value	Material	Thickness (cm)	R value
air space in walls	2-10	1.0	inside air film	film	0.7
brick (clay or shale)	10	0.4	insulating fibreboard	2.5	2.4
cinder block	10	1.1	outside air film	film	0.2
	20	1.7	plywood	1.5	0.8
	30	1.9	softwood (pine, fir, spruce)	2.5	1.3
concrete block	10	0.7	stucco	1.0	0.04
	20	1.1	Western cedar	2.5	1.6
	30	1.3	window glass	single pane	1.0
gypsum board	1.25	0.3		double pane	2.0
hardwood (maple, oak)	2.5	0.9		triple pane	3.0

Calculating the Total R Value of a Wall

The total R value of a wall is found by adding the R values for each layer. Thin air films cling to the inner and outer surfaces. These act as insulation. They must be included when calculating total R values.

Example Problem

Calculate the total R value of a wall of frame and brick construction

(Fig. 37-5). The wall has 10 cm of brick, 2 cm air space, 1.5 cm plywood, 9.0 cm air space, and 1.25 cm of gypsum board.

Given
Fig. 37-5

Required
Find the total R value of the wall.

Analysis
List the individual R values from Table 37-2. Sum the values.

Solution

Item	R value
inside air film	0.7
1.25 cm gypsum board	0.3
9 cm air space	1.0
1.5 cm plywood	0.8
2 cm air space	1.0
10 cm brick	0.4
outside air film	0.2
Total	4.4

Statement
The total R value of the wall is 4.4.

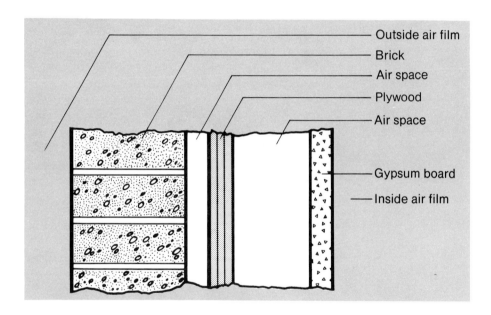

Fig. 37-5 Cross section of a wall of frame and brick construction

Section Review

1. Define R value.
2. What is the R value of 1 cm of vermiculite?
3. What is the R value of a fibreglass batt 5 cm thick?
4. Complete the following table.

Insulating material	R value/cm	Thickness (cm)	Total R value
wood shavings		3	
loose fibreglass		15	
styrofoam			5.4
cellulose fibre			3.0
		2	4.4

Section 37.2

5. The cross section of a cottage wall is shown in Figure 37-6. Find the total R value of the wall.

37.3   ACTIVITY   The Cost and Characteristics of Insulation

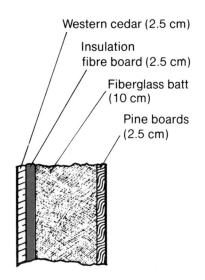

Fig. 37-6 Cross section of a cottage wall

Several factors should be considered when selecting insulation for a job. One is the cost of covering a square metre of surface to the desired R value. Others are the fire hazard, settling, and whether it harbours vermine. You will obtain information about these factors for three kinds of insulation in this activity.

Problem

Which insulation would you select: cellulose fibre, fibreglass batt, or vermiculite?

Procedure

a. Copy Table 37-3 into your notebook.
b. Choose a volunteer to phone a supplier of insulation and obtain the information needed to complete the table.
c. Find out other factors that are considered when choosing insulation.

Table 37-3   Cost and Characteristics of Insulation

Kind of insulation	Unit (bag, bale, etc.)	Unit cost	Area covered by the unit to an R value of 10 (m^2)	Cost to cover 1 m^2 to an R 10	Characteristics		
					Fire hazard	Settling	Habitat for Vermine
Cellulose fibre							
Fibreglass batt							
Vermiculite							

Discussion

1. Which material costs the least to cover 1 m^2 to an R value of 10?
2. Which material has the best characteristics?
3. What other factors besides fire hazard, settling and vermine should be considered when choosing insulation?

37.4 Retrofitting and Heating a House

Retrofitting [RE-tro-fit-ing] a building is improving its ability to retain heat. This involves adding insulation to ceilings, floors, and walls. It includes sealing cracks with weather stripping and caulking. And it means adding storm windows and doors. These can reduce fuel bills by 20% or more. Retrofitting is a good investment. Energy costs money. As energy costs go up, so do heating costs. The better the retrofit, the lower the heating costs. But retrofitting also costs money. What factors should be considered when deciding how much money to spend to conserve energy?

The Zone in Which you Live

It pays to add more insulation where the winters are longer and colder. Figure 37-7 shows four heating zones in Canada. Zone A, containing Vancouver, is the warmest. Zone D, containing Churchill, is the coldest. It takes over twice as much energy to heat a home for a year in Zone D as in Zone A. Thus it pays to use more insulation in Zone D than in Zone A.

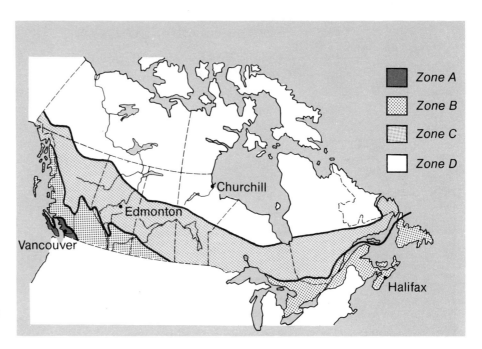

Fig. 37-7 A map showing the heating zones for Canada

Recommended Minimum Insulation Values

Table 37-4 shows the minimum insulation levels by zone for existing buildings. If possible, the owner should aim higher. With rising energy costs, a larger investment now will pay dividends later. Unfortunately, it may not even be possible to go as high as this minimum if walls and/or attic space is limited.

Table 37-4 Recommended Minimum Insulation Levels for Existing Buildings (In R Values)

	Zone A	Zone B	Zone C	Zone D
Walls	R 14	R 17	R 20	R 21
Basement walls	R 12	R 12	R 12	R 12
Roof or ceiling	R 30	R 32	R 36	R 40
Floor (over unheated spaces)	R 28	R 28	R 28	R 28

The Style of Home

The amount and location of the insulation required depends on the style of home. A detached home uses the most energy for heating. A semi-detached home of the same size uses about 30% less. A townhouse uses 45% less. Since these homes have common walls, a smaller surface area is exposed to the outside.

A single storey house loses a greater part of its heat through the roof than a two storey house does. Its roof has a greater surface area for the same floor area. So when adding insulation you have to decide where it does the most good.

Surface Being Insulated

The thickness of insulation depends on whether the ceiling, floor, or walls are being insulated. Ceilings need the most insulation; above ground walls need less and floors need the least. Hot air rises by convection. Therefore ceilings get warmer than walls or floors. The greater the temperature difference between the inner and outer surface, the faster the heat loss. Ceilings account for about 20% of the heat loss from most buildings.

The Area of Windows and Doors

Windows and doors account for about 20% of the heat loss from buildings. Glass conducts heat energy about 40 times better than trapped air. Double and triple glazed windows have layers of air in them (Fig. 37-8). The air cuts down on conduction and the space lessens convection.

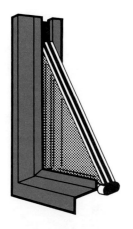

Fig. 37-8 Triple glazed windows are three times as effective as single glazed windows at decreasing heat loss.

Section Review

1. What is meant by retrofitting a house?
2. What factors affect the amount of fuel needed to heat a home?
3. **a)** Which zone has the most severe winter?
   b) Why is it economical to use more insulation in Zone D than in Zone B?
4. Why does a townhouse lose less heat than a detached house?
5. Why is more insulation needed in the ceiling than in the walls or floor of a home?

37.5 ACTIVITY Retrofitting an Old House

In this activity you will prepare a plan to retrofit a house. You will first analyze the R value of the wall, ceiling, and floor. Then you will decide how to improve the house to approach the minimum insulation standards for Zone C (Table 37-4).

Problem

How do you retrofit a house?

Table 37-5 House Description

Part	Description	Existing R value	How to improve the insulation to the recommended R value
Wall	outside air film 1 cm stucco 10 cm brick 2 cm air 3 cm soft wood 10 cm air space 1.25 cm gypsum board inside air film 　　　　Total	_____	
Ceiling	outside air film 2 cm gypsum board 5 cm loose rock wool inside air film 　　　　Total	_____	
Floor	outside air film 3 cm hardwood inside air film 　　　　Total	_____	

Procedure

a. Study Figure 37-9. Then copy Table 37-5 into your notebook.
b. Refer to Tables 37-1 and 37-2. Record the R value for each part of the wall, ceiling, and floor in your table.
c. Determine the total R value of the wall, ceiling, and floor. Record this in the table.
d. Refer to Table 37-3. Record the R value that the wall, ceiling, and floor should have for Zone C.
e. Decide how this R value can be achieved. Consider costs, availability, ease of installation, settling, vermine, and safety.
f. Prepare to discuss your results with others in the class.

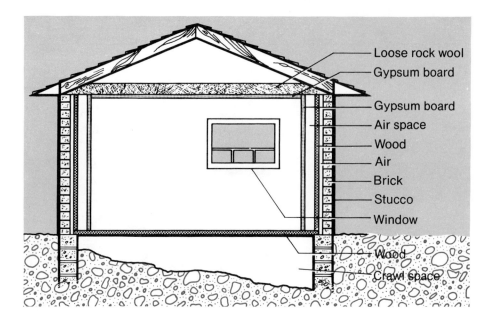

Fig. 37-9 Cross section of a house

Discussion

1. What total R value did you get for the wall? the ceiling? the floor?
2. What insulating material did you plan to use for the wall? the ceiling? the floor? Why?
3. How much insulation did you plan to use in the wall? the ceiling? the floor?

Main Ideas

1. Energy transfer occurs by conduction, convection, and radiation.
2. Air is a good insulator. But it transfers heat by convection.
3. Insulation traps air and reduces conduction and convection.
4. Insulation materials and reflectors decrease the transfer of heat by radiation.
5. The ability of different materials to limit heat transfer is compared using R values.
6. The R value of a material depends on the kind of material and the thickness.
7. The total R value of a wall is found by adding the R values of the different layers.
8. Retrofitting is improving the ability of a building to prevent heat transfer.
9. Insulation is more economical in severe climates.
10. Ceilings need more insulation than walls or floors.

Glossary

insulation	in-su-LAY-shun	any material which lessens heat transfer
R value		a measurement of the resistance of a material to heat transfer
retrofit	RE-tro-fit	improving the ability to prevent heat transfer

Study Questions

A. True or False

Decide whether each of the following sentences is true or false. If the sentence is false, rewrite it to make it true. (Do not write in this book.)

1. Conduction is reduced by using air as an insulator.
2. Small air pockets are made in insulation to reduce radiation losses.
3. Wood shavings have a higher R value than styrofoam.
4. The colder and longer the winters, the higher the economical R values.
5. A unit area of a wall loses more heat than a unit area of floor.

B. Completion

Complete each of the following sentences with a word or phrase which will make the sentence correct. (Do not write in this book.)

1. Air molecules which move from place to place transfer heat by ▨▨▨ .
2. The air in a house should be exchanged with the outside air every 5-6 h to prevent ▨▨▨ and ▨▨▨ from building up.
3. The total R value of any thickness of insulation is found by multiplying its thickness by the ▨▨▨ .
4. The total R value of a wall is found by adding the ▨▨▨ .
5. A detached house uses more energy for heat than a semi-detached house of the same size because ▨▨▨ .

C. Multiple Choice

Each of the following statements or questions is followed by four responses. Choose the correct response in each case. (Do not write in this book.)

1. What percentage of the heat that escapes from a house is caused by convection?
    a) 15%   b) 25%   c) 35%   d) 45%

2. Caulking and weather stripping are used to decrease
   a) conduction
   b) convection
   c) conduction and convection
   d) radiation
3. The humidity is too high in a house in winter if
   a) the temperature of the air next to the wall is 4-5°C colder than the air in the middle of the room
   b) the air is free to move between the inside and outside
   c) heavy condensation forms on the inside of a double glazed window
   d) the R value of the walls, ceiling, and floor is too large
4. For an average house the region with the highest heating cost is
   a) Zone A   b) Zone B   c) Zone C   d) Zone D
5. Which of the following partitions in a building should have the largest R value?
   a) ceilings   b) floors   c) walls   d) windows

D. Using Your Knowledge

1. Explain how electrical outlet boxes in outside walls increase heat transfer.
2. Turning the thermostat down at night to 15°C can save 15% on the fuel bill. However the temperature of the house has to be increased in the morning. Why, then, does turning down the thermostat at night save energy?
3. How are sleeping bags made to reduce heat transfer?
4. Note that the recommended minimum R values in Table 37-3 for walls and ceilings increase in going from Zone A to Zone D. The R values for basement walls and floors stay the same. Why do you think this is the case?

E. Investigations

1. Research the different ways to insulate windows. Find out what it costs to insulate a window 80 cm x 120 cm using each method. Then find out how much energy and money is saved during a winter by insulating the windows in an average house.
2. Research the construction, insulation, and caulking of your home. Calculate its R values. Find out what zone you are in. Prepare a plan to minimize heat transfer in your home.
3. Research ways to use landscaping to reduce the energy used for heating and cooling a house.

38 Conserving Electricity and Gasoline

38.1 Measuring and Conserving Electric Energy
38.2 Activity: Conserving Electric Energy
38.3 Conserving Energy for Transportation
38.4 Activity: Planning to Buy a New Car
38.5 Conservation or Crisis: The Decision is Ours

An active person needs about ten million joules of energy a day to survive. Canadians use about one hundred times this amount. This is because we waste electric energy and gasoline.

How is electric energy measured? What are some ways to conserve electricity and gasoline? You will find answers to these and other questions in this chapter.

38.1 Measuring and Conserving Electric Energy

Measuring Electric Energy

There are several ways to measure the electric energy we use.

A. Using the Power Rating. Some electrical devices have their power rating marked in **watts** (Fig. 38-1). To find out how much electricity the device uses, multiply the power rating in watts by the time the device is used in seconds. **Energy (joules) = power rating (watts) x time (seconds)**. Use the equation $E = Pt$. The power ratings of some common appliances is shown in Table 38-1. The approximate energy used by each appliance is also shown.

Fig. 38-1 This light bulb has a power rating of 100 W. Unfortunately, only 5% of the energy from the bulb becomes light. The rest is wasted as heat.

Table 38-1 Power Ratings for Some Common Appliances

Appliance	Average power (W)	Approx. yearly energy used (MJ)	Appliance	Average power (W)	Approx. yearly energy used (MJ)
clothes dryer	4356	3600	TV (black & white)	55	400
dishwasher	1200	1300	TV (colour)	200	1600
range and oven	12 200	4200	washing machine	512	400
refrigerator/ freezer	615	6600	water heater	2475	15 000

B. Using Electric Current and Electric Potential. Some devices do not have the power rating indicated. Instead they show the **electric current (I)** the device draws when connected to an **electric potential (V)** (Fig. 38-2). Most devices in the homes are connected to an electric potential of 120 V. Some appliances, such as stoves, water heaters, and clothes dryers, however, are connected to 240 V. To find the electric energy used in joules, multiply the electric potential in volts by the electric current in amperes by the time in seconds. **Energy (joules) = potential (volts) x current (amperes) x time (seconds).** $E = VIt$

Fig. 38-2 This power sander draws a current of 3 A when connected to a standard wall outlet.

Example Problem

Find the electric energy used by a drill that is used for 5 min. The drill is connected to a 120 V outlet. It draws a current of 4 A.

Given
$t = 5$ min
$V = 120$ V
$I = 4$ A
Required
E

Analysis
Change the time to seconds.
Then use the expression: $E = VIt$
Solution

$t = 5 \text{ min} \times 60 \frac{\text{s}}{\text{min}} = 300 \text{ s}$

$E = VIt$
$= 120 \text{ V} \times 4 \text{ A} \times 300 \text{ s}$
$= 144\,000 \text{ J} = 144 \text{ kJ}$

Statement
The drill used 144 kJ of electric energy.

C. Using the Electric Meter. The total electric energy used in the home is measured using an electric meter. The meter records the energy used in the same way as the odometer of a car records the distance travelled. Newer electric meters are digital. However some older meters have four dials. Their reading must be multiplied by 10. Others have five dials. Figure 38-3 shows a four dial meter. The dial reading in this meter is 8 0 1 6. Multiplying this by ten gives an actual reading of 8 0 1 6 0 units. The meter records energy in kilowatt hours (kW•h). Therefore the reading on the meter is 80 160 kW•h. (After Canada changes completely to SI, the unit will no longer be kilowatt hours.)

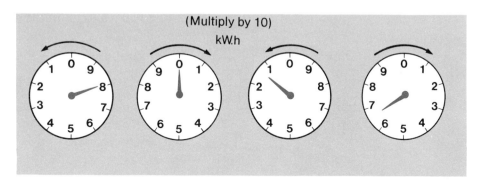

Fig. 38-3 A four dial electric meter. Notice the legend "Multiply by 10" and the unit kW/h.

Ways to Conserve Electric Energy in the Home

Some ways to conserve electric energy in the home are given in Table 38-2.

Table 38-2 Ways to Conserve Electric Energy in the Home

1. Turn off lights when they are not needed.
2. Replace large wattage incandescent bulbs with the smallest possible wattages for the job.
3. Thaw frozen foods before putting them in the oven.

4. Cook the entire meal in the oven instead of using many burners.
5. Don't peek while foods are cooking. Each peek costs 20% of the energy in the oven.
6. Don't boil any more water than you need for tea, coffee, or hot chocolate.
7. Cook vegetables in the least amount of water possible. Then turn down the heat when boiling starts.
8. Cook concentrated foods in a microwave oven.
9. Don't bother preheating the oven if the cooking time is more than an hour.
10. Turn off the oven before cooking time is up. The heat in the oven will finish the job.
11. Defrost manual refrigerators when the ice is 5 mm thick.
12. Make sure refrigerators and freezers are properly sealed.
13. Place your refrigerator well away from the stove, direct sunlight, and heating vents.
14. Leave at least 10 cm between the wall and back of a refrigerator.
15. Clean the condenser coils on the back of the refrigerator regularly.
16. Use the clothes washer only for a full load.
17. Use as few hot cycles as possible when washing clothes.
18. On sunny windy days, dry clothes on the yard line instead of in the dryer.
19. Use the clothes dryer only for a full load.
20. Set the thermostat on the hot water tank as low as possible (say 60°C).
21. Wrap the hot water tank in an extra blanket of insulation.
22. Wrap hot water pipes with insulation (Fig. 38-4).
23. Fix leaking water taps.
24. Wash dishes by hand and only when there is a full load.
25. Wash only a full load of dishes in the dishwasher.
26. Let dishes air dry instead of using the dryer cycle on a dishwasher.
27. Take quick hot showers rather than deep hot baths and wash and shave using the minimum amount of hot water.
28. Turn off the TV when you are not watching it.
29. Turn off the electric power to the wall plug for "instant-on" TVs.
30. Use hand tools rather than power tools for small repair jobs.

Fig. 38-4 Hot water pipes being wrapped with insulation.

Section Review

1. How do you find the energy used by an appliance if the power rating is given?
2. A dishwasher has a power of 1200 W. Suppose it runs for 10 min. How much energy will it use?
3. Suppose you know the electric potential, current, and time. How can you calculate the electric energy?
4. A food processor draws 6.5 A on a 120 V circuit. How much energy will this processor use in 8 min?
5. What is the reading on each of the electric meters shown in Figure 38-5?

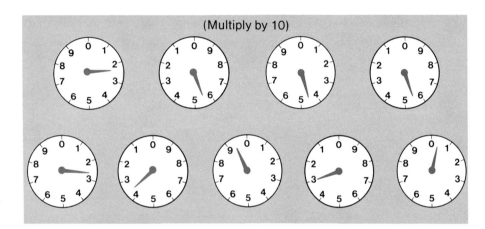

Fig. 38-5 Readings on a four and a five dial electric meter

38.2 ACTIVITY Conserving Electric Energy

In this activity you will measure the electricity used in a home during a normal week. Then you will measure it during a conservation week. Finally you will calculate the energy and money saved as a result of conservation.

Problem

How much energy and money can conservation save?

Procedure

a. Copy Table 38-3 into your notebook.
b. Choose two weeks close together with similar conditions.
c. Read the electric meter at a specific time.
d. Use electricity in the normal way for the next week.
e. During this normal week plan ways with your family to conserve electricity during the coming week. Refer to Table 38-2 for ideas, but include your own.
f. Read the electric meter at the end of one full normal week.

Section 38.2 475

g. Ask everyone to co-operate to save electricity during the conservation period.
h. Read the electric meter at the end of the full conservation week.
i. Calculate the electric energy used during the normal week and during the conservation week.
j. Find the cost of electricity by consulting your local hydro office.
k. Calculate the energy and money saved during the conservation week.

Table 38-3 Conserving Electric Energy

Description	Data	
	Normal week	Conservation week
Meter reading at end		
Meter reading at start		
Energy used (kW•h)		
Cost/kW•h		
Total cost		

Discussion

1. What conservation measures did you put into practice?
2. How much electric energy was saved?
3. How much money was saved?
4. Compare the conservation measures taken by different members in the class.

38.3 Conserving Energy for Transportation

The Efficiency of the Car

Transportation uses about one-quarter of all the energy used in Canada. Over half of this goes to moving people by car. Cars change the chemical energy stored in the fuel to bulk kinetic energy. Table 38-4 shows what happens to 100 J of chemical energy originally in the ground. Less than 10 J out of every 100 J ends up as bulk kinetic energy to move the passengers. The rest is lost forever to the environment as waste heat.

Table 38-4 Efficiency of an Average Car

Step	Energy used/wasted (J)	Usable energy remaining (J)
Crude oil in the ground	0	100
Removal of crude oil from ground	4	96
Refining of crude oil into gasoline	12	84
Transportation of gasoline to gas station	3	81
Conversion of chemical energy into mechanical energy in car's engine	60	21
Transfer of mechanical energy from engine to car's wheels	10	11
Frictional losses due to motion	3	8
Energy originally in crude oil which becomes kinetic energy		8

Decreasing Fuel Consumption in Cars

A. Driving Smaller Cars. Engines can be made more efficient. This will decrease fuel consumption. But the basic problem is one of mass. Most cars are too heavy. For each 100 kg of additional mass, fuel consumption increases by 6%. Figure 38-6 shows a graph of gasoline consumption plotted against mass. Note how much more gasoline it takes to go 100 km in a heavy Ford LTD than in a light Volkswagen Rabbit. Look up the mass of your family car. Where does it fit on the graph? If your car uses more gas than it should, perhaps it needs a tuneup. Or perhaps you drive or accelerate too quickly.

B. Driving Slowly and Steadily. Figure 38-7 shows how gasoline consumption increases as speed increases. Driving more slowly saves gas and lives! Driving in the city uses more energy than driving on the highway because there are more starts and stops. This is compounded by stop signs on residential streets. Also, rapid increases in speed and idling a car too long gobble up energy. In 45 s an idling car uses more energy than it takes to start it again.

Using Other Means of Transportation

Energy is saved by leaving the car at home. If you drive, don't travel alone; use car pools. For travel between cities, use a passenger train or a bus. A passenger train uses one sixth as much energy per person as a car carrying one person. A bus uses half the energy. For travel within the city, use trolleys and subways. They are three and a half times as efficient as a car. A bus is twice as efficient. Better

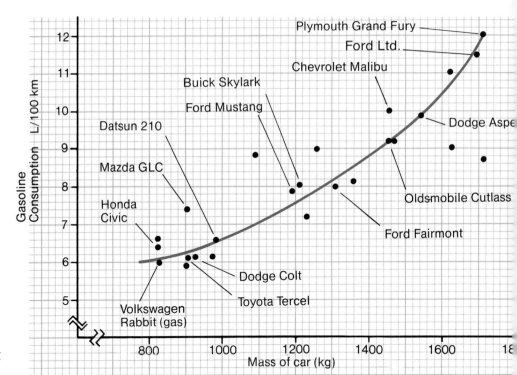

Fig. 38-6 Gasoline consumption against mass for some 1980 cars travelling at 90 km/h

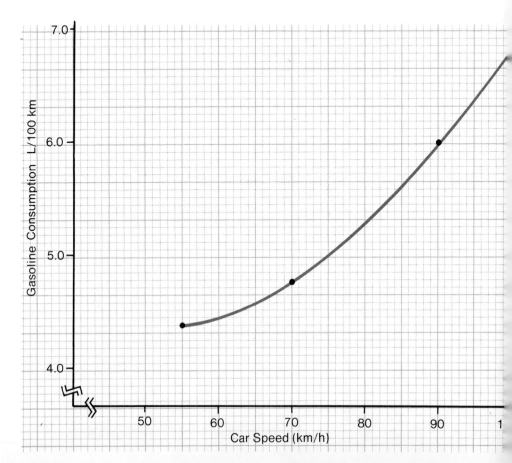

Fig. 38-7 Gasoline consumption against speed for a Volkswagen Rabbit (1980). The Rabbit is one of the best streamlined cars.

still, use a bicycle. It carries a passenger over forty times as far as a car for the same energy. And it keeps you fit into the bargain. If you can't pedal, ride a moped.

Section Review

1. What fraction of Canada's energy is used for transportation?
2. What fraction of the energy used for transportation is used by cars?
3. a) Of 100 J of chemical energy stored in crude oil in the ground, how many joules become bulk kinetic energy in the car?
   b) What gobbles up most of the difference?
4. How does the mass of a car effect gasoline consumption?
5. Describe how the speed of a car effects gasoline consumption.
6. Describe how to conserve energy in travelling between cities.
7. Describe how to conserve energy in travelling in a city.

38.4 ACTIVITY Planning to Buy a New Car

Assume that your family plans to buy a new car in the near future. In this activity you will research different cars. You will determine the most cost efficient car to meet your family's needs.

Problem

Which car should your family buy?

Materials

Consumer Reports Buying Guide
Fuel Consumption Guide
Advertising Brochures (car dealers)
Financial Post Magazine
New Car Guide
Popular Science

Procedure

a. Copy Table 38-5 into your notebook.
b. Meet with your family. Discuss the project. List the characteristics of the car that meets your family's needs.
c. Visit different car dealers to find 3 or 4 cars with these characteristics.
d. Research the literature and complete the table.
e. Discuss the findings with your family. Together decide which make and model is most cost efficient. Perhaps you can test drive the car.
f. Share your findings with your peers.

Table 38-5 Planning to Buy a New Car

Desirable characteristics	Cars with these characteristics	Basic cost	Mass (kg)	Fuel consumption (L/100 km)
	1.			
	2.			
	3.			
	4.			

Discussion

1. To conserve energy, what factors should be considered before buying a new car?
2. What did you learn about buying a new car?

38.5 Conservation or Crisis: The Decision is Ours

Over the years our energy demands have grown at the rate of about 7% per year. As a result, the demand for energy has doubled every ten years. What happens when energy demand doubles every ten years? Suppose your community needs one thermal generating plant today to meet its electrical needs. Ten years from now it would need two plants. In twenty years it would need four plants. In thirty years it would need eight plants. In seventy years your community would need 128 large electric power plants! This is shown graphically in Figure 38-8. Seventy years is about one human life span. This kind of

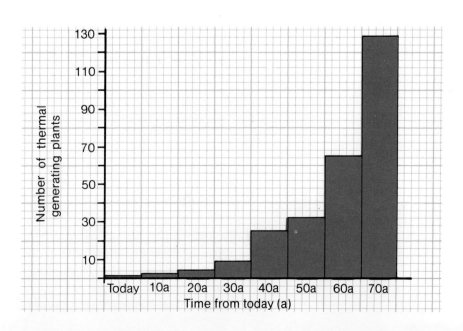

Fig. 38-8 Graph of the number of thermal generating plants against time for a growth rate of 7.0% per year

growth is called **exponential growth** [ex-po-NEN-shall]. Exponential growth of energy consumption cannot be allowed to continue.

Causes of Increased Demand

What factors cause the increased demand? Some of the demand is caused by an increasing population. Some is caused by new high energy industries. But much of the increased demand results from wasteful energy habits. Each of us is using more energy today than ten years ago. Many people drive bigger cars than they need. Heavy cars waste fuel. Also, lights are left on when not needed. Homes are poorly insulated. Television sets and radios are left running unattended. Ovens are used to heat small dishes of food. Full kettles of water are boiled to produce a single cup of coffee.

Results of Increased Use of Energy

Our increasing use of energy is causing heat pollution. The average city uses six times as much energy as reaches it by solar radiation. Waste heat is put into the atmosphere. Also lakes and rivers near electrical generating stations are warmed. Some scientists say that, if we continue to increase the use of energy, the temperature of the earth will increase. This could change weather patterns and cause flooding. Other scientists claim that the increased pollution in the atmosphere decreases the sunlight which reaches the earth. This could bring on another ice age. Neither result is desirable.

North America is having difficulty keeping pace with the yearly increase in demand for electric energy. More fossil and nuclear fueled electric generating stations are planned. These are costly. They are also harmful to the environment. Today's energy comes largely from concentrated sources such as fossil fuels and uranium. These were stored in the earth over thousands of millions of years. But oil and gas are running out; coal pollutes the atmosphere; and a safe way has yet to be found to store nuclear waste. Renewable energy sources could fill the gap in the years ahead. In Canada, wind and solar energy are predicted to increase from 1.5% to 4.5% by the year 2000. Biomass (especially wood products) is expected to rise from 3.5% to 10% by the year 2000. Researchers are even looking for ways to harness the hydrogen in water. But money and especially time are needed to bring these about. If the human race is to survive as we know it, we must learn to conserve energy. We are going to have to curb our demand for energy. We can no longer afford the luxury of waste, no matter how small the waste may be.

Section Review

1. What is exponential growth? Give an example.
2. Name three factors that cause an increase in demand for energy.
3. What undesirable effects result from using too much energy?

4. Describe two reasons why energy conservation is essential.

Main Ideas

1. The electric energy used by an appliance is found by multiplying the power rating by the time.
2. Electric energy used in an appliance is also found by multiplying current by electrical potential by time.
3. An electric meter measures the total electric energy used.
4. Electric appliances that produce heat use large amounts of energy.
5. A car is less than 10% efficient at moving people.
6. Fuel consumption is decreased by driving lighter cars more slowly.
7. Several factors have caused the increase in energy use. They are: more people, high energy industries, and wasteful energy habits.
8. We must learn to conserve energy to prevent a crisis.

Glossary

efficiency	e-FISH-en-see	the ratio of the effective work to the energy used in doing it
electric meter		a device for measuring and recording the total electric energy used
power rating		how fast in watts a device uses energy

Study Questions

A. True or False

Decide whether each of the following sentences is true or false. If the sentence is false, rewrite it to make it true. (Do not write in this book.)

1. The electric energy used by an appliance is found by dividing the power rating by the time.
2. To determine the energy an appliance uses, multiply the potential by the current.
3. An electric meter measures total electric energy used in joules.
4. A refrigerator with an ice build up in the freezer compartment uses more energy than one which is defrosted.
5. As the speed of a car increases, the distance it will go on a litre of gasoline increases.

B. Completion

Complete each of the following sentences with a word or phrase which will make the sentence correct. (Do not write in this book.)

1. A black and white TV viewed for the same time as colour TV uses ▓▓▓ energy.
2. The equation for finding the electric energy in terms of electric potential, current, and time is ▓▓▓ .
3. Taking a long bath rather than a short shower uses ▓▓▓ energy.
4. As the mass of a car increases the gasoline consumption ▓▓▓ .
5. For travel between cities, the bus is ▓▓▓ efficient than the car.

C. Multiple Choice

Each of the following statements or questions is followed by four responses. Choose the correct response in each case. (Do not write in this book.)

1. A washing machine has a power of 500 W. A black and white TV has a power of 50 W. Which will use more energy?
   a) the black and white TV
   b) the washing machine
   c) they will consume the same energy
   d) there is not enough information to tell
2. The greatest loss in converting energy in the oil in the ground to kinetic energy of a car occurs in
   a) refining the crude oil into gasoline
   b) transporting the gasoline to the gas station
   c) the car's engine
   d) the car's driveshaft
3. An electric meter records the electric energy in
   a) W   **b)** kW•h   **c)** J   **d)** MJ
4. A power saw connected to 120 V draws a current of 10 A. How much energy does it use in 2 min?
   a) 600 J   **b)** 2400 W   **c)** 2400 J   **d)** 144 000 J
5. What factor does *not* cause an increasing demand for energy?
   a) increasing population   **c)** poor energy use habits
   b) energy conservation   **d)** conversion to electricity

D. Using Your Knowledge

1. Some electric appliances such as kettles are used as sources of heat. Others such as electric sanders have electric motors. Which kind uses energy faster?
2. A dishwasher has a power of 1200 W. A colour TV has a power of just 200 W. Yet the colour TV uses more electric energy in a year than the dishwasher. Why?
3. It takes more hot water to get a bath to a comfortable temperature if the thermostat is set at 60°C rather than at 70°C. Why,

then, does it save energy to set the thermostat on the hot water tank as low as possible?
4. A bus is much heavier than a car. Why, then, does it save energy to take a bus rather than to drive a car to work?

E. Investigations

1. The higher the coefficient of performance of a refrigerator, the more heat it removes from inside for a given amount of electric energy. Research the coefficient of performance of various makes of refrigerators. Try to get the local newspaper to publish your results.
2. Find out what a heat pump is. How is it used to heat and cool a home? Why does a heat pump powered by electricity use less energy than the usual electric heaters?
3. Take a bath one night. Take a shower another night. Measure the quantity of water used in each case. See which saved energy and money.

Index

A
Acid rain 217, 230
Air pollution 214-24
Area 45-48
   definition **45**
   measuring 46-47
   units, uses 45-46
Asbestos 224
Atomic
   mass 199
   number 198
   particles 193
Atoms 163, 198
   radioactive 194

B
Batteries
   storage 398-400
Bohr, Neils 192
Boiling **81**
   of water 84-85
Boiling point
   effect of antifreeze on 111-12
   effect of pressure on 99-100
Brownian motion 87

C
Carbon dioxide
   in air 218-19
   in water 229-30
Carbon monoxide 219-20
Catalyst **183**-84
Cells
   and batteries 398-400
   dry 395-96, 411-13
   lead storage 396-98
   voltaic 392-95
Chemical change 174-75
Chemical properties **71**, 173-74
Chemical symbols 164-67
Circuits
   parallel 408- **410**, 412-13
   series 407-8, **410**-12, 413
   simple 404-7
Combustion of fuels 220-21
Compounds **163**-64
   and chemical formulas 164-66
Condensation **81**
Constant Composition, Law of **190**
Crystallization 101-2
Curie, Pierre and Marie 194
Current electricity **391**-94
   sources of 394-96

D
Dalton, John 190-91
Decimal System 36-37
Decomposition 179-81
Democritus 188
Diffusion 100-1
Distillation **141**-46
Dry Cells 395-96, 411-13

E
Electric
   charges 381-84
   circuits 404-14
   current 391-94, 472
   potential **391**-92, 472
Electrical symbols 407
Electricity 378
   conserving 471-76
   current **391**-401
   generation of 441-42
   measurement of 471-73
   static **380**-81
Electroscope 384-88
Elements **163**, 166
   chemical symbols of 164-67
Energy **70**-71, **267**-68, 418
   characteristics of 428-29
   content of food 372-75
   conversions 425-28
   forms of 420-24
   kinetic 423-24
   Law of Conservation of **426**
   non-renewable sources of 432-43
   nuclear 438-41, 442
   potential 422-23
   renewable sources of 445-55
   solar 445-40
Energy conservation 480-82
   in heating 458-68
   in transportation 476-80
   of electricity 471-76
Evaporation **81**
   effect on temperature of 106-7
   of water 107-8

F
Filtration 136-40
Flame tests 202-4
Floc formation 150-51, 153, 154-55
Food energy 372-75
Force **248**-50
   balanced 253-54
   gravity-mass ratio 252-53
   measuring 250-51
   of gravity 251-53
   unbalanced 254
Fossil fuels 432-38, 441-42
Freezing **81**
Freezing point
   effect of antifreeze 110-11
Friction 255-57

G
Gaseous
   state of matter 75-76
Gases
   and vapour **76**
   effect of temperature on pressure of 96-97
   effect of temperature on volume of 92-93
   properties of 75-77
   relationship between pressure and volume of 97-99
Gravity 61, 251-53

H
Heat
   and changes of state 367-75
   and energy 342
   and food energy 372-75
   and temperature 355-56
   conservation of 458-68
   kinetic molecular theory of 350-52
   latent **367**-72
   specific capacities of 357-64
   specific latent 367-71
   transfer 342-46, 458-60
Heterogeneous 121, 160, **161**-62, 167
Homogeneous 120, **161**
Hydrocarbons 222-23
Hypothesis 7, 14-15

I
Ice
   melting temperature 83-84
Insulation 460-68
Ions 200-1
Isotopes **198**-200

J
Joule **265**

K
Kilogram, standard 62
Kilopascal 259
Kinetic energy **423**-24
Kinetic molecular theory 350-52

L
Latent heat 367-68, 371-72
Lead
  in air 223-24
  storage cell 396-98
Lever 279-84
Liquid
  effect of temperature on
    volume 93-94
  properties 74-75
  state of matter 74-75
Liquefaction **81**

M
Machines 274-87
  efficiency 286-87
  mechanical advantage 285-86
Mass **61**-67
  atomic 192-93
  conservation of 188-90
  measuring 63-65
  number 199
  units, uses 62-63
Matter **70, 160**
  classification of 160, 168-69
  heterogeneous 160, **161**-62
  homogeneous 160, **161**, 168
  particle theory of 86-87
  properties of 71, 173-74
  states of 71-77
Measurement
  of area 45-47
  of length 36-41
  of mass 61-65
  of volume 49-58
  need for 32-34
Mechanical advantage 285
Melting **81**
  of ice 82-84
Melting point
  effect of salt on 109-10
Mendeleef, Dmitri 205
Metals 202
Metre, definition 34-35
Metric system
  history 34-35
  prefixes 35-36
Mixtures 121-23, 167-68
  distillation of 140-46
  filtration of 135-40
  floc formation in 150-51, 153, 154-55
  sedimentation in 150-55

Models 22-26
  atomic 191-92, 194
Molecules 163, **208**

N
Newton 251-52
Nitrogen
  in water 232
  oxides 221-22
Noise pollution 332-36
Non-metals 202
Nuclear
  fission 438-40, 442
  fusion 440-41

O
Observations
  direct and indirect 21-22
  qualitative 8, 32-33
  quantitative 8, 33, 41
Oxygen
  in water 227-29
Ozone 223

P
Parallax 39, 42
Particle theory 86-88, 187
Particles, airborne 215-17
Periodic table 205-8
pH 229-31
Phase 120
Phosphorus, in water 232
Physical changes **174**
Physical properties **71**, 173
Pollution
  air 214-24
  noise 332-36
  water 227-35
Potential energy 422-23
Power **269**-71
  electrical 471-72
Pressure **257**-60
  effect on boiling point 99-100
Prefixes, metric 35-37
Primary unit
  of area 45
  of length 37
  of mass 62
Properties
  chemical **71**, 173-74
  physical **71**, 173
Pulleys 284-85
Pure substances 162-64

Q
Qualitative characteristics 173
Quantitative characteristics 173

R
R values 460-68
Refrigerator 108-9
Retrofitting 465-68
Rutherford, Ernest 192-93

S
Scientific method 4, 6-9
  conclusions 6, 8-9
  experiments 7-8, 15-17
  hypothesis 7, 14-15
  observations 8, 13-14
  use by consumers 240-43
Sedimentation 150-55
Solar systems 445-50
Solids
  effect of temperature on
    volume 94-96
  properties 71-72
  state of matter 71-72
Solidification **81**
Solubility **129**
  factors affecting 130-31
Solute **121, 168**
Solutions 120-**121**, 168
  classes of 127-28
  concentration of 127-29
  rate of dissolving 126-27
  saturated **128**
  supersaturated **128**-29
  temperature changes during
    dissolving 124
  unsaturated **128**
  volume changes during
    dissolving 125
Solvent **121, 168**
Sound
  and communication 292-93
  and matter 308-10
  and pitch 327-31, 336
  and vibrations 295-302
  characteristics of 324-32, 336
  detectors 319-321
  effects of 333-35
  frequency 298-99, 303-5, 336
  intensity 324-27, 336
  quality 331-2, 336
  range of 302-5
  speed of 316-19
  sources of 293-95
  transmission of 308-21
  waves 310-13
Specific heat capacity 357-64, **359**
  measuring 357-58, 362-64
Specific latent heat 367-71
Standard chemical tests 176-79
States of matter 71-72
  changes of 80-89
  gaseous 75-76

liquid 74-75
solid 73
Sublimation **81**
of iodine 85-86
Sulfur dioxide 217-18

T
Temperature 355-56
measuring 346-49
Thomson, J.J. 191, 193
Tyndall effect 122

U
Ultrasonic vibrations 303-5

V
Vapour, see Gases
Vapourization 81
Variable 7
Vibrations
and sound 295-97
longitudinal **299**-302
transverse **297**-99
ultrasonic 303-5
Viscosity 112
experiments 112-14
Voltaic cells 392-95
Volume **49**-58
measuring 52-58
units, uses 49-51

W
Water
boiling temperature 84-85
effect of salt on freezing
point 109-10
pollution 227-35
Watt **269**
Wave energy 313-16
Waves
longitudinal 312-13
transverse 310-12
Weight 59
Word equations 175-76
Work **263**-66